T0174573

New Concepts in Polymer Science

Structure of the Polymer Amorphous State

New Concepts in Polymer Science

New Concepts in Polymer Science

Structure of the Polymer Amorphous State

G.V. Kozlov* and G.E. Zaikov**

* *Kabardino-Balkarian State University, Nalchik, Russia*
** *N.M. Emanuel Institute of Biochemical Physics, Moscow, Russia*

CRC Press
Taylor & Francis Group
Boca Raton London New York

CRC Press is an imprint of the
Taylor & Francis Group, an **informa** business

First published 2004 by VSP Publishing

Published 2018 by CRC Press
Taylor & Francis Group
6000 Broken Sound Parkway NW, Suite 300
Boca Raton, FL 33487-2742

© 2004 by Taylor & Francis Group, LLC
CRC Press is an imprint of Taylor & Francis Group, an Informa business

First issued in paperback 2019

No claim to original U.S. Government works

ISBN 13: 978-0-367-44658-1 (pbk)
ISBN 13: 978-90-6764-401-3 (hbk)

This book contains information obtained from authentic and highly regarded sources. Reasonable efforts have been made to publish reliable data and information, but the author and publisher cannot assume responsibility for the validity of all materials or the consequences of their use. The authors and publishers have attempted to trace the copyright holders of all material reproduced in this publication and apologize to copyright holders if permission to publish in this form has not been obtained. If any copyright material has not been acknowledged please write and let us know so we may rectify in any future reprint.

Except as permitted under U.S. Copyright Law, no part of this book may be reprinted, reproduced, transmitted, or utilized in any form by any electronic, mechanical, or other means, now known or hereafter invented, including photocopying, microfilming, and recording, or in any information storage or retrieval system, without written permission from the publishers.

For permission to photocopy or use material electronically from this work, please access www. copyright.com (http://www.copyright.com/) or contact the Copyright Clearance Center, Inc. (CCC), 222 Rosewood Drive, Danvers, MA 01923, 978-750-8400. CCC is a not-for-profit organization that provides licenses and registration for a variety of users. For organizations that have been granted a photocopy license by the CCC, a separate system of payment has been arranged.

Trademark Notice: Product or corporate names may be trademarks or registered trademarks, and are used only for identification and explanation without intent to infringe.

Visit the Taylor & Francis Web site at
http://www.taylorandfrancis.com

and the CRC Press Web site at
http://www.crcpress.com

If you perceive the Proper Way in the morning, you can die at night politely with the awareness of completed duty.

Confucius,
Chinese philosopher, 500 B.C.

The person did not life up to the hilt, who has not known poverty, love and war.

Mark Twain,
The American writer, XIX century

TABLE OF CONTENTS

PREFACE

Presently, the problem of quantitative description of the polymer amorphous state structure (i.e. the structure of both amorphous glassy-like polymer and non-crystalline zones in amorphous-crystalline polymers) is one of the most actual problems of the polymer physics. This point is important by itself: our understanding of the mentioned state shall not be complete without quantitative structural model. Therefore, the much more important this problem appears in another aspect: no theoretical justified structure – properties relationships can be obtained without development of such a model, which is the basic objective of the polymer physics. Despite the seeming simplicity, anyway, compared with the problems for biopolymers and others possessing a complex chemical structure, to this time, this problem is not solved even for such relatively simple and studied in detail polymers as polyethylenes or polycarbonate. This is the more amazing in the presence of the well-developed experimental capabilities for studies of the polymer structures, which also includes computerized simulation widely spread in recent years.

Not the minor role in determination of the polymer structure is played by its specificity. To a vivid expression by Academician Kargin, the amorphous polymer structure is coded at molecular level and is realized at the supermolecular one. That is why any description of the polymer amorphous state structure must include a characterization of the basic polymer element – the polymeric chain. As a consequence, when applied to the objective mentioned, the modern physical concepts (for example, fractal analysis) must be somewhat modified with respect to specificity of the polymer state of the substance.

At the present time, practically all investigators in the branch of polymer physics accept the presence of a definite part of local (short-range) order zones in the polymer amorphous state structure. However, the quantitative identification of these zones and similar classification of the crystalline (long-range) order zones in amorphous-crystalline polymers, which is put to the basis of the latter structure, are absent now.

To the above-said, the cluster theory suggested in the monograph is based on the revised understanding of experimental data on the local order zone (cluster) as the main structural unit. This allowed clear quantitative identification of clusters, i.e. determination of their relative part, size, number of segments per single cluster, number of clusters per specific volume, etc. Calculations of these basic characteristics were proved experimentally.

The important role in detecting cluster structure – polymer properties relations was played by principally new definition of linear structural defect (dislocation analogue) in polymers, based on the diametrical opposition of ideal (defect-free) structures of amorphous polymers and crystalline solids. Such definition has initiated quite unexpected result: equations of the dislocation theory without any changes can be applied to description of polymer properties. This suggests that the notion of defect by itself possesses much broader physical meaning than it was accepted before.

At present, there are many concepts describing or capable of somehow description of the polymer structure. One of the advantages of the model suggested is its *self-consistency*, i.e. quantitative correspondence to these concepts.

The authors are aware of that the suggested monograph represents the unique generalization of the cluster model and that complexity and variety of the polymer state of the matter may require various modifications of this structural model. They will be thankful to the readers for any notes and advises to the general contents of the monograph and details of the questions under discussion.

Dr. Kozlov Georgy Vladimirovich
Kabardino-Balkarian State University, Nalchik, Russia

Prof. Zaikov Gennady Efremovich,
N.M. Emanuel Institute of Biochemical Physics, Russian Academy of Sciences, Moscow Russia

INTRODUCTION

The basic problem of polymer physics is obtaining of the "structure – properties" correlations for their future application for practical purposes. However, they cannot be obtained without development of a quantitative model of the polymer structure, which is not yet solved completely due to the complexity of the task.

One of possible ways of solving the mentioned problem is application of ideas on the local order of the polymer amorphous state structure, for which the structure of both amorphous polymer and non-crystalline zone in amorphous-crystalline polymer are taken. This problem is actively investigated during last 45 years, which results in obtaining a great amount of experimental (mostly indirect) proofs of existence of the local order in the polymer amorphous state. It is also confirmed by many general statements such as the Ramsey theorem well-known in the theory of numbers, etc. Therefore, the time has come for creating a structural model of polymer basing on the local order ideas. The cluster model of the polymer amorphous state structure represents the realization of such attempts.

Development of the cluster model is based on the well-known experimental observation: the behavior of glassy polymers in the area of stimulated high-elasticity plateau is described in the framework of the rubber elasticity concepts. Thus this gives an opportunity to present the local order (cluster) zone as a multifunctional entanglement of the physical network consisting of several collinear closely packed segments of different macromolecules (the amorphous analogue of crystallite with extended chains) and surrounded by a packless matrix. An independent method for assessing the local order zone fraction in the structure has been elaborated. Segment length in the cluster equals the length of the chain statistical segment that gives a correlation between molecular and structural parameters of the polymer.

Application of the cluster model allowed description and obtaining of analytical "structure – properties" correlations for many processes proceeding in polymers: elasticity, yielding, degradation, transport, some thermodynamic processes, structural relaxation, plasticization, structural stabilization at thermooxidative degradation, etc.

The correspondence between the cluster model and some modern physical concepts, for instance, fractal analysis, fluctuation free volume kinetic theory, percolation theory, irreversible colloidal aggregation models, shall be outlined.

A separate Chapter is devoted to the features of formation and control of the relative part of local order zones in non-crystalline domains of semicrystalline polymers and their connection to the structure of crystallites.

Chapter 1.

Brief historical review of opinion development on the local (short range) order in polymers

---

The structure of the polymer amorphous state represents one of the most important and disputable problems of the polymer physics. Before 1957, the idea dominated that chain macromolecules in the polymer amorphous phase display a random distribution and that "spaghetti" or "unwoven felt" models (the statistic coil) correctly describe the structure (or, more precisely, the absence of it) in this phase. For the first time, existence of the short range (or local) order in the polymers was suggested in ref. [1]. The hypothesis was based on comparison of the segmental volume and density of the amorphous phase, the effect of solid-phase crystallization, etc.

Henceforth, this hypothesis was widely discussed; hence, opinions of the followers of the two mentioned concepts were in conflict. There are, at least, two reasons for such increased attention to the problem of the local order in amorphous polymers. The first is principal, because obviously the presence of the local (short range) order must cause an effect on properties of polymers similar to that of crystallinity, caused on properties of amorphous-crystalline polymers. The second reason is of the methodological character, because at present there are no direct experimental techniques able to solve this perplexity unambiguously.

The results of small angle neutron diffraction are considered the most weighty argument to the advantage of the concept of the statistic coil. In accordance with these results, the inertia radii, $\left\langle R_g^2 \right\rangle$, of the macromolecule in θ-solvent and in bulky polymer are approximately the same [2]. However, there are different opinions, too: Boyer [3] has achieved a tentative scheme (Figure 1.1a), in accordance to which the presence of the local order does not necessarily change the inertia radius of the macromolecule. Moreover, the experiments on small angle neutron diffraction have indicated that polyethylsiloxane which forms the mesophase displays a noticeable deviation between chain conformations in solution and the condensed state [4]. For crystallizing high molecular poly(ethylene terephthalate), this difference is less expressed giving ~15% [5].

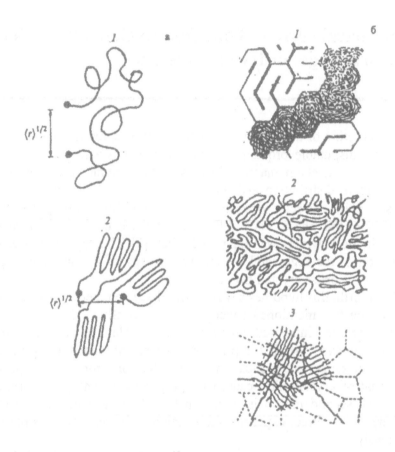

Figure 1.1. Schematic (simulated) representation of a macromolecular coil (a): 1 - statistic coil; 2 statistic coil with local order zones [3]; amorphous state structures (b): 1 – the meander model, 2 – the micellar model with folded chains, 3 – micellar domain structure [12]

It may be simply assumed that conformational changes in chains during formation of the local order will be much lower, than during formation of the mesophase or crystalline zones. That is why in the first case $\left\langle R_g^2 \right\rangle^{1/2}$ will be changed by several percents only, which coincides with the experimental accuracy (~8%) [6], and can hardly be determined. It will be shown below that invariability of $\left\langle R_g^2 \right\rangle^{1/2}$ value at transition from solution to the condensed state makes no obstacles to formation of the local (short range) order zones.

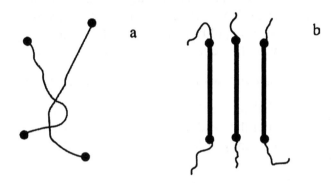

Figure 1.2. Schematic representation of macromolecular "hooking" (a) and cluster (b)

There are two main models of the local order zones in the amorphous state of glassy polymers. Primarily, it has been assumed that such zone is similar to a crystallite with folded chains (CFC) [7]. Certainly, this point of view was strongly affected by discovery of the crystallite structure, made by Fisher [8], Keller [9] and Till [10], according to which they are composed of monocrystallite lamellas with folded chains. The Pechhold "meander" model [11] does also approach the latter. The second model representing the local order zones is the so-called micellar domain structure [12] representing a sort of crystallite analogue with extended chains (CEC) (Figure 1.1b).

The most weighty arguments to the advantage of the local order existence in the amorphous polymer can be obtained from X-ray diffraction patterns [13 – 16]. It has been observed [14] that the shape of amorphous halo in wide-angle X-ray diffraction patterns qualitatively characterizes the presence of one or another constant of the local order, which becomes the interplanar distance for polymer crystallites, i.e. already the parameter of the long range order. Of importance is that the number of these constants, i.e. the totality of reflexes which form the amorphous halo may be both the property of the material and the individual characteristic of the sample, which gives an opportunity to judge about degree and character of the amorphous phase regularity. Application of this general principle to experimental confirmation of the cluster model of the polymer amorphous state will be shown below. Indirect proofs of the local order existence in the amorphous state of glassy polymers were also obtained with the help of other modern experimental techniques: nuclear magnetic resonance [17 – 19], IR-spectroscopy [20], thermal analysis [21, 22], electron diffraction [23], small angle neutron diffraction [24], and thermal expansion analysis [25]. Simulation of the polymer structure applying the Monte-Carlo method has confirmed the possibility of local regularity basing on consideration of even the packing density only [26]. Despite the above-mentioned variability (anyway, within the experimental uncertainty) of $\left\langle R_g^2 \right\rangle^{1/2}$ value [3], a series of theoretical studies was carried out to indicate the possibility of realization of the local order in the solid polymer amorphous state, which also explained constancy of the glass transition and melting temperatures of polymers [27].

As concerns the size of the local order zones, one should note that their primary extent was suggested to be quite high (about several hundred nanometers [28 - 30]). However, at the later time evolution of this opinion has turned it towards decrease of this value, and Boyer following Robertson has suggested the assumption that the local order zones are sized below 5.0 nm, and the long-range order is above 10.0 nm [3]. Certainly, size is not the unique difference of the local and the long-range order zones. For the latter, the necessary condition is spatial regularity [31].

A definite stage in development of views on the origin of the local order in glassy polymer was presented in the review by Fisher and Dettenmaier [32]. Primarily, it has been shown that the size of the local order zones can hardly exceed 1.0 nm (in accordance with Vendorff estimation - 2.0 nm [33]). Secondly, the local order in the glassy state is of thermofluctuation origin or, to put it differently, its exponent will be the temperature function. That is why, when the term "the order frozen below the glass transition temperature, T_g," is used, one should understand it as the constancy of the exponent parameter (free volume, for example [32]) at $T < T_g$. Contrary to low-molecular liquids, where the local order is of the dynamic origin, "freezing up" of the local order in the glassy state is determined by high viscosity of the structure in this state [3]. It will be shown below that the cluster model fully corresponds to these conditions.

At the present time, the investigators justify existence of the local order in the amorphous phase of glassy polymers. Eventually, this was justified even by the most consistent opponent of this concept Flory [2], though under a reservation that the exponent of this order in polymers is not greater than in low-molecular liquids. That is why the basic problem is accurate structural identification of the local order zones and creation of a quantitative concept on this basis, which is the cluster model.

One more important aspect of the mentioned approach is determination of a structural defect for the glassy polymer amorphous state. The influence of thoroughly developed theory of dislocations on development of the metal science is commonly known [34]. That is why many attempts were made to apply disclination-dislocation analogies to description of amorphous polymer properties [35 – 40]. However, absence of a quantitative concept of the structure induced either molecular origin of the structural defects [35, 36, 39] or their irrelevance to the structure of polymers [37, 38, 40]. Presented below is the principally new interpretation of the structural defect for polymers.

There are many different interpretations of the local order and experimental results assuming one or another quantitative assessment of it. The present monograph presents examples displaying both qualitative and quantitative correspondence of the cluster model to previously obtained data. This proves that the cluster model does not conflict with the results already obtained in the physics of polymers, but represents the basis for uniting them.

As the cluster model is the quantitative interpretation, it is purposed for analytical description of the polymer structure - properties relations. This has required revision of some approaches to such phenomena as elasticity, yielding properties, degradation, etc. Application of the cluster model to description of various properties (mechanical, thermodynamic, transfer, etc.) of polymers of different classes will also be demonstrated below.

It should be noted as well that consideration of the local order as dissipative structures can become one of the prospective directions of investigations [41 – 44]. Such approach allows application of the mathematical apparatus of synergism [45, 46] to description of the polymer structure. The principal advantage of this new approach to the structure of the amorphous phase of polymers is the possibility of studying interrelations between parameters of the initial structure of solid polymer and the one derived as a result of any impact (for example, mechanical). The same principle of investigations was used in works [47, 48], in which the Grüneisen parameter characterizing the anharmonicity degree of interatomic bonds was used as the polymer structure variation index. In this connection, the aim of the present monograph is analysis of data on the conditions of views on the local order, structure and properties of solid polymers using the cluster model of the polymer amorphous state structure [49].

REFERENCES

1. Kargin V.A., Kitaigorodski A.I., and Slonimski G.L., *Colloid. Zh.*, 1957, vol. **19**(2), pp. 131 – 132. (Rus)
2. Yoon D.Y. and Flory P.J., *Polymer*, 1977, vol. **18**(5), pp. 509 - 513.
3. Boyer R.I., *J. Macromol. Sci.-Phys.*, 1976, vol. **B12**(2), pp. 253 - 301.
4. Fischer E.W., Maerz K., Willenbacher N., Ballauff M., and Stamm M., *33rd IUPAC Int. Symp. Macromol.*, Montreal, July 8 - 13, 1990: *Book Abstr.*, Montreal, 1990, p. 335.
5. Crist B., Tanzer J.D., and Graessley W.W., *J. Polymer Sci.: Polymer Phys. Ed.*, 1987, vol. **25**(3), pp. 545 - 556.
6. Callea K.P., Schultz J.M., Gardner K.H., and Wignall G.D., *J. Polymer Sci.: Polymer Phys. Ed.* 1987, vol. **25**(3), pp. 651 - 661.
7. Yiekh G.S., *Vysokomol. Soedin.*, 1979, vol. **21A**(11), pp. 2433 – 2446. (Rus)
8. Fischer E.W., *Z. Naturforsch.*, 1957, Bd. **122**(9), S. 753 - 754.
9. Keller A., *Phil. Mag.*, 1957, vol. **2**(21), pp. 1171 - 1175.
10. Till P.H., *J. Polymer Sci.*, 1957, vol. **24**(106), pp. 301 - 306.
11. Pechhold V.W., *Kolloid-Z. Z. Polymere*, 1968, Bd. **228**(1÷2), S. 1 - 28.
12. Kaush H.H., *Polymer Fracture*, New York, Springer-Verlad, 1978, 440 p.
13. Antipov E.M., Ovchinnikov Yu.A., Markova G.S., and Bakeev N.F., *Vysokomol. Soedin.*, 1975, vol. **17B**(3), pp. 172 – 174. (Rus)
14. Poddubni V.I. and Lavrentiev V.K., *Vysokomol. Soed.*, 1990, vol. **32B**(5), pp. 354 – 356. (Rus)
15. Ovchinnikov Yu.K., Kuzmin N.N., Markova G.S., and Bakeev N.F., *Vysokomol. Soedin.*, 1978, vol. **20A**(8), pp. 1742 - 1753. (Rus)
16. Antipov E.M., Ovchinnikov Yu.K., Rebrov A.V., Belov G.P., Markova G.S., and Bakeev N.F., *Vysokomol. Soedin.*, 1978, vol. **20A**(8), pp. 1727 - 1735. (Rus)
17. Kashaev R.S., Prokopiev V.P., Deberdeev R.Ya., Valeev I.I., Stoyanov O.V., and Shmakova O.P., *Vysokomol. Soedin.*, 1978, vol. **26A**(2), pp. 115 - 117. (Rus)
18. Fedotov V.D. and Kadievski G.M., *Vysokomol. Soedin.*, 1978, vol. **20A**(7), pp. 1565 - 1574. (Rus)
19. Fedotov V.D., Ovchinnikov Yu.K., Abdrashidova P.A., and Kuzmin N.N.,

Vysokomol. Soedin., 1977, vol. **19A**(2), pp. 327 - 330. (Rus)

20. Rashchupkin V.P., Goncharov T.K., Karapetyan Z.A., Dzhavadyan E.A., Rozenberg B.A., and Korolev G.V., *Vysokomol. Soedin.*, 1972, vol. **14B**(7), pp. 484 - 485. (Rus)

21. Grubb D.T. and Yoon D.Y., *Polymer Commun.*, 1986, vol. **27**(3), pp. 84 - 88.

22. Jäger E., Müller J., and Yungnickel B., *J. Progr. Colloid and Polymer Sci.* 1985, vol. **71**(2), pp. 145 - 153.

23. Ovchinnikov Yu.K., Markova G.S., and Kargin V.A., *Vysokomol. Soedin.*, 1969, vol. **11A**(2), pp. 329 - 348. (Rus)

24. Červinka L. and Fischer E.W., *J. Non-Cryst. Solids*, 1985, vol. **75**(1-3), pp. 63 - 68.

25. Takeuchi Y., Yamamoto F., and Shuto Y., *Macromolecules*, 1986, vol. **19**(7), pp. 2059 - 2061.

26. Muzeau E. and Johari G.P., *J. Chem. Phys.*, 1990, vol. **149**(1-2), pp. 173 - 183.

27. Bouda V., *Polymer Degrad. Stab.*, 1989, vol. **24**(4), pp. 319 - 326.

28. Kargin V.A., Berestneva Z.Ya., Bogdanov M.E., and Efendiev A.A., *Doklady AN SSSR*, 1966, vol. **167**(2), pp. 384 - 385. (Rus)

29. Kargin V.A., Berestneva Z.Ya., and Kalashnikov V.T., *Uspekhi Khimii*, 1967, vol. **36**(2), pp. 203 – 216. (Rus)

30. Nadezhdin Yu.S., Sidorovich A.V., and Asherov B.A., *Vysokomol. Soedin.*, 1976, vol. **18A**(12), pp. 2626 - 2630. (Rus)

31. Kitaigorodski A.I., Mixed Crystals, 1983, Moscow, Nauka, 276 p. (Rus)

32. Fischer E.W., and Dettenmaier M., *J. Non-Cryst. Solids*, 1978, vol. **31**(1-2), pp. 181 - 205.

33. Wendorff J.H., *Polymer*, 1982, vol. **23**(4), pp. 543 - 557.

34. Honeycombe R.W.K., *The plastic deformation of metals*, Cambridge: Edward Arnold (Publishers) Ltd., 1968, 408 p.

35. Argon A.S., *Phil. Mag.*, 1973. vol. **28**(4), pp. 839 - 865.

36. Argon A.S., *J. Macromol. Sci.-Phys.*, 1973, vol. **8B**(3-4), pp. 573 - 596.

37. Bowden P.B. and Raha S., *Phil. Mag.*, 1974, vol. **29**(1), pp. 149 - 165.

38. Escaig B., *Ann. Phys.*, 1978, vol. **3**(2), pp. 207 - 220.

39. Pechhold W.R. and Stoll B., *Polymer Bull.*, 1982, vol. **7**(4), pp. 413 - 416.

40. Sinani A.B. and Stepanov V.A., *Mechanics of Composite Materials*, 1981, vol. **16**(1), pp. 109 – 115. (Rus)

41. Aliguliev R.M., Khiteeva D.M., and Oganyan V.A., *Vysokomol. Soed.*, 1988, vol. **30B**(4), pp. 268 – 272. (Rus)

42. Aliguliev R.M., Khiteeva D.M., Oganyan V.A., and Nuriev R.A., Doklady

AN AzSSR, 1989, vol. 45(5), pp. 28 – 31. (Rus)

43. Perepechko I.I. and Maksimov A.V., *Vysokomol. Soed.,* 1989, vol. **31B**(1), pp. 54 – 57. (Rus)
44. Kozlov G.V., Beloshenko V.A., and Varyukhin V.N., *Applied Mechanics and Technical Physics*, 1966, vol. **37**(3), pp. 115 - 119. (Rus)
45. Ivanova V.S., Balankin A.S., Bunin I.Zh., and Oksogoev A.A., *Synergetics and Fractals in Material Science*, 1994, Moscow, Nauka, 383 p. (Rus)
46. Ivanova V.S., *Synergetics. Strength and Destruction of Metal Materials*, 1992, Moscow, Nauka, 160 p. (Rus)
47. Kozlov G.V. and Sanditov D.S., *Anharmonic Effects and Physicomechanical Properties of Polymers*, 1994, Novosibirsk, Nauka, 261 p. (Rus)
48. Sanditov D.S. and Kozlov G.V., *Physics and Chemistry of Glass*, 1995, vol. **21**(6), pp. 547 – 576. (Rus)
49. Kozlov G.V. and Novikov V.U., *Uspekhi Fizicheskikh Nauk*, 2001, vol. **171**(7), pp. 717 – 764. (Rus).

Chapter 2.

Physical grounds for cluster model of polymer amorphous state structure

2.1. FUNDAMENTALS

The well-known experimental observations have become the prerequisites for creation of the cluster model of the polymer amorphous state structure. As shown by Haward *et al.* [1, 2], at deformation of glassy polymers outside the yielding range (on the plateau of induced rubbery state) they obey the regularities of the rubbery elasticity theory. In this case, the polymer behavior under high deformations is described by either the Langevin equation [3, 4] or the Gaussian interpretation [5], when the polymer chain does not approach completely stretched state, and correlation between the real stress, σ^{real}, and the stretch degree, λ, for the axial tension is presented as follows [5]:

$$\sigma^{real} = G_p\left(\lambda^2 - \lambda^{-1}\right),\tag{2.1}$$

where G_p is the so-called strain hardening modulus.

Formally, G_p value allows determination of macromolecular entanglements network frequency, v_{ent}, in accordance with the well-known formulae of the rubber elasticity [5]:

$$M_{ent} = \frac{\rho KE}{G_p},\tag{2.2}$$

$$v_{ent} = \frac{\rho N_A}{M_{ent}},\tag{2.3}$$

where ρ is the polymer density; R is the universal gas constant; T is the test temperature; M_{ent} is the molecular mass of chain segment between entanglements; N_A is the Avogadro number.

However, the attempts to assess M_{ent} (or v_{ent}) via G_p values, determined from the equation (2.1), have given incredibly low values of M_{ent} (or unrealistically high v_{ent}), which conflict with the requirements of the Gaussian statistics. This statistics suggests the presence of, at least, 13 units in the chain section between entanglements [6]. Possible reasons for such inconsistency under consideration of entanglements as traditional macromolecular "backlashes" (Figure 1.2a) are discussed in detail in the work [5]. The principally different solution of this problem, suggested in ref. [7], indicates that besides the network of the above-mentioned backlashes, there is another type of entanglements in the polymer glassy state, the structure of cross-link points in which is analogous to crystallites with extended chains (CEC), schematically depicted in Figure 1.2b. Such cross-link point possesses quite high functionality, F (by the cross-link point functionality the number of chains coming out of it is meant [8]). Hereinafter, this cross-link point will be named the cluster.

The cluster consists of segments of different macromolecules, and the length of each segment is postulated equal to the statistical segment length, l_{st}, of the chain "rigidity segment" [9]. In this case, the effective (real) molecular mass of the chain segment between clusters, M_{cl}^{eff}, can be calculated as follows [8]:

$$M_{cl}^{eff} = \frac{M_{cl}F}{2},\qquad(2.4)$$

where M_{cl} is the molecular mass of the chain segment between clusters, calculated by the equation (2.2).

Obviously, at quite high values of F one can obtain reasonable values of M_{cl}^{eff} meeting requirements of the Gaussian statistics. Further on, for the purpose of distinguishing parameters of the cluster entanglement network and macromolecular hooking network indices "cl" and "h" will be used, respectively. Thus, as assumed in the model suggested in the work [7], the structure of the polymer amorphous state represents areas consisting of collinear closely packed segments, composed of different macromolecules (clusters), submersed into a packless matrix. Simultaneously, the clusters play the role of multifunctional cross-link points of physical entanglements. The value of F can be estimated (back again in the framework of the rubber elasticity concept) as follows [10]:

$$F = \frac{2G_\infty}{kT\nu_{cl}} + 2, \tag{2.5}$$

where G_∞ is the equilibrium shear modulus; k is the Boltzmann constant.

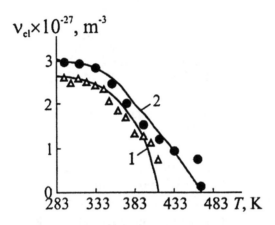

Figure 2.1. Dependencies of macromolecular entanglement cluster network frequency, ν_{cl}, on temperature, T, for PC (1) and PAr (2) [7]

Figure 2.1 shows $\nu_{cl}(T)$ dependencies for polycarbonate, synthesized from bisphenol A (PC) and polyarylate (PAr). These dependencies indicate ν_{cl} decrease with T increase that suggests thermofluctuational origin of clusters (the local order zones). Moreover, the mentioned dependencies display two characteristic temperatures. The first of them is the glass transition temperature of the polymer, T_g. It determined complete decay of clusters that corresponds to an inflection on $\nu_{cl}(T)$ curves and is located approximately by 50 K below T_g. Recently, in the framework of the local order concepts also, it has been shown [11, 12] that temperature T_g is associated with defrosting of the segmental mobility in packless zones of the polymer. This means that in the framework of the cluster model T_g can be associated with devitrification of the packless matrix. For the same polymers, $F(T)$ dependencies are of similar shape (Figure 2.2).

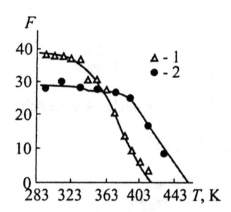

Figure 2.2. Dependencies of cluster functionality, F, on temperature, T, for PC (1) and PAr (2) [7]

Two basic models of the local order zones in polymers (with folded and extended chains, refer to Figure 1.1) possess the point of coincidence: they play the role of cross-link points of macromolecular entanglements physical network [7, 13]. However, their response to mechanical deformation should be significantly different: at high deformations the zones with folded chains ("packs") are capable of unfolding and forming stretched conformations, and clusters are incapable of doing that and polymers can be deformed by stretching "tie" (cluster linking) chains, i.e. by their orienting in direction of applied stress. Turning back to the analogy with crystalline morphology of polymers, let us note that high deformations of amorphous-crystalline polymers (especially for polyethylenes, which give 1,000 – 2,000%) are realized due to unfolding of crystallites [14]. That is why using values of border deformations of polymers one can obtain arguments for the benefit of one or another type of the local order zones in the polymer amorphous state that was performed in the work [15].

It was suggested [33] that the "packs" are capable of delivering parts of chains to "interstices" (i.e. to inter-pack zones) by unfolding, thus acting similarly to crystallites with folded chains (CFC). On this evidence, of special attention is a great difference between maximum stretch degrees, λ_{max}, of amorphous elastic polymers and such amorphous-crystalline polymers as high density polyethylene (HDPE) and polypropylene (PP), for which unfolding of crystallite wrinkles at high deformations was proved in experiments ($\lambda_{max} \approx 1.6$ for PC, $\lambda_{max} \approx 6$ for PP, and $\lambda_{max} \approx 13$ for HDPE at room temperature [16]). Explanation of this difference requires quantitative assessments [15].

Gent and Madan [16] have suggested that stretching proceeds due to extension of crystalline and amorphous chain sequences. In this case, the value λ_{max} can be expressed via f, the number of times a molecule passes through the same crystallite (or "pack"):

$$\frac{1}{\lambda_{max}} = \frac{K}{f} + \frac{(1-K)^{1/2}}{n_{st}^{1/2}}, \tag{2.6}$$

where K is the crystallinity degree; n_{st} is the number of equivalent statistical units between entanglements in the polymer melt. Usually, n_{st} varies within the range of $100 - 300$; as this value is not sufficient for assessment results, for all polymers used in the work [15], it was accepted equal 225.

Clearly the value f determines the number of folds formed by a macromolecule in CFC (or the "pack"). Formation of folds requires meeting the condition $f \geq 2$. Assessment results of f for HDPE and PP are shown in Table 2.1. They have expectedly indicated that macromolecules of the former compound are folded over 50 times, and the latter ~5 times that correlates well with the known data [16].

For amorphous glassy polymers the equivalent of the crystallinity degree, K, is the relative part of clusters, φ_{cl}. The value φ_{cl} can be assessed as follows. Total length of macromolecules, L, per specific polymer volume under their dense packing equals [17]:

$$L = S^{-1}, \tag{2.7}$$

where S is the macromolecule cross-section.

The length of statistical segment, l_{st}, is assessed as follows [18]:

$$l_{st} = l_0 C_\infty, \tag{2.8}$$

where l_0 is the skeletal bond length in the backbone; C_∞ is the characteristic relation, which is the statistical flexibility index of the macromolecule [19].

Total length of segments in the clusters, L_{cl}, per specific volume of the polymer is assessed as follows [20]:

$$L_{cl} = l_{st} v_{cl}, \tag{2.9}$$

and φ_{cl} is assessed by the relation:

$$\varphi_{cl} = \frac{L_{cl}}{L}.$$

(2.10)

Table 2.1

Assessment of folding parameter f for amorphous and amorphous-crystalline polymers [15]

Polymer	T, K	λ_{max}	K	φ_{cl}	f
HDPE	293	13	0.69	-	53
PP	293	6	0.50	-	4.7
PC	333	1.91	-	0.33	0.70
	353	2.23	-	0.29	0.73
	373	2.15	-	0.24	0.57
	393	2.36	-	0.19	0.52
	413	2.75	-	0.11	0.35
PAr	293	1.66	-	0.50	0.89
	313	1.67	-	0.38	0.68
	333	1.66	-	0.37	0.66
	353	1.76	-	0.31	0.59
	373	1.66	-	0.25	0.45
	393	1.70	-	0.19	0.35
	413	1.75	-	0.16	0.30
	433	1.80	-	0.13	0.25
	453	1.86	-	0.11	0.22
	473	1.97	-	0.01	0.02

As combined the equations (2.7) – (2.10) give the final formula for φ_{cl} calculation [20]:

$$\varphi_{cl} = Sl_0 C_\infty v_{cl}.$$

(2.11)

The values of φ_{cl} for PC and PAr, calculated by equation (2.11), are shown in Table 2.1. Using experimental values of λ_{max}, these results indicate that in all cases $f < 1$ (Table 2.1). This proves the absence of macromolecule folding in the local order zones (macromolecular entanglement points of physical network) of these polymers. Note also that unfolding in crystallites of amorphous-crystalline polymers is initiated at deformations of about 50 – 100% [21]. Substituting $\lambda_{max} \sim 2$ and $\varphi_{cl} \approx 0.20$ for amorphous zones of HDPE into equation (2.6), we obtain $f < 1$. This means that in amorphous-crystalline polymers macromolecules are folded in the crystalline zones only.

Thus calculations executed in the work [15] have shown that the crystallite analogue with stretched chains – the cluster, is the most probable type of supersegmental structures in the polymer amorphous state (Figure 1.2b).

The well-known inhomogeneity of elastic deformation of amorphous glassy polymers [22, 23] allows their acceptance as heterogeneous systems. This statement is also true for the amorphous phase of amorphous-crystalline polymers [24, 25]. Nevertheless, the behavior of polymers of both classes is successfully described by both continual models (remind that primarily the known Dagdale model was developed for metals [26]) and molecular concepts. In this connection, the question about the scale which may be considered to be the lower border of continual models' applicability is raised.

One more problem is what the probability of an inhomogeneous molecular system (which unambiguously the amorphous glassy polymer is) consideration as a two-phase system is. If there is such a probability, this will allow application of the so-called "composite" models to description of the amorphous polymer behavior. These models are well developed and successfully used, for example, for description of artificial two-phase systems, which also include the filled ones. These two problems were discussed in the work [27].

Fellers and Huang [28] have applied the fluctuation statistical theory to description of crazing in amorphous polymers. They have deduced an expression for assessing the polymer volume, V_0, in which the fluctuation probability is of the unit value:

$$\frac{\sigma_c V_0^{1/2}}{(2kT_0 B)^{1/2}} = 3.87, \qquad (2.12)$$

where σ_c is the crazing stress; T_0 is the equilibrium temperature, the lower border of which range is the glass transition temperature, T_g; B is the polymer volumetric modulus bound to the Jung modulus, E, by the following relation [29]:

Kozlov G.V. and Zaikov G.E.

$$B = \frac{E}{3(1-v)},$$ (2.13)

where v is the Poisson index.

The distance between clusters, R_{cl}, can be assessed using a simple formula as follows [30]:

$$R_{cl} = 18\left(\frac{2v_{cl}}{A}\right)^{-1/3}, \text{Å}.$$ (2.14)

Table 2.2 shows comparison of R_{cl} and linear dimension L_0, at which the fluctuation probability is of the unit value ($L_0 = V_0^{1/3}$) for 5 amorphous glassy polymers. As indicated by the Table data, parameters R_{cl} and L_0 are close by both the absolute values and variation tendencies. This means that in the framework of characteristic dimensions of the cluster structure amorphous polymer (or amorphous phase of amorphous crystalline polymer) may be considered as the inhomogeneous system [27].

Table 2.2

Comparison of characteristic dimensions of the fluctuation theory, L_0, and the cluster model, R_{cl}, for amorphous glassy polymers [27]

Polymer	L_0, Å	R_{cl}, Å
Polystyrene	76.4	36.1
Poly(methyl methacrylate)	31.7	31.6
Polyvinylchloride	54.0	27.1
Polycarbonate	39.7	31.1
Polysulfone	36.4	25.0

Katsnelson [31] has suggested the following definition of the substance phases: they are "…states of the substance able, being in touch, to exist simultaneously in equilibrium with one another. Obviously different properties are corresponded to different phases. Hence, it should be taken into account that by different phases … parts of a body are meant, related to the solid phase, but possessing different structure and properties". Clusters and packless matrix which in accordance with the cluster model [7] are the main structural elements of the polymer amorphous state meet the above definition, at least, partly. It is known that these elements possess different mechanical properties [32] and different glass transition temperatures [33]. All these facts together give an opportunity to consider the amorphous state of a polymer to be a quasi-two-phase state, disclaiming full strength of the definition [27].

Let us now consider applicability of models, developed for two-phase filled polymers, for describing mechanical behavior of amorphous glassy and amorphous-crystalline polymers. By the analogy with the dispersion theory of strength the shear stress, τ_a, of the composite flow behavior (or for the case under consideration, the amorphous polymer structure) is presented by the equation as follows [34]:

$$\tau_a = \tau_m + \frac{Gb}{R_{cl}}, \qquad (2.15)$$

where τ_m is the shear flow stress of a packless matrix; b is the Burgers vector.

For polymers $\tau_m = 0$ [32, 35], and in the case of aggregation of the filler particles (i.e. association of segments to clusters) the equation (2.15) for the polymer amorphous state can be presented as follows [34]:

$$\tau_a = \frac{Gb}{k(d)R_{cl}}, \qquad (2.16)$$

where $k(d)$ is the aggregation parameter.

The equation (2.16) can be used for description of the temperature dependence of τ_a displaying one principal difference from filled polymers. As G value for amorphous and amorphous-crystalline polymers the macroscopic shear modulus should be used instead of its value for packless matrix. This is explained so that, contrary to τ_a value, the value of G for the polymer amorphous state is determined by the structure of both quasi-phases [32]. The plain truth is that application of G value to the equation (2.16) for packless matrix only would mean determination of the cluster property (τ_a) from properties of another component of the structure only, which is physically meaningless.

Figure 2.3 compares experimental and calculated (by the equation (2.16)) temperature dependencies of τ_a (under the condition $k(d) = $ const for every polymer). This comparison indicates good coincidence of the results thus proving consistency of equation (2.16) use for description of amorphous and amorphous-crystalline polymer properties.

Kozlov G.V. and Zaikov G.E.

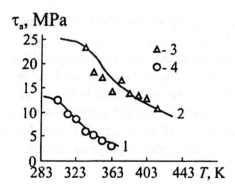

Figure 2.3. Experimental (1, 2) and calculated by equation (2.16) temperature dependencies of the shear flow stress, τ_a, for HDPE (1, 3) and PC (2, 4) [27]

The aggregation parameter $k(d)$ value was assessed by equation (2.16) for 8 polymers at $T = 293$ K. Obviously the physical meaning of it must be analogous to functionality, F, for the cluster structure of polymers. The relation between $k(d)$ and F values is shown in Figure 2.4, from which the expected correspondence follows. This observation and the fact that $k(d)$ is independent of temperature induce an assumption that for the polymer amorphous state $k(d)$ represents some qualitative measure of the polymer tendency to cluster formation at the segmental level [27].

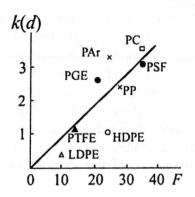

Figure 2.4. The relation between aggregation parameter, $k(d)$, and cluster functionality, F, for amorphous and amorphous-crystalline polymers [27]

For calculating filled polymer strength, σ_d, a series of empirical equations was deduced, for example [36]:

$$\sigma_d = a - c\varphi_{fl}, \tag{2.17}$$

where a and c are constants; φ_{fl} is the volumetric content of the filler.

Figure 2.5 shows dependencies of the degradation stress, σ_d, on φ_{cl} for three polymers (PC, PAr, and HDPE; for the latter results were obtained by the blow tests [37]). The Figure indicates that strength of the mentioned polymers is described by a simple equation as follows [27]:

$$\sigma_d \approx 119\varphi_{cl} \text{ (MPa)}, \tag{2.18}$$

which at $a = 0$ and $c = -119$ MPa is analogous to the equation (2.17).

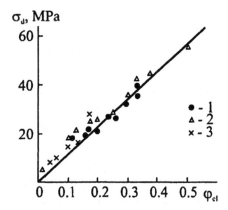

Figure 2.5. Destruction stress, σ_d, dependence on relative fracture of clusters, φ_{cl}, for PC (1), PAr (2) and HDPE (3) [27, 37]

Thus, under some conditions, the "composite" models can be successfully applied to description of amorphous and amorphous-crystalline polymers' behavior in the framework of the quasi-two-phase cluster model. In this case, obvious possibility for obtaining quantitative correlation between the theory and the experiment exists under the condition of heretophase structure of the amorphous state and consistency of the cluster model [38].

To finishing this Section, let us note that in the presence of the local order in the polymer amorphous state (irrelatively to particular model of its zones) there are the most general strict mathematical proofs. In accordance with the Ramsey theorem proved in the theory of numbers, any quite great amount of numbers, points or objects, $i > R(i, j)$, (in the case under consideration, statistical segments) necessarily contains highly oriented subsystem from $N_j \leq R(i, j)$ such segments. That is why the absolute irregularity of large systems (structures) is impossible [39, 40].

2.2. THERMODYNAMICS OF THE LOCAL ORDER FORMATION

As shown in the present Section, cluster structure formation is the integral part of much more general concepts, for example, the theory of hierarchy systems evolution [41 – 46]. Correlation between the specific Gibbs function of intermolecular interactions, $\Delta \overline{G}^{im}$ (*im* symbol means intermolecular or, in the present case, intersegmental type of interactions; symbol "‒" indicates the specific type of the value; symbol "~" outlines heterogeneous type of the system; hereinafter, for the sake of simplicity it will be denoted as $\Delta \widetilde{G}^{im}$) and the melting temperature, T_m, [43, 45, 46] was chosen for experimental validation of the physicochemical theory of chemical systems evolution [41]. Selection of these parameters is stipulated in the works [43 – 46].

First of all, the fundamentals stated in ref. [78], necessary for better understanding of the material below should be reminded.

It is common knowledge that the Gibbs-Helmholtz equation is valid for processes proceeding in simple closed systems:

$$\left[\frac{\partial(\Delta G/T)}{\partial(1/T)} \right]_p = \Delta H, \qquad (2.19)$$

where ΔG and ΔH are changes of the Gibbs function and enthalpy during the process, respectively; T and p are temperature and pressure, respectively.

If it is accepted that in a definite temperature range ΔH is independent of T, then for a non-equilibrium phase transition (self-assembly of an individual substance) at temperature T the following relation is valid:

$$\Delta G^{im} = \left(\Delta H_m^{im}/T_m\right)\left(T_m - T\right) = \left(\Delta H_m^{im}/T_m\right)\Delta T = \Delta S_m^{im}\Delta T, \qquad (2.20)$$

where ΔG^{im} is the Gibbs function change during crystallization (self-assembling) of the studied substance from overcooled state at $T = T_m - \Delta T$, ΔH_m^{im} is the enthalpy change during crystallization (solidification); ΔS_m^{im} is the crystallization entropy (the change of entropy at the phase transition).

It has been suggested [41, 43, 45] to use the equation (2.20) for open systems, the composition and T_m values of which vary negligibly. Further on, the possibility of application of this equation was displayed for various chemical compounds melting at $T_m < 373$ K and condensing at constant standard temperature $T = T_0 = 298$ K [43 – 46]. At the extended approach, for these cases the equation (2.20) should be represented as follows:

$$\Delta G_i^{im} = \left(\Delta H_{mi}^{im}/T_{mi}\right)\left(T_{mi} - T_0\right) = \Delta S_{mi}^{im}\Delta T, \qquad (2.21)$$

where index $i = 1, 2, ..., n$ indicates different substances. In shape, the equation (2.21) represents the analogue of correlation (2.20). Simultaneously, these equations are principally different in the following. In equation (2.20) ΔG^{im} is a variable characterizing non-equilibrium transition of an individual substance in the system at any temperature $T < T_{mi}$. Values of ΔH_m^{im}, ΔS_m^{im} and T_m belong to this individual substance and accepted as constant values. As a whole, the equation (2.20) represents the functional dependence $\Delta G^{im} = f(T)$.

In equation (2.21) ΔG^{im} is the variable related to non-equilibrium transitions of different compounds with different melting temperatures, T_{mi}, at a standard (constant) temperature, T_0. In this case, the equation (2.21) represents the function $\Delta G_i^{im} = f(T_{mi})$ [45]. The technique for calculating $\Delta \widetilde{G}^{im}$ for polymers is described in the work [47].

Figure 2.6 shows $\Delta \widetilde{G}^{im}$ dependence on $\Delta T = T_g - 293$ K for 15 amorphous glassy, amorphous-crystalline polymers and polymer networks [47]; $\Delta \widetilde{G}^{im}$ value is given in kcal/mol. As might be expected, linear decrease of $\Delta \widetilde{G}^{im}$ with ΔT (or, intrinsically, T_g) increase is observed. Yet more important fact is that the straight line plotted in Figure 2.6, which approximates well the results obtained, corresponds to the data of the works [43 – 46] shown in $\Delta \tilde{G}^{im}$ - ΔT coordinates for absolutely different chemical compounds, but at 10:1 scale by $\Delta \tilde{G}^{im}$ axis.

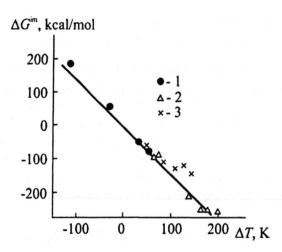

Figure 2.6. Dependence of the Gibbs specific function of non-equilibrium phase transition, $\Delta \tilde{G}^{im}$ (kcal/mol), on $\Delta T = T_g - 293$ K for amorphous-crystalline (1) and amorphous glassy (2) polymers, and polymer networks (3). The straight line is plotted in accordance with ref. [43] at 10:1 scale by $\Delta \tilde{G}^{im}$ axis [47]

The latter circumstance is provided by, approximately, the order of magnitude difference between molar volumes of the segments (which are also kinetically independent fragments) in compounds, used in the work [47], which is proved by the plot in Figure 2.7 ($\Delta \tilde{G}^{im}$ is measured in cal/g).

As follows from this graph, $\Delta \tilde{G}^{im}(\Delta T)$ dependence for the mentioned polymers corresponds, both qualitatively and quantitatively, to the data shown in refs. [43, 45, 46]. Principally, this allows calculation of the segment dimension, which is different for different polymers. Deviation from the graph plotted for polymers with high T_g indicates the features of supersegmental structure of these substances [43]. Data from Figures 2.6 and 2.7 indicate that the cluster model postulated in ref. [7], based on the existence of the local order in the amorphous state of the polymers, qualitatively and quantitatively conforms with yet more general macrothermodynamic hierarchy model [41 –

46], occupying the corresponded energetic niche in the hierarchy of the real world structure. Graphs in Figures 2.6 and 2.7 demonstrate direction of the polymer structure evolution during its physical aging. Striving of the polymer structure to equilibrium means $\Delta \widetilde{G}^{im}$ striving to the minimum (i.e. $\Delta \widetilde{G}^{im}$ shift towards lower negative values) and increase of the local order degree, respectively, which is accompanied by T_g increase [48]. Polymer "rejuvenation", which is the process opposite by its thermodynamic directivity to the above-considered one, is also possible. In practice, this is realized by "injection" of energy (for example, mechanical) to the polymer [49].

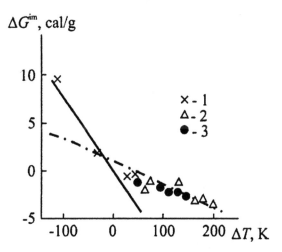

Figure 2.7. Temperature dependence of the Gibbs specific function $\Delta \widetilde{G}^{im}$ (cal/g) for amorphous-crystalline (1), amorphous (2) polymers and polymer networks (3) [47]. The straight line is plotted with respect to data from ref. [43]

Regularities (the Gibbs-Helmholtz-Gladyishev equation), shown in Figures 2.6 and 2.7, are true for both different polymers and a single polymer at varying temperature, T (the Gibbs-Helmholtz equation). Figure 2.8 shows the $\Delta \widetilde{G}^{im}(\Delta T)$ dependence for PC, which also quantitatively coincides with the one shown in Figure 2.6. Therefore, equations (2.20) and (2.21) are fulfilled for polymers simultaneously, i.e. formation of supersegmental structures represents a non-equilibrium transition resulting in formation of non-equilibrium structures. Typically, the beginning of their formation corresponds to glass transition, i.e. to the transition from equilibrium devitrified state to low non-equilibrium glassy-like one.

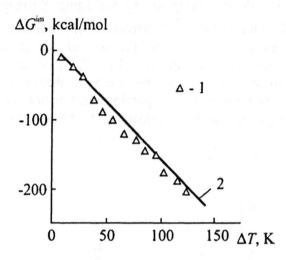

Figure 2.8. Dependence of the Gibbs specific function of non-equilibrium phase transition, $\Delta\widetilde{G}^{im}$, on $\Delta T = T_g - T$ for PC (1). Line (2) is plotted in accordance with Figure 2.6 [47]

Finally, it should be noted that $\Delta\widetilde{G}^{im}$ values "regulating" formation of supersegmental structures in polymers are definitely associated with molecular characteristics of the latter. Since polymer is the solid consisting of long chain macromolecules, it should be expected that the most important (or at least one of the most important) property is flexibility of the polymer chain, which can be expressed with the help of characteristic relation C_∞ [19, 50]. That is why Figure 2.9 shows $\Delta\widetilde{G}^{im}$ (C_∞) dependence, which clearly displays the tendency to $\Delta\widetilde{G}^{im}$ increase (and consequently, T_g decrease) with the chain flexibility. The unique noticeable deviation, detected for polystyrene may be stipulated by the well-known specificity of its chemical structure [51]. Correlation of $\Delta\widetilde{G}^{im}$ (C_∞) dependence fully conforms with the previously postulated [51] T_g increase with the polymer chain rigidity (which is now thermodynamically substantiated).

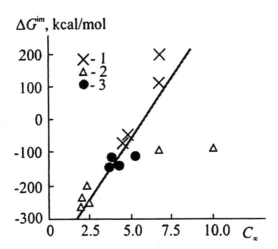

Figure 2.9. Dependence of the Gibbs specific function, $\Delta\widetilde{G}^{im}$, of non-equilibrium phase transition on characteristic relation C_∞ for amorphous-crystalline (1), amorphous glassy (2) polymers and polymer networks (3) [47]

Thus $\Delta\widetilde{G}^{im}(\Delta T)$ dependencies for supersegmental structure of polymers, obtained in the framework of macrothermodynamic hierarchy model, qualitatively and quantitatively conform to previously obtained analogous correlations for a broad selection of substances [43, 45, 46, 52]. This proves the reality of these structures for the polymer amorphous state. Equations (2.20) and (2.21) are equally applicable to description of thermodynamic behavior of these structures and can be used for their quantitative simulation. If more rigid calculations are required, corrections for heat capacity change at phase transitions must be introduced [52].

2.3. POLYMER STRUCTURE REGULARITY AND CLUSTER MODEL

It should be expected that formation of the local order zones will affect the general regularity of the polymer amorphous state structure. At the present time, there are some methods allowing characterization of order (or disorder) of the polymer structure, and they will be compared below with parameters characterizing the cluster structure of these materials.

The possibility of using the Mooney-Rievlyn equation constants for characterization of the local order; therefore, the principal difference from the methods previously used for rubbers only is application of this approach to solid polymers [53]. The simplest form of the Mooney-Rievlyn empirical equation is as follows [55]:

$$f^* = 2C_1 + 2C_2\lambda^{-1}, \tag{2.22}$$

where f^* is the reduced stress; $2C_1$ and $2C_2$ are equation constants; λ is the stretch degree.

The value of f^* is determined by the follow relation [56]:

$$f^* = \frac{\sigma}{\lambda - \lambda^{-2}}, \tag{2.23}$$

where σ is the nominal stress, i.e. the one calculated using the initial cross-section of the sample.

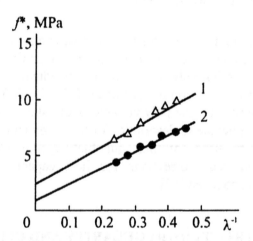

Figure 2.10. Dependencies of reduced stress, f^*, on the strain degree, λ, for PC (1) and HDPE (2) [53]

The equation (2.22) is widely applied to the study of mechanical properties of rubbers. Basing on works by different authors, Boyer [56, 57] has made an assumption that the relation $2C_2/2C_1$ can be the measure of the short range order in cross-linked rubbers and has presented the summary Table for a series of polymers in the rubbery state, which proves this assumption.

Figure 2.10 shows typical dependencies corresponded to equations (2.22) and (2.23) for PC (test temperature $T = 403$ K) and HDPE ($T = 293$ K). Clearly the Mooney-Rievlyn equation is applicable to both amorphous glassy PC and amorphous crystalline HDPE and gives reasonable $2C_1$ and $2C_2$ values. The latter statement is based on the following observation. It is known [55] that $2C_1$ constant can be expressed as follows:

$$2C_1 = \frac{A\rho RT}{M_{ent}},\qquad(2.24)$$

where A is a coefficient determined by functionality of the entanglement network cross-link points.

The equation (2.24) gives a possibility to estimate M_{ent} with respect to the known $2C_1$ values. The estimation results displayed good correspondence of the values obtained to analogous ones, calculated by the equations (2.2) and (2.3). It is typical that for macromolecular overwhelms' network, M_{cl} but not M_{ent} values (which are by one or two orders of magnitude greater than M_{cl}) conform to $2C_1$ values [53].

Figure 2.11 shows M_{cl} dependence on $2C_2/2C_1$ relation value, obtained from the Mooney-Rievlyn graphs. Data from this Figure display good linear correlation between the mentioned parameters that confirms the Boyer assumption about the possibility of $2C_2/2C_1$ relation use as the measure for the short range (local) order in polymers. However, Boyer has also assumed [56] that increase of the absolute $2C_2/2C_1$ value displays growth of the short range order degree in rubbers. For studied polymers [53, 54], M_{cl} increase with $2C_2/2C_1$ means decrease of the entanglement network frequency (equation (2.3)), decrease of the number of segments in clusters and, consequently, reduction of the local order degree (equation (2.11)). To put it differently, for amorphous glassy and amorphous-crystalline polymers, the increase of $2C_2/2C_1$ reflects the effect opposite to that observed in rubbers. Such deviation is not accidental and reflects the difference in structures for these classes of polymers. If density of the cross-linked carcass and the possibility of chain segments packing between cross-link points in rubbers correspond to different structural elements and display opposite tendencies of variation [56], then for studied polymers both the entanglement network frequency and increase of the local order degree possess symbate tendencies that follows from the cluster model

[38]. To put it differently, analogies between structural and mechanical properties of absolute cross-linked rubbers and linear polymers, studied in the work [53] are true to some extent only.

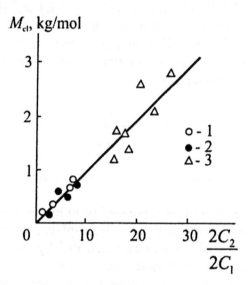

Figure 2.11. Dependence of the molecular mass, M_{cl}, of chain fracture between clusters on the relation of Mooney-Rievlyn equation constants for PC (1), poly(arylate sulfone) (PAS) (2), and HDPE (3) [53]

In the framework of phenomenological theory of the second-order transition by Landau [58], the order parameter ψ_0, unambiguously associated with one of the most important thermodynamic properties which is entropy change, ΔS, is determined as follows:

$$\psi_0 = (a/2C)^{1/2}(T_{tr} - T)^{1/2}, \qquad (2.25)$$

where a and C are parameters; T_{tr} is the transition temperature.

The experimental proof of the Landau theory applicability is correspondence of the temperature dependence of ψ_0 to $(T_{tr} - T)^{1/2}$ shape [58]. Graphs in Figure 2.12 indicate the same shape of v_{cl} temperature dependence for amorphous-crystalline HDPE [59]. In this case, the glass transition temperature T_g (as usual for the Landau theory [58]), the melting temperature T_m and temperature of "liquid 1 – liquid 2" transition T_{ll} [60] were accepted for T_{tr}. Temperature of "liquid 1 - liquid 2" transition can be estimated as follows [9]:

$$T_{ll} = (1.20 \pm 0.05)T_m. \tag{2.26}$$

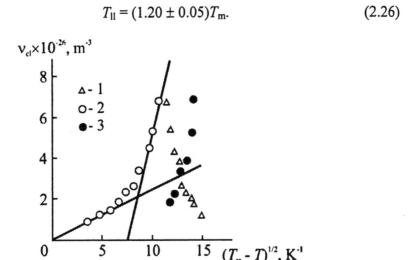

Figure 2.12. Dependencies of cluster entanglement network frequency ν_{cl} on $(T_{tr} - T)^{1/2}$ parameter for HDPE, corresponded to the Landau equation for the second-order phase transitions. The transition temperature T_{tr} is as follows: T_g (1), T_m (2), and T_{ll} (3) [59]

Data from Figure 2.12 give a possibility to assume that the cluster entanglement network frequency, ν_{cl}, is analogous to the order parameter ψ_0 and, consequently, characterizes local ordering degree in non-crystalline zones of polyethylenes. Two interesting features of dependencies shown in Figure 2.12 should be displayed. First of all, on the dependence $\nu_{cl}(T_{tr} - T)^{1/2}$ at $T = 333$ K an inflection is observed. As is known [61], a relaxation transition of polyethylenes, which Boyer named the "glass transition I", was detected at this temperature. Secondly, the condition $\psi_0 = 0$ at $T = T_{tr}$ is assumed in the equation (2.25). The graph in Figure 2.12 indicates that the identity $\nu_{cl} = 0$ is reached at $T_{tr} = T_m$, but not at $T_{tr} = T_g$. To put it differently, temperature T_m is corresponded not only to melting of crystallites, but also to "melting" of segments, and above T_m a network from macromolecular backlashes remains only [30]. This situation can be explained by the specificity of the local order formation in non-crystalline zones of amorphous-crystalline polymers similar to polyethylenes, for which this process is of "induced" type due to chain tensioning in the amorphous phase during crystallization [62, 63]. Note also that similarity of ψ_0 and ν_{cl} parameters eliminates consideration of entanglement cross points as traditional backlashes [30] and supposes that the local order zones (clusters) play their role [59].

Obviously the structure of amorphous polymers (or amorphous phase of amorphous-crystalline polymers) gives grounds to assume the presence of a definite chaos degree in it. It is also absolutely obvious that the chaotic character of the amorphous phase structure represents an important parameter determining structural characteristics and, consequently, properties of the polymer. That is why the question about interconnection of these indices is brought up. The second important problem is the physical origin of chaos in polymers: if it is random (and unpredictable) or deterministic chaos.

Among possible measures of the chaos degree in dynamics of a system (in this case, in polymer structure), the common one is Lyapunov's index, λ_L [64]. It represents the measure of exponential rate of neighboring trajectories divergence or convergence in the phase space along the current axis of coordinate. Chaotic processes are characterized by exponential divergence of neighboring trajectories, which gives, at least, one positive Lyapunov's index. The technique for λ_L assessment is described in the work [65]. Because the basic distinguishing characteristic of a polymer is its composition from long chain macromolecules, and the basic characteristic of the latter is their flexibility, the interconnection between Lyapunov's index λ_L and characteristic relation C_∞ has been studied [65]. There are two reasons for choosing C_∞ parameter for the flexibility measure. The first reason is that C_∞ is determined more precisely compared with other similar parameters [66]. The second one is that it can be estimated basing on chemical structure of the polymer macromolecule only [67]. Primarily, it should be noted that the presence of positive λ_L [65] indicates chaos existence of in the polymer structure. Further on, a systematic increase of λ_L (i.e. chaos intensification) with the chain flexibility is observed. Obviously C_∞ increase means growth of the chain mobility and, as a consequence, intensification of chaotic processes in the system (Figure 2.13). Two polymers are outside of the general dependence: PS and PMMA, which was frequently observed before [66]. This suggests that C_∞ value is not always the unique molecular parameter that determines structural characteristics and properties of polymers, as suggested in ref. [67]. It is a matter of familiar experience [68] that both PS and PMMA have side groups, which are the reason for sharp increase of the cross-section, S, of their macromolecules. That is why one may suggest that applying S as the normalizing factor will improve $\lambda_L(C_\infty)$ correlation. Actually, the dependence $\lambda_L(C_\infty/S)$ does not give fall out results (Figure 2.14) and, despite a definite data scattering, can be approximated by a straight line. The most important result following from the above-mentioned data is regular variation of the chaos, λ_L, exponent for the structure of polymers possessing molecular characteristics C_∞ (or C_∞/S), which presumes deterministic (predictable) chaos [65].

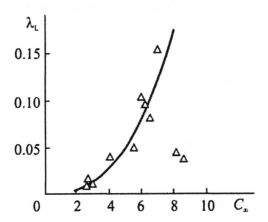

Figure 2.13. Dependence of Lyapunov's index, λ_L, on characteristic relation, C_∞, for amorphous glassy and amorphous-crystalline polymers [65]

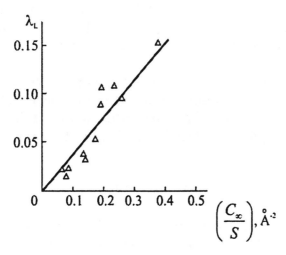

Figure 2.14. Dependence of Lyapunov's index, λ_L, on molecular parameter (C_∞/S) for amorphous glassy and amorphous-crystalline polymers [65]

Figure 2.15 shows dependence of macromolecular entanglement cluster network, v_{cl}, on λ_L. As might be expected, chaos intensification (λ_L increase) decreases v_{cl} value, i.e. the local ordering degree in the structure of polymer amorphous state [65]. More precise interpretation of the chaotic character of the polymer structure within the framework of multifractal formalism is given below.

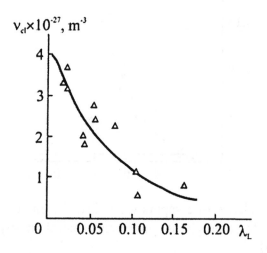

Figure 2.15. Dependence of macromolecular entanglement cluster network frequency, v_{cl}, on Lyapunov's index, λ_L, for amorphous glassy and amorphous-crystalline polymers [65]

Finishing this Section, let us discuss from positions of thermodynamics the interconnection between disorder degree of the polymer structure and the local regularity degree. One of possible characteristics of disorder can be the random free volume, f_f [69], which is associated with entropy variation, ΔS, as follows [70]:

$$\Delta S = (3 \div 5)Rf_f \ln f_f. \tag{2.27}$$

One more quantitative interpretation of disorder is suggested in ref. [71], in which the disorder parameter, δ, associated with thermal mobility of macromolecules in liquid at the melting point T_m is determined as follows:

$$S = \frac{P_i\left(\widetilde{V}_m - 1\right)}{p^*}, \tag{2.28}$$

where P_i is the internal pressure; $\widetilde{V}_m$ is the reduced molar volume at the melting point; p^* is the characteristic pressure.

The value of P_i can be determined as follows [71]:

$$P_i = \frac{p^*}{\widetilde{V}_m^2}, \tag{2.29}$$

and finally, after substitution of equation (2.29) into (2.28), the equation for δ determination is obtained [72] as follows:

$$\delta = \frac{\widetilde{V}_m - 1}{\widetilde{V}_m^2}. \tag{2.30}$$

Taking into account successful application of liquid theories to description of amorphous polymer behavior [9], the range of equation (2.30) application was extended upon the glassy state of these polymers (by substituting $\widetilde{V}_m$ by $\widetilde{V}$) [72].

Obviously the problem of quantitative assessment of δ is reduced now to determination of $\widetilde{V}$ that for the polymer may be implemented by using the equation as follows [73]:

$$\frac{\left(\widetilde{V}^{1/3} - 1\right)}{\left(4 - 3\widetilde{V}^{1/3}\right)} = \alpha T, \tag{2.31}$$

where α is the heat linear expansion coefficient.

Temperature dependence of α can be assessed by different methods. To the first approximation, α value may be accepted constant, and direct experimental measurements (which is intrinsically the most accurate method) or the known Barker equation can be applied [74]:

$$\alpha^2 E = C_1, \tag{2.32}$$

where E is the elasticity modulus; C_1 is the coefficient varying in the range of $7.5 – 24$ N/m²K² giving, on average, 15 N/m²K² [74].

The equation (2.32) gives an opportunity of quite simple assessment of α by mechanical test results. Variation of C_1 does not change the quality of δ assessment results, but gives a possibility of improving quantitative correspondence of parameters, calculated in ref. [72], by $15 – 20\%$, approximately. Preliminary assessments indicate the best correspondence at C_1 equal 7.5, 15 and 24 N/m²K² for PAr, PC and PMMA, respectively [72].

Relative random free volume, f_f, represents the function of Poisson coefficient, ν, expressed by the following equation [65]:

$$f_f = C_2\left(\frac{1+\nu}{1-2\nu}\right). \qquad (2.33)$$

Constant C_2 value may be varied, but this causes no effect on the quality of f_f assessment; thus it was accepted [29, 75, 76] that:

$$C_2 = 0.017. \qquad (2.34)$$

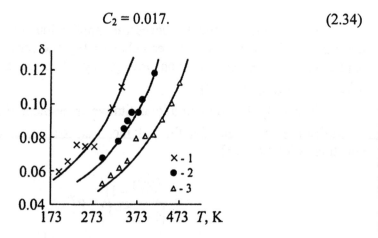

Figure 2.16. Temperature dependencies of disorder degree δ for PMMA (1), PC (2), and PAr (3) [72]

Figure 2.16 shows temperature dependencies of disorder parameter δ, assessed by the above-considered method, for three glassy polymers. Clearly dependencies $\delta(T)$ are symbate and indicate the increase of disorder degree with temperature. It should be noted that this result is not trivial. There are several concepts considering the disorder degree as the "frozen" one at transition of amorphous polymer to the glassy state. According to equation (2.27), the condition f_f = const also proves this point of view.

The symbate type of $\delta(T)$ curves (Figure 2.16) indicates the possibility of their superimposition by shifting along the temperature axis (i.e. applying the principle of temperature superimposition). Actually, the dependence $\delta(\Delta T)$, where ΔT represents the difference between the glass transition temperature T_g and the test one, T, indicated the reality of such superimposition (Figure 2.17). Practically, this means that the disorder degree in amorphous glassy polymers is independent of their chemical structure and is determined by temperature only (more specifically, by the difference between T_g and test temperature). This gives grounds to an assumption that disorder in the structure of amorphous polymers, determined by the above-mentioned method, is of thermofluctuation origin only (that follows from Sharma's determination of δ [71]).

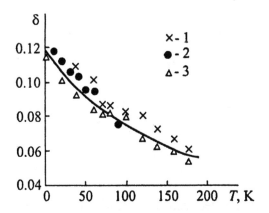

Figure 2.17. Dependence of disorder degree (δ) on the temperature difference ($\Delta T = T_g - T$) for PMMA (1), PC (2), and PAr (3) [72]

Figure 2.18 shows comparison of δ and f_f values, calculated for PMMA, PC and PAr. This comparison displays approximate equality of the mentioned parameters, i.e. random free volume is also the disorder index for amorphous glassy polymer [72]. Thus as v_{cl} determines the number of closely packed segments in the local order zones and, consequently, is the index of this order, inversely proportional correlation between δ and v_{cl} must be observed. Obviously for the case of "backlash" network, none of such correlations is expected, because it is more probable that "backlashes" will open the structure of amorphous polymer [77], i.e. will intensify disorder.

Figure 2.19 shows the relation between δ and φ_{cl} (i.e. between disorder and order parameters, respectively). As might be expected, inversely proportional correlation approximated well by a straight line is observed for these parameters. The line is composed basing on the ideas as follows. Obviously at $\varphi_{cl} = 1.0$ the system possesses the ideal ordering and $\delta = 0$, whereas at $\varphi_{cl} = 0$, δ is accepted equal f_f at T_g with respect to the data in Figure 2.18 (in the present case, $f_f = 0.113$ is used [78]). The unambiguously indicated result favorably testifies the cluster interpretation of macromolecular entanglement cluster network.

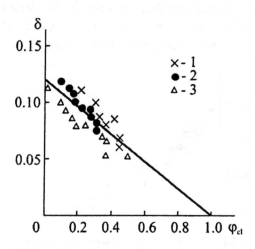

Figure 2.19. Correlation between the disorder parameter (δ) and the relative part of clusters (φ_{cl}) for PMMA (1), PC (2), and PAr (3) [72]

For linear polymers δ and f_f are approximately equal (Figure 2.18), and for glassy polymer networks $\delta > f_f$ [69]. This difference will be explained in Chapter 5.

Simultaneously, data from Figure 2.19 induce an assumption that there is a tight connection between the cluster model [7] and the random free volume theory [75, 76]: separation of a segment from a cluster means formation of the free volume microscopy, whereas addition of a segment to a cluster means its "implosion" or collapse. If it is true, the entropy variation, ΔS_f, due to random free volume formation (equation (2.27)) must be equal to entropy change, ΔS_{cl}, at partial thermofluctuational decay of clusters. The value of ΔS_{cl} can be assessed in the framework of the Forsman theory [79] by the equation as follows:

$$\Delta S_{cl} = \frac{Rc_3\varphi_{cl}}{M_0}\left[\frac{m-1}{m}\ln\left(\frac{c_3\rho\varphi_{cl}}{m}\right) - \frac{(m-1)+\ln m}{m}\right], \tag{2.35}$$

where c_3 is the polymer concentration in the system (in the case under consideration, $c_3 = 1$); M_0 is the molecular mass of the repeating unit; m is the number of repeating units in the cluster; ρ is the polymer density.

Figure 2.20 compares ΔS_f and ΔS_{cl} for PC and PAr and indicates an interesting detail. Let us note first that the change of $\Delta S_f(T)$ curve slopes correspond to devitrification temperature of the packless polymer matrix, T_g' [12]. Thus in the case of glassy packless matrix, the disorder increase due to partial cluster decay is transformed into disorder of random free volume microcavity formation. For devitrified packless matrix ΔS_{cl} is partly transformed into ΔS_f and partly increases disorder in accordance with other mechanisms (possibly, due to conformational changes [80]). As a consequence, strictly saying, δ value (and parameters f_f, v_{cl}, and φ_{cl} associated with it) characterize thermofluctuational "structural" disorder only, but not the entire spectrum of it [72].

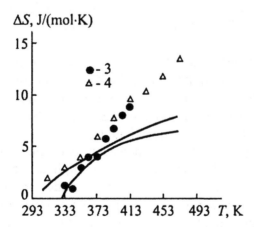

Figure 2.20. Temperature dependencies of entropy change stipulated by formation of the random free volume, ΔS_f (1, 2), and cluster decay, ΔS_{cl} (3, 4), for PC (1, 3) and PAr (2, 4) [72]

It should be noted that the study of disorder degree of amorphous polymer structure is not of the empirical significance only. Figure 2.21 shows dependence of the offset yield stress of three studied polymers on reverse δ value. Clearly this dependence is of the simplest shape: it is linear and passes through the origin of coordinates. That is why it is quite suitable for forecasting mechanical behavior of polymers. Similar dependencies were obtained for the yield strain, ε_y, and elasticity modulus, E, deformation [72].

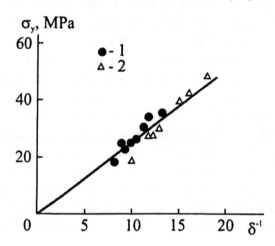

Figure 2.21. Dependence of offset yield stress σ_y on disorder parameter δ for PC (1), PAr (2), and PMMA (3) [72]

The above-shown results allow making the following conclusions. First of all for amorphous glassy polymers, the disorder degree is the function of temperature. Dependence of δ on temperature only suggests purely thermofluctuational origin of this disorder. Secondly, for amorphous polymers, the random free volume, f_t, is the quantitative parameter of the disorder degree. Thirdly, cross-link points of macromolecular entanglement cluster network represent the local ordering degree of the amorphous polymer structure, as postulated in the cluster model. Fourthly, disorder in the systems under consideration is not necessarily of the thermofluctuational origin, but the latter component of it defines mechanical behavior of the polymer (Figure 2.21) [72].

2.4. THERMOFLUCTUATIONAL ORIGIN OF CLUSTERS

A set of parameters indicating thermofluctuational origin of the local order zones (clusters) was mentioned above, for example, in equation (2.12), Figures 2.8 and 2.19. In the present Section, this point will be discussed in more detail. First of all, the interconnection between parameters of the entanglement cluster network and density fluctuations should be studied. Density fluctuations $\langle \Delta \rho / \rho \rangle^2$ represent the measure of disorder in polymers and are determined as follows [81, 82]:

$$\langle \Delta \rho / \rho \rangle^2 = \frac{\left\langle \left(N - \langle N \rangle \right)^2 \right\rangle}{\langle N \rangle}, \qquad (2.36)$$

where N is the number of electrons in randomly selected volume; $\langle N \rangle$ is the mean value of N.

For a liquid in the equilibrium state, statistical mechanics gives the following expression for $\langle \Delta \rho / \rho \rangle^2 (V)$, where V is the standard volume, in the limit $V \to \infty$ (i.e. in thermodynamic limit) [82]:

$$\langle \Delta \rho / \rho \rangle^2 (\infty) = \rho k T \chi_T, \qquad (2.37)$$

where χ_T is the isothermal compressibility.

Equation (2.37) shows that density fluctuations are stipulated by heat mobility of atoms with energy kT, but limited by volumetric rigidity $(\chi_T)^{-1}$.

Based on the data from literature for ρ and χ_T, assessment of $\langle \Delta \rho / \rho \rangle^2 (\infty)$ at T_{ll} or T_{ll}' temperatures (T_{ll}' is the analogue of T_{ll} for amorphous-crystalline polymers associated with T_m similar to equation (2.26)) has shown that the value mentioned is approximately constant at these temperatures. This observation suggests that as some critical value of $\langle \Delta \rho / \rho \rangle^2$ is reached, formation of the local order zones, i.e. thermofluctuational cluster network of macromolecular entanglements, is impossible due to high heat mobility of macromolecules. Vice versa, at $\langle \Delta\rho/\rho \rangle^2$ values below the critical ones the local order zones (clusters) in the polymer melt are formed, which according to the current interpretation are identified as cross-link points of macromolecular entanglement network, i.e. a state is formed, which was defined by Boyer as a "liquid with fixed structure" [56].

Basing on the above-discussed arguments, critical temperatures, T_{cr}, at which critical density fluctuations are reached, were calculated. For the latter, $\langle \Delta\rho/\rho \rangle^2$ value for PS at 433 K was accepted. Calculation results are shown in Table 2.3, from which it follows that T_{cr} and T_{ll} (T_{ll}') for a series of amorphous and amorphous-crystalline polymers are very close: maximal deviation of these temperatures does not exceed 6% [84, 85]. Note that in the case under consideration we are dealing with dynamic (short-living) local order, which is "frozen" below T_g (T_m) [9].

Table 2.3

Calculated temperatures T_{ll} (T_{ll}'), T_{cr} and their relative deviation Δ [85]

Polymer	T_{cr}, K	T_{ll} (T_{ll}'), K	Δ, %
Poly(methyl methacrylate)	398	415	4.1
Polystyrene	462	433	6.2
Polycarbonate	488	502	2.8
Low density polyethylene	488	480	1.7
High density polyethylene	496	480	3.3
Polypropylene	510	528	3.4
Polyamide-6	635	600	5.8

Sanditov and Bartenev [75] have shown that $\langle \Delta\rho/\rho \rangle^2$ value is associated with the relative random volume by the relation as follows:

$$\langle \Delta\rho/\rho \rangle^2 = f_f(V_h/V_{at}), \qquad (2.38)$$

where V_h is the microcavity volume of the random free volume; V_{at} is the atomic volume.

The value of $\langle \Delta\rho/\rho \rangle^2$ can also be determined using the Poisson coefficient, ν [75]:

$$\langle\Delta\rho/\rho\rangle^2 \approx \frac{(1-2v)^3}{6(1+v)^2}. \qquad (2.39)$$

Figures 2.22 and 2.23 show temperature dependencies of $\langle\Delta\rho/\rho\rangle^2$ for high and low density polyethylene (HDPE and LDPE, respectively), Calculated by equations (2.38) and (2.39). Despite different absolute values of density fluctuations, calculated by the above-mentioned equations, shapes of their temperature dependencies correlate with the modern data [9, 82]. Typically, similar to Figure 2.12, dependencies $\langle\Delta\rho/\rho\rangle^2(T)$ do again display bendings at ~333 K. Higher $\langle\Delta\rho/\rho\rangle^2$ values for LDPE are induced by higher V_h value for this polymer [87].

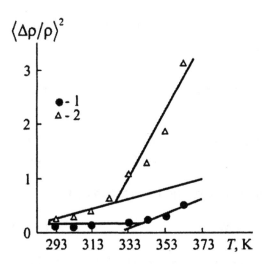

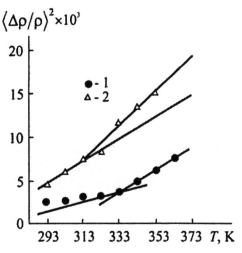

Figure 2.22. Temperature dependencies of density fluctuation $\langle\Delta\rho/\rho\rangle^2$, calculated by equation (2.38), for HDPE (1) and LDPE (2) [86]

Figure 2.23. Temperature dependencies of density fluctuation $\langle\Delta\rho/\rho\rangle^2$, calculated by equation (2.39), for HDPE (1) and LDPE (2) [86]

Data in Figure 2.24 prove existence of inversely proportional correlation between order (v_{cl}) and disorder $\left(\langle\Delta\rho/\rho\rangle^2\right)$ indices for $\langle\Delta\rho/\rho\rangle^2$ values calculated by the equation (2.38). Correlation $\langle\Delta\rho/\rho\rangle^2\left(v_{cl}^{-1}\right)$ is one more reason for the benefit of the thermofluctuational origin of local order zones (clusters) in the polymer amorphous state [86].

Thermofluctuational origin of entanglement cluster network allows theoretical assessment of temperature variation of its density v_{cl}. For this purpose, the authors of ref. [88] have used the model from [89], in which the following expression for relaxation time, τ_r, determining the speed of macromolecule fracture motion, N_s segments long, on the test temperature in the range of $T_s < T < T_m$ was used:

$$\tau_r = \tau(T)F(N_s), \qquad (2.40)$$

where $\tau(T)$ dependence is described by the Frenkel-Eiring-Arrhenius formula [89]:

$$\tau(T) = \tau_0 \exp\left(\frac{\Delta U}{RT}\right), \qquad (2.41)$$

where τ_0 is the constant; ΔU is the potential barrier of segment transition from one quasi-equilibrium state to another; the activation energy of viscous flow, U_{fl}, is suitable for estimating this value [89].

For function $F(N_s)$ the following scaling relation was accepted [89]:

$$F(N_s) = N_s. \qquad (2.42)$$

Following the general way of deducing the model [89], it has been suggested [88] that at all temperatures the size of macromolecule segment between cross-link points of macromolecular entanglement network, N_{ent} (expressed by the number of statistical segments), equals N_s, at which $\tau_r(T, N_s)$ becomes equal some value, τ_{cl}. The latter parameter can be interpreted as time required for packing a segment of the chain possessing a definite axial collinearity with the ones already precipitated to the cross-link point (by analogy with crystallization), to an entanglement cross-link point. Clearly in this case, the cross-link point of thermofluctuational network of macromolecular entanglements is considered as a closely packed cluster of macromolecule segments with approximately parallel disposition. Solving the following equation for N_{ent}:

$$\tau_{cl} = \tau_r(T, N_{ent}),\qquad(2.43)$$

we obtain that

$$N_{ent} = \left(\frac{\tau_{cl}}{\tau_0}\right)^{1/x} \exp\left(-\frac{\Delta U}{RTx}\right).\qquad(2.44)$$

This dependence reflects the fact that at temperature decrease the mobility type becomes lower scale; in this case, N_{ent} can be considered the effective parameter characterizing the chain mobility dormancy or probability of its fluctuations [88].

Because at present the relation τ_{cl}/τ_0 can be hardly subject to reasonable theoretical and experimental assessment, the conditions mentioned in the work [89] were followed in further calculations. Taking $N_{ent} = 1$ at $T = T_g$ and $x = 3.3$, ΔU (or U_{fl}) was assessed in accordance with the data from ref. [90]. For PC, molecular mass of a segment, M_s, equals 726 g/mol, i.e. it is suggested that statistical segment of PC includes two monomeric units. Then $M_{ent} = N_{ent}M_s$. Figure 2.25 shows the comparison of M_{ent}, calculated by the above method, and experimental M_{cl} values of the chain segment molecular mass between entanglement cross-link points for PC as temperature function, where good coincidence is indicated. Such coincidence can be accepted to be the proof of the condition $N_{ent} = 1$ validity at $T = T_g$, shown in ref. [89].

Fluctuation theory of glass transition allows assessment of the volume V_a of so-called cooperatively rearranging areas (CRA) [8, 91]:

$$V_a = kT_g \frac{\Delta\left(C_V^{-1}\right)}{\rho_g\left(\delta T\right)^2},\qquad(2.45)$$

where $\Delta\left(C_V^{-1}\right)$ is the reverse heat capacity jump at the glass transition; ρ_g is the polymer density at T_g; δT is the average temperature fluctuation per one CRA.

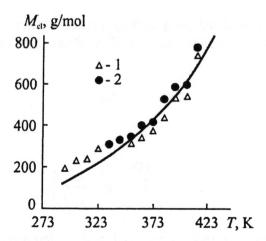

Figure 2.25. Theoretical (1) and experimental (2, 3) temperature dependencies of molecular mass M_{ent} (M_{cl}) of the chain segment between entanglement cross-link points for PC: initial (2) and annealed at 393 K (3) [88]

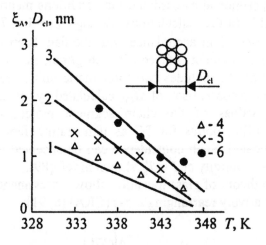

Figure 2.26. Temperature dependencies of CRA characteristic length, ξ_A, (1 – 3) and cluster diameter, D_{cl}, (4 – 6) for PMMA (1, 4), PVC (2, 5) and PS (3, 6). Insertion shows schematic image of the cluster cross-section [94]

Bittrich [91] has assessed CRA characteristic length, ξ_A, which varies in the range of 0.2 – 2.5 nm in the temperature interval of 323 – 348 K and declines with temperature increase. Since both CRA [92] and clusters [38] are postulated to be of the thermofluctuational origin, consistency of two above-mentioned models should naturally be determined by comparing temperature dependencies of ξ_A and the size of clusters for considered polymers [93, 94]. As mentioned above, the cluster model suggests decrease of the number of segments in clusters with temperature increase [38]. As the cluster is considered the analogue to crystallite with stretched chains consisting of several collinear closely packed segments, decrease of the number of segments in the cluster with temperature increase induces the cross-section decrease, which is schematically shown in Figure 2.26 (insertion), and correspondingly, the cluster diameter, D_{cl}. Temperature dependencies of ξ_A and D_{cl} as characteristic parameters of microheterogeneity of the structure in both models were compared [93, 94].

Figure 2.26 shows comparison of the mentioned dimensions, where ξ_A values are taken from the work [91] and D_{cl} calculation technique from the work [94]. It is clear that temperature dependencies of these structural characteristics display good coincidence. This gives grounds for an assumption that CRA and clusters might be identical structural elements and proves thermofluctuational origin of clusters [93, 94].

To conclude this Section, let us discuss temperature stability of the cluster structure. A simple empirical approximation connecting ν_{cl} to the Poisson coefficient, ν, was deduced as follows [95]:

$$\nu_{cl} \approx 2.38 \times 10^{27}(1 + \nu)^2, \, \text{m}^{-3}. \tag{2.46}$$

In accordance with the Le Chatelier-Brown principle (the principle of least constraint) ν value determining stability of a solid falls within the range as follows [39]:

$$0 \leq \nu \leq 0.5. \tag{2.47}$$

Combination of equations (2.46) and (2.47) gives the conditions of cluster structure stability for the polymer glassy state [95]:

$$2.38 \times 10^{27} \leq \nu_{cl} \leq 5.34 \times 10^{27}, \, \text{m}^{-3}. \tag{2.48}$$

Figure 2.27 shows temperature dependence of ν_{cl} for polyhydroxyester (PHE), where horizontal dashed line marks the lower border $\left(\nu_{cl}^{min}\right)$ of the cluster structure stability. Good concordance of v_{cl}^{min} and v_{cl} values is observed, at which accelerated thermofluctuational decay of clusters is initiated. Similar dependencies were obtained for PC and polysulfone (PSF) [95].

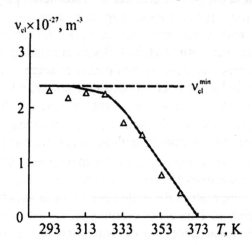

Figure 2.27. Temperature dependence of entanglement cluster network frequency, v_{cl}, for polyhydroxyester; horizontal dashed line marks the lower border $\left(v_{cl}^{min}\right)$ of the cluster stability [95]

As follows from the equation (2.46), v values become negative at $v_{cl} < v_{cl}^{min}$ (e.g. $T \geq T_g'$). The classic theory of elasticity assumes negative v values limiting the range of their variation as follows [39]:

$$- 1 < v \leq 0.5. \tag{2.49}$$

It is absolutely obvious that according to the equation (2.46) at v approaching -1, v_{cl} approaches zero. Negative values of v observed for some substances near phase transitions testify about a change of their volumes. Making definite comparisons, note that the glass transition is also accompanied by discrete change of the polymer volume [75].

Maximum possible value of v_{cl}, permitted by molecular structure of the polymer $\left(v_{cl}^{str}\right)$, can be determined by the following procedure. The length of macromolecules, L, in the specific volume of polymer is determined from the equation (2.7), and the segment length in cluster, taken equal to statistical segment length, l_{st}, according to the equation (2.8). Thus v_{cl}^{str} equals [95]:

$$v_{cl}^{str} = \frac{L}{l_{st}} = (SC_\infty l_0)^{-1}. \qquad (2.50)$$

For PHE, PC and PSF, v_{cl}^{str} values estimated by the above method, approximately, equal 8.45×10^{27}, 10.78×10^{27} and 11.0×10^{27} m^{-3}, respectively. According to (2.48), these values are much higher than the upper border of v_{cl}. This means that the condition of the cluster structure stability does not allow complete packing of polymer chain segments in the local order zone (cluster). Other criteria of this effect are discussed below.

The results obtained [95] allow an assumption of the following variation of macromolecular entanglement cluster network frequency with temperature (Figure 2.27). As the lower border of the clusters' stability $\left(v_{cl}^{min}\right)$ is reached, their accelerated decay is initiated, which leads to much more abrupt decrease of v_{cl}. It is suggested [38] that a definite part of clusters with low functionality, F (e.g. with small number of segments in the cluster), is present in a packless matrix. These clusters preserve the matrix in the glassy state. This supposition is confirmed by experiments of polyarylate annealing, in which increase of v_{cl} and decrease of F were observed simultaneously. When T_g' (or v_{cl}^{min}) is reached, clusters with low F (thus thermodynamically unstable ones) decay completely, and the packless matrix devitrifies promoting an abrupt F decrease of stable clusters due to high molecular mobility of the environs. Abrupt decrease of v_{cl} is finished by complete thermofluctuational decay of the clusters "frozen" in the glassy state at $T = T_g$ [38].

2.5. FUNCTIONALITY OF CLUSTERS AND ASSESSMENT METHODS

As mentioned above, clusters consist of several collinear closely packed segments of different macromolecules and are considered as multifunctional cross-link points of macromolecular entanglement random network [38]. By functionality, F, of such cross-link point the number of chains in it is usually meant [8]. Thus full characterization of the polymer amorphous state requires obtaining of two parameters: macromolecular entanglement cluster network frequency, v_{cl}, and the number of segments, n_{seg}, in a single cluster (local characteristic). Because cluster represents an amorphous analogue of crystallite with extended chains, $n_{seg} = F/2$. Several possible techniques for F (or n_{seg}) value assessment were considered [96].

The first of possible methods of n_{seg} assessment is of a semi-quantitative type. It is supposed [9] that the volume of local order zones in polymers equals about 10 nm^3. Dividing this value by the statistical segment volume we obtain the following values: $n_{seg} \approx 88$ for PC and $n_{seg} \approx 40$ for HDPE. These values display the proper order (several ten segments in cluster) and can be the upper assessment border.

One more assessment method for F involves equation (2.5), suggested by Graessley [10].

Temperature dependencies of the cluster functionality F, calculated for HDPE and PC by the Graessley equation, are shown in Figures 2.28 and 2.29, respectively. As indicated by graphs in these Figures, the number of segments in a single cluster varies from 7 to 12 units for HDPE and from 3 to 18 for PC. As should be expected [7], F value is decreased with temperature increase.

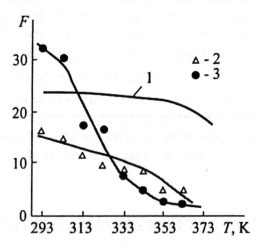

Figure 2.28. Temperature dependencies of the cluster functionality, F, for HDPE calculated as follows: 1 – by equation (2.5); 2 – by equation (2.51); 3 – by equation (2.52) [96]

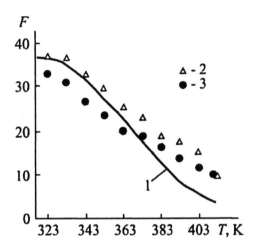

Figure 2.29. Temperature dependencies of the cluster functionality, F, for PC calculated as follows: 1 – by equation (2.5); 2 – by equation (2.51); 3 – by equation (2.52) [96]

There are two more approximate methods of F assessment. Several conditions, well approved at present, must be observed for application of any model of macromolecular entanglement network (cluster or "backlash" [18]). One of the conditions is fulfillment of the Gaussian statistics of polymer chains. Fulfillment of this condition requires the presence of, at least, 20 repeating units between cross-link points of the entanglement network [79]. Table 2.4 shows molecular masses of the chain segments between entanglement cross-link points in the "backlash" network alternative for HDPE and PC, accepted from references [17, 18, 30, 68, 97 – 99] or calculated in accordance with them. Though different authors have presented significantly M_{ent} values, on the whole, it is observed that the Gaussian statistics condition for chains is fulfilled. For multifunctional cross-link points, Flory [8] has displayed a relation between M_{ent} detected during experiment and effective molecular masses, M_{ent}^{eff}, of the mentioned chain fracture, expressed by equation (2.4). For entanglement cluster network M_{cl} is estimated in accordance with the equation (2.2), and M_{ent} is taken for M_{cl}^{eff}. Taking into account that the cross-link point of "backlash" network is of the four-functional type, it is finally obtained [96] that

$$F = \frac{4M_{eng}}{M_{cl}}.$$ (2.51)

Table 2.4

Molecular masses M_{ent} for chain segments between cross-link points of the "backlash" network for HDPE and PC in accordance with data from ref. [96]

Polymer	M_{ent}, g/mol			
HDPE	1,900 [17]	2,936 [68]	950÷2,140 [98]	737 [30]
PC	1,780 [18]	2,426 [99]	2,453 [98]	3,214 [97]

Figures 2.28 and 2.29 also show F values, estimated by the equation (2.51), and indicate their good coincidence with the values, calculated by the Greassley equation for PC and a bit poorer one for HDPE. Nevertheless, F magnitude and temperature dependence run conform in both cases.

It is common knowledge [8] that macromolecular coils in θ-solvent and in condensed state of polymers possess equal shapes, which is expressed by approximately equal mean-square distances between ends of the macromolecule. Boyer [56] has schematically indicated that this circumstance does not inhibit formation of local order zones (refer to Figure 1.1a). More stricter this conclusion is proved in the work [96]. Obviously this experimentally proved condition should also be true for the cluster model. For assessment of n_{seg} in polymer clusters, Forsman [79] has suggested the following equation:

$$n_{seg}^{1/2} = \frac{\theta_2}{4A}\left\{1 - \left[1 - \frac{16A[1 - \ln(c\rho\varphi_{cl})]}{\theta_2^2}\right]^{1/2}\right\},$$ (2.52)

where θ_2 is a dimensionless (normalized by kT) value characterizing energy of the segments' interaction, the technique of determination of which is given in the next Chapter; c is the polymer concentration; ρ is the polymer density.

The value A describes variation of the macromolecule size before and after cluster formation and is determined from the following equation [79]:

$$A = \frac{\alpha^4 - \alpha^2 + 1}{\alpha^2},$$ (2.53)

where parameter α is given as follows [79]:

$$\alpha = \frac{\left\langle r^2 \right\rangle}{\left\langle r_0^2 \right\rangle}.$$ (2.54)

Here r_0 and r are the mean-square distances between the ends of the macromolecule before and after cluster formation, respectively.

If the distance between ends of the macromolecule before and after cluster formation is the same ($r = r_0$), $\alpha = 1$ and assessment results by the equation (2.52) are also shown in Figures 2.28 and 2.29. Similar to the previous case, despite a definite quantitative inconsistency for HDPE, they give proper magnitude of F and expected shape of the temperature dependence.

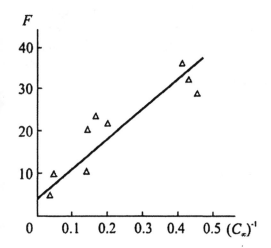

Figure 2.30. The relation between cluster functionality, F, and characteristic relation, C_∞, for 9 amorphous and amorphous-crystalline polymers [96]

Strictly saying, F value is not the exponent of the local regulation of the polymer structure, because clusters are formed from segments of different macromolecules, it can be the exponent of mutual penetration of macromolecular coils. As shown in refs. [19, 97], the same role can be played by the characteristic relation C_∞. If this assumption is correct, a definite correlation between F and C_∞ must be observed. Data from the Figure 2.30 show that such correlation is really observed for 9 amorphous and amorphous-crystalline polymers (F value is calculated for $T = 293$ K) [96].

As a consequence, the number of segments in the cluster with respect to characteristics of the polymer and temperature can vary from several ten segments. It is suggested that the most accurate assessment of F is given by the Greassley equation (equation 2.5). This conclusion follows from the fact, for example, that it describes decrease of v_{cl} at the glass transition temperature of packless matrix, T_g' (Figures 2.28 and 2.29). For equations (2.51) and (2.52), equal accuracy should not be expected, because the data of the former are highly scattered, and the latter represents an approximation.

These conclusions were proved [100] on the example of two series of cross-linked epoxidiane resin ED-20, hardened by 3,3'-dichloro-4,4'-diaminodiphenylmethane (EP-1 composite) and isomethyltetrahydrophthalic anhydride (EP-2) with variable curing agent : oligomer ratio. Figure 2.31 shows the comparison of clusters' functionalities, F_1 and F_2, calculated by equations (2.4) and (2.52), respectively, for epoxy-polymers EP-1 and EP-2. Clearly a good conformity between F_1 and F_2 values is obtained, which again indicates that invariability of sub-chains' statistics in epoxy-polymers during their cross-linking does not hinder formation of the local order zones (clusters) [100].

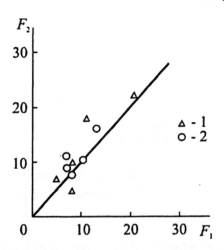

Figure 2.31. Correlation between cluster functionalities, F_1 and F_2, calculated by equations (2.4) and (2.52), respectively, for epoxy-polymers EP-1 (1) and EP-2 (2) [100]

Of special attention is much worse conformity of F values, calculated by different methods for amorphous-crystalline polymers in contrast with amorphous ones (Figures 2.28, 2.29 and 2.31). This nonconformity is explained by different mechanisms of the cluster formation for the mentioned classes of polymers. Since clusters in amorphous glassy polymers display thermofluctuation origin and their relative part, φ_{cl}, is the function of temperature [38], for non-crystalline zones of such amorphous-crystalline polymers as polyethylenes and polypropylene the situation is somewhat more difficult. As is known [48], clusters represent a "frozen" local order of the glassy state, and at temperatures above the glass transition temperature of the polymer their complete decay is observed. Nevertheless, for polyethylenes and polypropylene at about room temperature or higher, i.e. at $T > T_g$ of the amorphous phase, the presence of the local order in this phase is assumed [101 - 103]. It is postulated that in the current case, the mechanism of cluster formation is associated with the polymer chain tension during crystallization [29, 104]. The tension mentioned can be assessed with the help of parameter β [105], which is determined from the relation as follows [29]:

$$E = \frac{\rho RT}{M_{cl}} \cdot \frac{\beta^2}{5(1-K)^3}, \qquad (2.55)$$

where E is the elasticity modulus; K is the crystallinity degree.

Figure 2.32 shows the dependence of functionality F on parameter β for HDPE and PP. Clearly this dependence is approximated well by a straight line passing through the origin of coordinates, general for both polymers. As a consequence, $F(\beta)$ correlation proves the supposition made that straining of amorphous parts of the chains during crystallization is the basic mechanism of local order zones' formation in devitrified amorphous phase of amorphous-crystalline polymers [106].

To complete this Section, let us consider one more principal point associated with formation of entanglement cluster network in amorphous-crystalline polymers. The equation (2.24) is valid for polymers, in which the entire volume of the compound is involved in formation of the carcass of macromolecular entanglements. In the case, if a part of polymer only with the volumetric fracture φ_{carc} is involved in the process, the equation (2.24) should be reduced to as follows [107]:

$$2C_1 = \frac{A\rho RT\varphi_{carc}^{1/2}}{M_{cl}}. \tag{2.56}$$

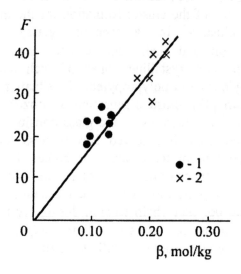

Figure 2.32. Dependence of the cluster functionality, F, on parameter β characterizing the chain tension during crystallization for HDPE (1) and PP (2) [106]

At the present time, there are two points of view on the carcass structure of macromolecular entanglements in amorphous-crystalline polymers. One of them [98] suggests that the cross-link points of macromolecular entanglement carcass in polyethylene are crystalline zones, which are lamellar crystallites. Another point of view [108] presumes that the cross-link points are concentrated in non-crystalline zones of the amorphous-crystalline polymer. Obviously the condition $\varphi_{carc} = 1.0$ (i.e. the entire polymer represents the carcass) is corresponded to the former opinion, and φ_{carc} equal the volumetric part of non-crystalline zones of the polymer, e.g. about 0.3 for HDPE and 0.5 for LDPE [83] is corresponded to the latter one.

Prior to assessment of correctness of one or another φ_{carc} value, one should understand the physical meaning of the front-factor A in equations (2.24) and (2.56). One alternative [109] provides for A variation within the range of $0.5 - 1.0$, where $A = 0.5$ is corresponded to the so-called "ghostable" or "phantom" carcass which displays the full freedom of fluctuation for cross-link points of macromolecular entanglements round their middle positions. Thus $A =$

1.0 is corresponded to an "affine" carcass, in which such freedom is completely suppressed. Another alternative [110] provides for quantitative relation between the factor A and functionality F of cross-link points of the macromolecular entanglement carcass:

$$A = 1 - \frac{2}{F}. \qquad (2.57)$$

Obviously the first alternative allows assessment of tightness of the carcass entanglement cross-link points basing on A value, calculated using experimental $2C_1$ value and the equation (2.56). The second alternative provides for qualitative assessment of functionality F [54].

Table 2.5 shows calculation results for A value under conditions as follows: $\varphi_{carc} = 0.3$ and 1.0 for HDPE and $\varphi_{carc} = 0.55$ and 1.0 for LDPE, whence it follows that use of the condition $\varphi_{carc} = 1.0$ gives $A = 0.33 - 0.42$, i.e. physically meaningless values, whereas $\varphi_{carc} = 0.30$ and $\varphi_{carc} = 0.55$ provide for A values in the expected interval of $0.5 - 1.0$. To put it differently, the calculation results indicate validity of the condition choice, in which the carcass is represented by the amorphous phase of amorphous-crystalline polymer and, consequently, there are no grounds for taking crystallites for cross-link points of the macromolecular entanglement carcass [54].

Table 2.5

Front-factor A and functionality F calculated for HDPE and LDPE [54]

Polymer	φ_{carc}	A	F
HDPE	0.30	0.75	8.0
	1.0	0.33	3.0
LDPE	0.55	0.64	5.6
	1.0	0.48	3.8

Figure 2.33 shows dependence of the front-factor A on the molecular mass M_{cl}. The graph shows that tightness of clusters' fluctuations is systematically varied from 1.0 at $M_{cl} = 0$ to 0.5 at $M_{cl} = 3,600$ g/mol. To put it differently, if the amorphous phase of amorphous-crystalline polymer represents an entire cluster (supercluster), fluctuations are completely suppressed, and its behavior corresponds to the "affine" model [107]. The value $M_{cl} = M_{ent} = 3,600$ g/mol corresponds to the melt of polyethylenes [17], where tightness applied by

crystallites is absent. Thus in this case, behavior of the entanglement carcass of polyethylenes corresponds to the "phantom" alternative [110].

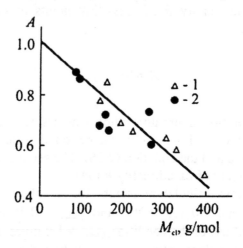

Figure 2.33. Dependence of front-factor A on molecular mass, M_{cl}, of a chain fracture between clusters for LDPE (1) and HDPE (2) [54]

Note also that in accordance with the equation (2.57) A increase is equivalent to increase of the cluster functionality F. As shown by Flory [111], the mean-square fluctuations of chain vectors, $\langle (\Delta r)^2 \rangle$, induced by joint fluctuations, is given by the following equation:

$$\langle (\Delta r)^2 \rangle = \frac{2\left(\langle r^2 \rangle_0 \right)}{F},$$ (2.58)

where $\langle r^2 \rangle_0$ is the mean-square distance between ends of the chain.

As follows from equation (2.58), fluctuations in the system must be suppressed with F increase, which is supposed by data from Figure 2.33. Note also that the condition $F = 2$ (or $n_{seg} = 1$), which determines the cluster decomposition, occurs at $\langle (\Delta r)^2 \rangle = \langle r^2 \rangle_0$, e.g. in the case of equality of mean-square fluctuations of the chain vectors and the distance between end of the chain. Essentially, this condition is the thermofluctuation criterion of the cluster formation, another alternative of which is discussed in Section 2.4 (refer to Table 2.3).

Thus results obtained in the framework of the rubber elasticity theory suggest that the carcass of macromolecular entanglements in amorphous-crystalline polymers is limited by noncrystalline zones, and the cross-link points of this carcass represent clusters consisting of 2 – 9 closely packed collinear segments [54].

2.6. THE METHOD OF ARTIFICIAL CONTROL OF THE CLUSTER STRUCTURE (ON THE EXAMPLE OF HDPE)

Introduction of low amounts of highly dispersed Fe – FeO (Z) mixture into HDPE causes significant changes in the structure and properties of this polymer [101, 103, 112, 113]. Table 2.6 illustrates these changes in properties of HDPE + Z composites on the example of blow viscosity, A_p, cracking resistance in aggressive media, τ_{50}, and gas penetration coefficient by nitrogen, P_{N_2}. Obviously so significant variations of mentioned (and some other [114]) properties suggest corresponded structural changes in the HDPE + Z composites compared with initial HDPE. Structural changes induced by introduction of Z into HDPE were studied [103] by classic methods for amorphous-crystalline polymers: differential scanning calorimetry (DSC), wide- and small-angle X-ray scattering, and IR-spectroscopy. Definite results were also obtained by mechanical tests, magnetic measurements, rheological studies, optic and electron microscopy, and gas penetration measurements.

Figure 2.34 shows dependencies of melting enthalpy, ΔH, and the melting temperature, T_m, determined by the DSC method, on Z concentration (C_Z) in HDPE + Z composites. As observed, minimal values of ΔH and T_m at C_Z = 0.05 wt.% correspond to the extreme variation of properties, shown in Table 2.6. This observation suggests decrease of the crystallite size at the mentioned concentration of Z [115]. Electron microscopy data prove this conclusion. Figure 2.35 shows electron micrographs of chemically etched initial HDPE samples and HDPE + Z composites. They visually illustrate changes in morphology of the samples studied. The initial HDPE possesses quite large lamellar crystallites with broad distribution by sizes. Moreover, the initial HDPE displays formation of larger morphological elements of the circular spherulite type. If just 0.1 wt.% of Z is injected to HDPE, circular structures disappear, and with further increase of Z concentration in the composites the size of crystallites is decreased and reaches its minimum in the HDPE + Z composite containing Z in concentration of 0.05 wt.% [116].

Table 2.6

Blow viscosity (A_p), cracking resistance (τ_{50}) and gas penetration coefficient $\left(P_{N_2}\right)$ for HDPE + Z composites [103]

Concentration of Z in HDPE + Z composites, wt.%	A_p, kJ/m²	τ_{50}, hour	$P_{N_2} \times 10^{-17}$, (mol/m)/(m²·s·Pa)
0	12.0	10	2.70
0.01	17.3	36	-
0.05	37.4	250	0.16
0.10	12.0	38	-
0.50	13.0	-	-
1.0	19.5	39	1.70

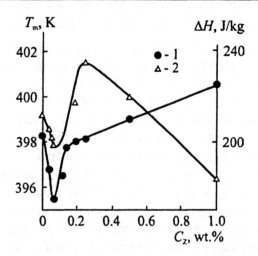

Figure 2.34. Dependencies of the melt enthalpy ΔH (1) and the melting temperature T_m (2) on Z concentration (C_Z) in HDPE + Z composite [103]

Since properties of melts of HDPE and its composite with Z are interrelated with the material morphology formed during its crystallization, the melt flow indices (MFI) of the mentioned polymers were measured. Dependence of the flow index of melt of Z concentration is shown in Figure 2.36: an extreme change of the MFI function on C_Z for HDPE + Z composite containing 0.05 wt.% of Z is observed. As shown in the works [101, 102], this increase of melt viscosity for HDPE + Z composite compared with initial HDPE should be related to the account of increase of entanglement cluster network frequency, v_{cl}, induced by Z introduction. In accordance with currently existing concepts of crystallization of polyethylenes [117], the increase of the number of cross-link points of macromolecular entanglements, rejected from crystallizing zones during crystallization, induces a break in the process continuity and appropriate decrease of the size of crystallites.

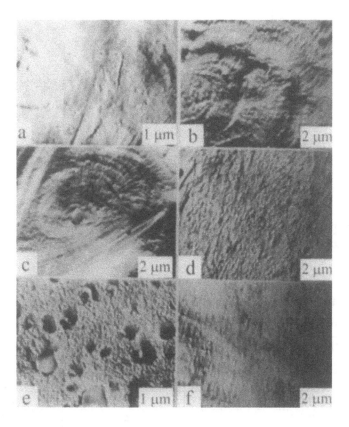

Figure 2.35. Electron micrographs of chemically treated surfaces of the initial HDPE (a – c) and HDPE + Z composites with Z concentration as follows: 0.01 (d), 0.05 (e) and 0.1 (f) wt.% [116]

One more structural feature of HDPE + Z composite containing 0.05 wt.% of Z is detected by small-angle X-ray scattering method. Study of the dependence of the part of oriented crystallites L_{or} on C_Z indicated its broad variation with the latter (Table 2.7). Clearly this variation is of the extreme type also. For initial HDPE, $L_{or} \cong 25\%$, and for $C_z = 0.05$ wt.% $L_{or} \cong 44\%$. It is assumed that L_{or} variation is induced by the melt viscosity increase (Figure 2.36), which determines an increase of shift stresses during polymer injection molding and orientation of crystallites [103].

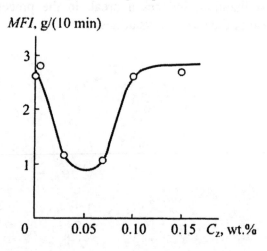

Figure 2.36. Dependence of the melt flow index *(MFI)* on Z concentration (C_Z) for HDPE + Z composites [102]

Table 2.7

The fracture of oriented crystallites (L_{or}) and normalized magnetic susceptibility $\left(\chi_m^n\right)$ for HDPE + Z composites [103]

Concentration of Z in HDPE + Z composites, wt.%	L_{or}, %	χ_m^n, Gs/g
0	25	-
0.01	27	0.30
0.05	44	0.25
0.10	36	-
0.15	25	-
0.20	24	0.18
0.50	27	-
1.0	25	0.14
12.0	-	0.17

Analysis of small-angle X-ray scattering patterns for the samples with different concentration of Z (Figure 2.37) indicates the broadest scattering in the samples with high concentration of oriented crystallites. For HDPE + Z composites with Z concentration of 0.05 wt.% the intensity of small-angle X-ray scattering reaches its maximum, and the pattern is maximum symmetrical and possesses the greatest half-width. All these observations indicate that samples of the current composite possess the most proper (regular) morphology with the minimum dispersion by sizes of both long period and crystalline and amorphous zones. These observations are in keeping with electron microscopy data (Figure 2.35). Such homogeneous structure results the oriented crystallization in HDPE + Z composite samples containing Z in concentration of 0.05 wt.%.

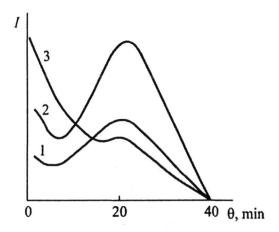

Figure 2.37. Small-angle X-ray scattering patterns recorded along the molding direction for initial HDPE (1) and HDPE + Z composite with Z concentration of 0.05 wt.% (2) and 1.0 wt.% (3) [101]

Completing the discussion on changes in crystalline morphology of HDPE + Z composites, one shall note that DSC, IR-spectroscopy, X-ray scattering and density measurement data indicated the absence of significant changes in the crystallinity degree, K, though its absolute values determined by DSC method were found minimal in the mentioned series of measurements [101].

To clear out differences in the structure of non-crystalline zones in HDPE + Z composites comparing with primary HDPE, it is suitable to search the relation of intensities (or extinction coefficients) of IR absorption bands at

1,303, 1,352 and 1,368 cm^{-1} using results obtained in the works [119, 120]. Figure 2.38 shows the change of extinction coefficient relations: $e_{1,303}/e_{1,352}$, $e_{1,303}/e_{1,368}$ and $e_{1,352}/e_{1,368}$, with respect to HDPE + Z composite density, ρ. This Figure also shows similar data from the work [119] for series of linear polyethylenes. Clearly good correspondence between these two series of data is observed, though for HDPE + Z composites the decrease of $e_{1,303}/e_{1,352}$ and $e_{1,303}/e_{1,368}$ relations with C_Z increase is much clearer [101, 121].

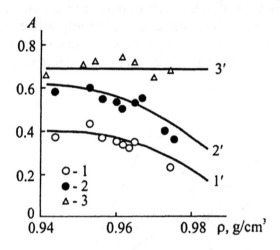

Figure 2.38. Dependencies of extinction coefficients' relations for IR-absorption bands, *A*, on density, ρ, for samples of HDPE + Z composites: 1, 1' - $e_{1,303}/e_{1,352}$; 2, 2' - $e_{1,303}/e_{1,368}$; 3, 3' - $e_{1,352}/e_{1,368}$; 1 - 3 – data from work [118], 1' - 3' - data from work [119]

Relations $e_{1,303}/e_{1,352}$ and $e_{1,303}/e_{1,368}$ were also estimated for paraffin melts [119]. They were found equal to the appropriate relations for polyethylene with the density below 0.960 g/cm^3. This induced a conclusion that chain units in gauche-conformations of these samples (possessing $\rho \leq 0.960$ g/cm^3) display the same structure (distribution of sequences) as in the primary melt at the same temperature. Decrease of the considered relations with ρ increase for $\rho > 0.960$ g/cm^3 was assigned [119] to higher regularity of the distribution of units in gauche-conformations. Essentially, analogous explanation of dependencies from Figure 2.38 is given in the work [120]. It is supposed that increase of the crystallinity degree provides for more compact packing of crystallites and, as a consequence, limitation of the selection of possible conformations, which can be realized in non-crystalline zones. For example, as the distance between crystallites decreases, chains in non-crystalline zones have to orient more parallel to crystallite surfaces (the so-called barrier effect).

Thus the above shown results of experimental studies allow the conclusions as follows. Introduction of low amounts of Z (up to 0.10 wt.%) into amorphous-crystalline HDPE induces significant structural changes which imply corresponded variations of physicomechanical properties of HDPE + Z composites. Mechanical and rheological test results induce an assumption that the starting point of the structural changes is increase of macromolecular entanglement cluster network frequency, v_{cl}, in the HDPE melt. This phenomenon predetermined occurrence of two important structural effects. First, a significant increase of the melt viscosity leads to consequent increase of shear stresses during processing of HDPE + Z composites and oriented crystallization. In turn, this induces morphology of HDPE + Z composite with 0.05 wt.% concentration of Z representing a selection of lamellar crystallites with low dispersion by sizes. Secondly, variation of frequency v_{cl} leads to a change of lamellar crystallite sizes and thickness of non-crystalline interlayers between them. This is accompanied by the change of conformational state of chain sequences in non-crystalline zones. The second effect is the basic one, because qualitatively identical change of properties was also observed for HDPE + Z composite samples, obtained by press molding.

It may be assumed that the most probable origin of forces stipulating entanglement network frequency increase at Z injection to HDPE is either adhesion by short-range van der Waals forces or long-range magnetic interaction between Z and the polymer matrix. Since particles of injected modifier display ferromagnetic origin and possess very high specific surface, both possibilities are equally permissible. However, injection of disperse magnetic powder possessing high coercive force and low specific surface into HDPE has indicated M_{cl} decrease from (771 ± 45) g/mol for initial HDPE to (688 ± 27) g/mol for the composite. Statistically in the latter case, a noticeable decrease of M_{cl} bespeaks for the magnetic origin of forces, which increase the entanglement cluster network frequency [122].

Another argument for the benefit of this assumption is comparison of action of identical amounts of Z and highly dispersed copper, injected as before, on PS and HDPE. In contrast with Z, the effect of which is described above, injection of dispersed copper does not practically change physicomechanical properties of the material. Both modifiers (Z and Cu) possess large specific surface, but contrary to Z, Cu is diamagnetic substance.

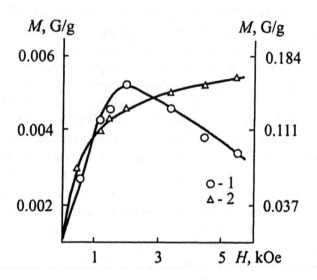

Figure 2.39. Dependencies of general magnetic moment M on external magnetic field intensity H for HDPE + Z composites containing 1 - 0.01 wt.% and 2 - 0.20 wt.% of Z

For further proofing of this assumption two series of experiments were performed [103, 122]. The first series represented measurements of general magnetic moment, M, as the function of external magnetic field intensity, H, for a series of HDPE + Z composites. The measurements were performed on magnetic-field meter MAGNET B-E15 "Bruker" at 295 K [122]. Figure 2.39 show magnetization characteristics (general magnetic moment M as the function of H) for composites containing 0.01 and 0.20 wt.% of Z. These curves allow estimation of general magnetic susceptibility, χ_m, of the samples as follows [123]:

$$\chi_m = \frac{M}{H}. \tag{2.59}$$

Table 2.7 shows χ_m values normalized for Z concentration in HDPE + Z composites, χ_m^n. These results, as well as the data from Figure 2.39 provide for some conclusions. For example, HDPE + Z composite containing 0.05 wt.% of Z displays $M(H)$ dependence analogous by shape to curve 1 in Figure 2.39 and composites containing 1.0 and 12.0 wt.% of Z to curve 2. After the saturation point, dependencies $M(H)$ for HDPE + Z composites containing 0.01 and 0.05 wt.% of Z are characterized by negative gradient that assumes a noticeable increase of diamagnetic contribution (by the absolute value) to general magnetic moment for these composites compared with others [123]. As is known [124], one of the reasons for such increase can be growth of the number of conductivity electrons or, to put it differently, relative increase of the part of free iron in Fe/FeO mixture, injected to the polymeric matrix. The data shown indicate the following dynamics of χ_m^n change with Z concentration increase for HDPE + Z composites: it is significantly increased first and then (at $C_Z \geq$ 0.20 wt.%) reaches asymptotic values [122].

Figure 2.40. Scanning electron photomicrography of the degradation surface of HDPE + Z composite with 0.05 wt.% concentration of Z [125]

Figure 2.40 shows scanning electron photomicrograph illustrating the appearance of Z particles. Figure 2.41 shows the dependence of the average size of these particles, R_{av}, on C_Z value. Anomalous behavior of composites containing 0.01 and 0.05 wt.%, for which R_{av} are much lower than for the rest composites under consideration, is again observed. This suggests higher relative content of Fe particles in the mentioned composites, because they are smaller than FeO particles [122].

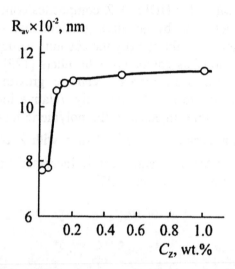

Figure 2.41. Dependence of the average size, R_{av}, of particles Z on C_Z in HDPE + Z composites [122]

The above stated results give an opportunity to suggest the following interpretation of experimentally observed extreme value of physicomechanical properties of HDPE + Z composites, which is associated with ν_{cl} increase, analogous to the one suggested for single-phase metallopolymers [126]. As Z is injected into HDPE using the technique described [127], iron is reduced from oxides by the known formula [101, 122] as follows:

$$\text{Fe}^{+2} \underset{\text{Red}}{\overset{O_2}{\rightleftarrows}} \text{Fe}^{+3}$$

(2.60)

where Red is the reducer, which can be HDPE oxidation products with hydroxyl and carbonyl groups, as well as mobile hydrogen. This assumption is proved by increase of diamagnetic contribution, decrease of Z particle size and higher χ_m^n values for composites with 0.01 and 0.05 wt.% concentration of Z. A part of Z fraction, due to high dispersion of Z particle size (refer to Figure 2.40), may fall within the range of sizes determining occurrence of super-paramagnetic domains [128] and, actually, the smaller is the average size of particles R_{av}, the greater part of them can become super-paramagnetic domains. The technique of metal injection to polymers used [126] suggests much smaller sizes of their particles and the absence of metal oxides. Nevertheless, general tendency in behavior of one-phase metallopolymers and HDPE + Z composites does really exist. In particular, global property of them is a significant decrease of the melt flow index [103].

Thus the following situation with formation of extreme value of properties for HDPE + Z composites can be suggested. In the case of low amounts of Z introduced into HDPE ($C_Z \leq 0.10$ wt.%), intensive reduction of Fe from its oxides proceeds in accordance with the reaction (2.60) and relatively large part of small Z particles (super-paramagnetic domains) is formed. These particles increase frequency of macromolecular entanglement network. Below $C_Z = 0.10$ wt.%, v_{cl} is increased due to simple increase of Z amount. At $C_Z = 0.10$ wt.% reducing abilities of the polymer matrix are exhausted and dispersion composition of Z with dominating content of larger particles with greater part of FeO is practically preserved. Though possessing high coercive force, these particles are incapable of forming super-paramagnetic domains due to purely dimensional restrictions. That is why they do not modify the network of macromolecular entanglements and, consequently, do not affect properties of HDPE + Z composites [103].

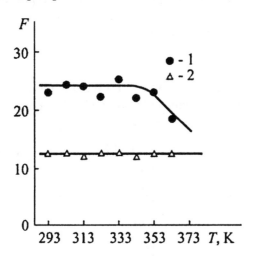

Figure 2.42. Temperature dependencies of cluster functionality F in the amorphous phase for HDPE (1) and HDPE + Z composite containing 0.05 wt.% of Z (2) [131]

Moreover, the above-described results also suggest that in non-crystalline zones of HDPE particles Z initiate entanglements of the type shown in Figure 1.2b (clusters), because magnetic field affecting polyethylene melt promotes ordering [129]. Thus it can be expected that local magnetic fields of super-paramagnetic particles will promote formation of the local order zones.

From these positions, τ_{50} increase and P_{N_2} decrease for HDPE + Z composites (Table 2.6) are simply explained, because increase of the amount of macromolecular backlashes must induce the opposite effect due to structure pack-off [130].

Figure 2.42 shows temperature dependencies of the cross-link point functionality, F, of macromolecular entanglement network for HDPE and HDPE + Z composite containing 0.05 wt.% of Z. The functionality was calculated by the equation (2.5). For these polymers $F \geq 11$, whereas for macromolecular backlashes $F = 4$ [18] which also testifies for the benefit of the cluster formation in HDPE non-crystalline zones [131]. Note also that for initial HDPE, F declines with temperature, whereas for HDPE + Z composite it remains practically constant. This observation allows a supposition that magnetic field of a super-paramagnetic domain is more effective for hampering variation of the number of segments in the cluster with temperature than tension of the chain amorphous areas. To put it differently, despite the smaller number of segments, clusters in HDPE + Z composite with 0.05 wt.% of Z are more stable than the ones in the initial HDPE [131].

Table 2.8

Comparison of parameters of macromolecular entanglement network obtained by different methods for PBTP + Z composites [134]

Z concentration, wt.%	M_{cl}^{m}, g/mol; equation (2.62)	$M_{cl}^{s.p.}$, g/mol; equation (2.2)
0	860	860
0.01	880	880
0.05	740	30
0.10	770	300
0.50	820	640
1.0	860	1,190

As is known [132], polymer melt viscosity is controlled by two parameters: molecular mass of the polymer, *MM*, and molecular mass of the chain segment between entanglements. General dependence of viscosity η of polymeric melt on the factors mentioned is of the following form [132]:

$$\eta \sim \frac{MM^2}{M_{cl}^3}. \qquad (2.61)$$

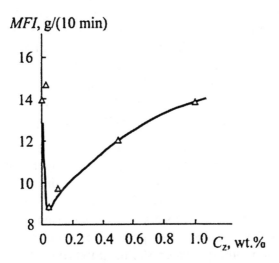

Figure 2.43. Dependence of the melt flow index (MFI) on Z concentration (C_Z) for PBTP + Z composite [134]

Correctness of the relation (2.61) was proved in the work [133], where for η the reverse melt flow index (MFI) was accepted for HDPE + Z composites. Figure 2.43 shows MFI dependence on Z concentration, C_Z, for poly(butylene terephthalate) composites with Z (PBTP + Z). Similar for HDPE + Z composites (Figure 2.36), the case under consideration displays a minimum of MFI (or a maximum of η) in the Z concentration range of $C_Z = 0.05 \div 0.10$ wt.%. Because there are no grounds for assuming changes of molecular mass of polymer at Z injection into PBTP [133], the relation (2.61) can be reduced to the form as follows [134]:

$$MFI = C_{ent} M_{cl}^r. \qquad (2.62)$$

Constant C_{ent} in this equation can be determined using parameters of the initial PBTP ($MFI = 14$ g/10 min, $M_{cl} = 860$ g/mol): $C_{ent} \approx 22$. Table 2.8 shows comparison of $M_{cl}^{s.p.}$ (equation (2.2)) and M_{cl}^{m} (equation (2.62)) values, where index "s.p." indicates the solid phase and "m" means melt. Remind that for HDPE + Z composites this comparison has given: $M_{cl}^{s.p.} \approx M_{cl}^{m}$. However, for PBTP + Z composite at $C_Z = 0\ 05$ and 0.10 wt.% a significant (approximately, 2.5-fold) difference in the indicated parameters is observed. This means that for HDPE + Z composite the cluster network is formed already in the melt, and during crystallization its cross-link points (clusters) are rejected from crystallizing zones to non-crystalline (mainly, interphase) ones [117]. For initial PBTP and PBTP + Z composites containing 0.01, 0.5 and 1.0 wt.% of Z (i.e. in cases, for which super-paramagnetic domains of Z are absent) the process of cluster network formation is fully analogous to the above-described one for HDPE + Z. Thus the condition $M_{cl}^{s.p.} = M_{cl}^{m}$ (Table 2.8) is met for them. For PBTP + Z composites containing 0.05 and 0.10 wt.% of Z, a cluster network with cross-link points (clusters) formed by molecular interaction forces is added to the one formed by interaction between polymer chains and Z in the melt at $T < T_g$. The type of the former network formation is, probably, caused by additional chain straining by clusters and suppression of fluctuations. The relation of frequencies of these two types of the cluster network is 1:2.2. This difference is typical of amorphous-crystalline polymers with devitrified (HDPE) and glassy-like (PBTP) amorphous phase [134].

Thus the above results induce the following conclusions:
1) cross-link points of macromolecular entanglement network in non-crystalline zones of amorphous-crystalline polymers represent the local order zones (clusters);
2) variation of entanglement cluster network frequency causes a strong effect on structures and properties of this class of polymers;
3) there is a possibility of purposeful control for v_{cl} and, consequently, properties of both amorphous-crystalline and amorphous polymers.

REFERENCES

1. Haward R.N., *J. Polymer Sci.: Part B: Polymer Phys.*, 1995, vol. **33**(8), pp. 1481 - 1494.
2. Haward R.N., *Polymer*, 1987, vol. **28**(8), pp. 1485 - 1488.
3. Boyce M.C., Parks D.M., and Argon A.S., *Mech. Mater.*, 1988, vol. **7**(1),

pp. 15 - 33.

4. Boyce M.C., Parks D.M., and Argon A.S., *Mech. Mater.*, 1988, vol. 7(1), pp. 35 - 47.
5. Haward R.N., *Macromolecules*, 1993, vol. 26(22), pp. 5860 - 5869.
6. Bartenev G.M. and Frenkel S.Ya., *Polymer Physics*, 1990, Leningrad, Khimia, 432 p. (Rus)
7. Belousov V.N., Kozlov G.V., Mikitaev A.K., and Lipatov Yu.S., *Doklady AN SSSR*, 1990, vol. 313(3), pp. 630 - 633. (Rus)
8. Flory P.Y., *Polymer J.*, 1985, vol. 17(1), pp. 1 - 12.
9. Bernstein V.A. and Egorov V.M., *Differential Scanning Calorimetry in Physicochemistry of Polymers*, 1990, Leningrad, Khimia, 256 p. (Rus)
10. Graessley W.W., *Macromolecules*, 1980, vol. 13(2), pp. 372 - 376.
11. Perepechko I.I. and Startsev O.V., *Vysokomol. Soedin.*, 1973, vol. 15B(5), pp. 321 - 322. (Rus)
12. Belousov V.N., Kotsev B.Kh., and Mikitaev A.K., *Doklady AN SSSR*, 1983, vol. 270(5), pp. 1145 – 1147. (Rus)
13. Arzhakov S.A., Bakeev N.F., and Kabanov V.A., *Vysokomol. Soedin.*, 1973, vol. 15A(5), pp. 1154 - 1167. (Rus)
14. Narisawa I., Polymer Material Strength, 1987, Moscow, Khimia, 400 p. (Rus)
15. Kozlov G.V., Sanditov D.S., and Serdyuk V.D., *Vysokomol. Soedin.*, 1993, vol. 35B(12), pp. 2067 - 2069. (Rus)
16. Gent A.N. and Madan S., *J. Polymer Sci.: Part B: Polymer Phys.*, 1989, vol. 27(7), pp. 1529 - 1542.
17. Graessley W.W. and Edwards S.F., *Polymer*, 1981, vol. 22(10), pp. 1329 - 1334.
18. Wu S.J., *J. Polymer Sci.: Part B: Polymer Phys.*, 1989, vol. 27(4), pp. 723 - 741.
19. Budtov V.P., *Physical Chemistry of Polymer Solutions*, 1992, Saint-Petersburg, Khimia, 384 p. (Rus)
20. Mashukov N.I., Vasnetsova O.A., Malamatov A.Kh., and Kozlov G.V., *Lakokrasochnye Materialy i Ikh Primenenie*, 1992, No. 1, pp. 16 – 17. (Rus)
21. Barinov V.Yu., *Vysokomol. Soedin.*, 1981, vol. 23B(1), pp. 66 - 68. (Rus)
22. Kambour R.P., *J. Polymer Sci.: Macromol. Rev.*, 1973, No. 7, pp. 1 - 54.
23. Kramer E.J., *J. Polymer Sci.: Polymeer Phys. Ed.*, 1975, vol. 13(2), pp. 509 - 516.
24. Lu X.-C. and Brown N., *J. Mater. Sci.*, 1986, vol. 21(11), pp. 4081 - 4092.
25. Pertsev N.A., Romanov A.E., and Vladimirov V.I., *J. Mater. Sci.*, 1981, vol. 16(8), pp. 2084 - 2093.

26. Dugdlave D., *J. Mech. Phys. Solids*, 1960, vol. **8**(2), pp. 100 - 104.
27. Kozlov G.V., Belousov V.N., and Mikitaev A.K., *Fizika i Tekhnika Vyisokikh Davleniy*, 1998, vol. **8**(1),pp. 101 – 107. (Rus)
28. Fellers J.F. and Huang D.C., *J. Appl. Polymer Sci.*, 1979, vol. **23**(8), pp. 2315 - 2326.
29. Kozlov G.V. and Sanditov D.S., *Anharmonic Effects and Physicomechanical Properties of Polymers*, 1994, Novosibirsk, Nauka, 261 p. (Rus)
30. Lin Y.-H., *Macromolecules*, 1987, vol. **20**(12), pp. 3080 - 3083.
31. Katsnelson A.A., *Introduction to Physics of Solids*, 1984, Moscow, Izd. IGU, 293 p. (Rus)
32. Shogenov V.N., Belousov V.N., Potapov V.V., Kozlov G.V., and Prut E.V., *Vysokomol. Soedin.*, 1991, vol. **33A**(1), pp. 155 – 160. (Rus)
33. Startsev O.V., Abeliov Ya.A., Kirillov V.N., and Voronkov M.G., *Doklady AN SSSR*, 1987, vol. **293**(6), pp. 1419 – 1422. (Rus)
34. Sumita M., Tsukumo Y., Miyasaka K., and Ishikawa K., *J. Mater. Sci.*, 1983, vol. **18**(5), pp. 1758 - 1764.
35. Mashukov N.I., Belousov V.N., Kozlov G.V., Ovcharenko E.N., and Gladyishev G.P., *Izv. AN SSSR, Ser. Khim.*, 1990, No. 9, pp. 2143 – 2146. (Rus)
36. Schrager M., *J. Appl. Polymer Sci.*, 1978, vol. **22**(8), pp. 2379 - 2381.
37. Malamatov A.Kh. and Kozlov G.V., *Doklady Adyigeiskoi (Cherkesskoi) Mezhdunarodhoi AN*, 1998, vol. **3**(2), pp. 78 – 81. (Rus)
38. Kozlov G.V. and Novikov V.U., *Uspekhi Fizicheskikh Nauk*, 2001, vol. **171**, pp. 717 – 764. (Rus)
39. Balankin A.S., *Synergysm of Deformable Body*, 1991, Moscow, MO SSSR, 404 p. (Rus)
40. Balankin A.S., Bugrimov A.L., Kozlov G.V., Mikitaev A.K., and Sanditov D.S., *Doklady RAN*, 1992, vol. **326**(3), pp. 463 – 466. (Rus)
41. Gladyshev G.P., *Thermodynamics and Macrokinetics of Natural Hierarchic Processes*, Moscow, Nauka, 1988, 290 p. (Rus)
42. Gladyshev G.P., *J. Theor. Biol.*, 1978, vol. **75**(4), pp. 425 - 444. (Rus)
43. Gladyshed G.P. and Gladyshev D.P., *On Physicochemical Theory of Biological Evolution* (Preprint), 1993, Moscow, Olimp, 24 p. (Rus)
44. Gladyshev G.P., *J. Biol. Systems*, 1993, vol. **1**(2), pp. 115 - 129.
45. Gladyshed G.P. and Gladyshev D.P., *Zh. Fiz. Khim.*, 1994, vol. **68**(5), pp. 790 – 792. (Rus)
46. Gladyshed G.P., *Izvestiya RAN., Ser. Biol.*, 1995, No. 1, pp. 5 – 14. (Rus)
47. Kozlov G.V. and Zaikov G.E., In Book: *Fractal Analysis of Polymers:*

From Synthesis to Composites, Ed. Kozlov G., Zaikov G., and Novikov V., New York, Nova Science Publishers, Inc., 2003, pp. 89 - 97.
48. Beloshko V.A., Kozlov G.V., and Lipatov Yu.S., *Fizika Tverdogo Tela*, 1994, vol. **36**(10), pp. 2903 – 2906. (Rus)
49. Matsuoka S. and Bair H.E., *J. Appl. Phys.*, 1977. vol. **48**(10), pp. 4058 - 4062.
50. Wu S., *J. Appl. Polymer Sci.*, 1992, vol. **46**(4), pp. 619 - 624.
51. Privalko V.P. and Lipatov Yu.S., *Vysokomol. Soedin.*, 1971, vol. **13**A(12), pp. 2733 – 2738. (Rus)
52. Gladyshev G.P., *J. Biol. Phys.*, 1994, vol. **20**(2), pp. 213 - 222.
53. Belousov V.N., Kozlov G.V., and Lipatov Yu.S., *Doklady AN SSSR*, 1991, vol. **318**(3), pp. 615 – 618. (Rus)
54. Kozlov G.V., Aloev V.Z., and Lipatov Yu.S., *Ukrainski Khimicheski Zhurnal*, 2001, vol. **67**(10), pp. 115 – 119. (Rus)
55. Mark J.E. and Sullivan J.L., *J. Chem. Phys.*, 1977, vol. **66**(3), pp. 1006 - 1011.
56. Boyer R.F., *J. Macromol. Sci.-Phys.*, 1976, vol. **B12**(2), pp. 253 - 301.
57. Boyer R.F., *Macromolecules*, 1992, vol. **25**(20), pp. 5326 - 5330.
58. White R.M. and Geballe T.H., *Long Range Order in Solids*, New York, Academic Press, 1979, 447 p.
59. Kozlov G.V., Milman L.D., and Mikitaev A.K., *Local Order in Amorphous-Crystalline Polyethylene*, Manuscript deposited to VINITI RAS, Moscow, February 26, 1997, No. 622-V97. (Rus)
60. Lobanov A.M. and Frenkel S.Ya., *Vysokomol. Soedin.*, 1980, vol. **A22**(5), pp. 1045 – 1057. (Rus)
61. Boyer R.F., *Polymer Eng. Sci.*, 1968, vol. **8**(3), pp. 161 - 185.
62. Mashukov N.I., Serdyuk V.D., Kozlov G.V., Ovcharenko E.N., Gladyshev G.P., and Vodakhov A.B., *Stabilization and Modification of Polyethylene by Oxygen Acceptors* (Preprint), 1990, Moscow, IKhF AN SSSR, 64 p. (Rus)
63. Belousov V.N., Kozlov G.V., and Mashukov N.I., *Doklady Adyigeiskoi (Cherkesskoi) Mezhdunarodnoi AN*, 1996, vol. **2**(1), pp. 74 - 82. (Rus)
64. Zhenyi M., Langford S.C., Dickinson J.T., Engelhard M.H., and Boyer D.R., *J. Mater. Res.*, 1991, vol. **6**(1), pp. 183 - 195.
65. Novikov V.U. and Kozlov G.V., *Materialovedenie*, 1999, No. 12, pp. 8 – 12. (Rus)
66. Boyer R.F. and Miller R.L., *Macromolecules*, 1977, vol. **10**(5), pp. 1167 - 1169.
67. Miller R.L. and Boyer R.F., *J. Polymer Sci.: Polymer Phys. Ed.*, 1984, vol. **22**(12), pp. 2043 - 2050.

68. Aharoni S.M., *Macromolecules*, 1985, vol. **18**(12), pp. 2624 - 2630.
69. Kozlov G.V., Beloshenko V.A., and Lipskaya V.A., *Ukrainsky Fizichesky Zhurnal*, 1996, vol. **41**(2), pp. 222 – 225. (Rus)
70. Matsuoka S., Aloisio C.J., and Bair H.E., *J. Appl. Phys.*, 1973, vol. **44**(10), pp. 4265 - 4268.
71. Sharma B.K., *Acoust. Lett.*, 1980, vol. **4**(2), pp. 19 - 23.
72. Kozlov G.V. and Zaikov G.E., *Izv. KBHTs RAN*, 2003, No. 9, pp. 132 – 140. (Rus)
73. Sharma B.K., *Acustica*, 1981, vol. **48**(2), pp. 121 - 122.
74. Barker R.E., *J. Appl. Phys.*, 1963, vol. **34**(1), pp. 107 - 116.
75. Sanditov D.S. and Bartenev G.M., *Physical Properties of Irregular Structures*, 1982, Novosibirsk, Nauka, 256 p. (Rus)
76. Sanditov D.S., In Coll.: *Nonlinear Effects in Degradation Kinetics*, 1988, Leningrad, FTI AS USSR, pp. 140 – 149. (Rus)
77. Privalko V.P. and Lipatov Yu.S., *Vysokomol. Soedin.*, 1974, vol. **16A**(7), pp. 1562 – 1568. (Rus)
78. Boyer R.F., *J. Macromol. Sci.-Phys.*, 1973, vol. **7B**(3), pp. 487 - 501.
79. Forsman W.C, *Macromolecules*, 1982, vol. **15**(6), pp. 1032 - 1040.
80. Hornboden E., *Int. Mater. Rev.*, 1989, vol. **34**(6), pp. 277 - 296.
81. Curro J.J. and Roe R.-J., *Polymer*, 1984, vol. **25**(10), pp. 1424 - 1430.
82. Rathje J. and Ruland W., *Colloid Polymer Sci.*, 1976, vol. **254**(3), pp. 358 - 370.
83. Kalinchev E.L. and Sakovtseva M.B., *Properties and Processing of Thermoplasts*, 1983, Leningrad, Khimia, 288 p. (Rus)
84. Mashukov N.I., Serdyuk V.D., Belousov V.N., Kozlov G.V., and Khatsukova M.A., *Fluctuation Network of Molecular Entanglements As The Percolation System*, Manuscript deposited to VINITI RAS, Moscow, June 20, 1994, No. 1537-B94. (Rus)
85. Belousov V.N., Kozlov G.V., and Mashukov N.I., *Doklady Adyigeiskoi (Cherkesskoi) Mezhdunarodnoi AN*, 1995, vol. **2**(2), pp. 76 – 80. (Rus)
86. Serdyuk V.D., Sanditov D.S., Mashukov N.I., Kozlov G.V., Novikov V.U., and Zaikov G.E., *Int. J. Polym. Mater.*, 1998, vol. **42**(1), pp. 65 - 73.
87. Afaunov V.V., Mashukov N.I., Kozlov G.V., and Sanditov D.S., *Izv. VUZov. Severo-Kavkazskii Region, Estestvennye Nauki*, 1999, No. 4(108), pp. 69 – 71. (Rus)
88. Mashukov N.I., Serdyuk V.D., Gladyshev G.P., and Kozlov G.V., *Voprosy Oboronnoi Tekhniki*, Ser. 15, 1991, Iss. 3, No. 97, pp. 11 – 13. (Rus)
89. Eliashevich G.K., *Vysokomol. Soedin.*, 1988, vol. **30A**(8), pp. 1700 – 1705. (Rus)

90. Kozlov G.V., Shogenov V.N., Kharaev A.M., and Mikitaev A.K., *Vysokomol. Soedin.*, 1987, vol. **29B**(4), pp. 311 – 314. (Rus)

91. Bittrich H.-J., *Acta Polymerica*, 1982, vol. **33**(12), pp. 741.

92. Adam G. and Gibbs J.H., *J. Chem. Phys.*, 1965, vol. **43**(1), pp. 139 - 146.

93. Kozlov G.V. and Lipatov Yu.S., *Vysokomol. Soedin.*, 2003, vol. **45B**, pp. 660 - 664. (Rus)

94. Kozlov G.V. and Lipatov Yu.S., In Coll.: *Perspectives of Chemical and Biochemical Physics*, Ed. Zaikov G., New York, Nova Science Publishers, Inc., 2002, p. 205-212.

95. Kozlov G.V., Belousov V.N., Serdyuk V.D., Mikitaev A.K., and Mashukov N.I., *Doklady Adyigeiskoi (Cherkesskoi) Mezhdynarodnoi AN*, 1997, vol. **2**(2), pp. 88 – 93. (Rus)

96. Kozlov G.V., Belousov V.N., Mikitaev A.K., and Mashukov N.I., *Doklady Adyigeiskoi (Cherkesskoi) Mezhdynarodnoi AN*, 1997, vol. **2**(2), pp. 94 – 98. (Rus)

97. Aharoni S.M., *Macromolecules*, 1983, vol. **16**(9), pp. 1722 - 1728.

98. Mills P.J., Hay J.N., and Haward R.N., *J. Mater. Sci.*, 1985, vol. **20**(2), pp. 501 - 507.

99. Prevorsek D.C. and De Bona B.T., *J. Macromol. Sci.-Phys.*, 1981, vol. **B19**(4), pp. 605 - 622.

100. Kozlov G.V., Novikov V.U., and Zaikov G.E., *Plasticheskie Massy*, 2002, No. 5, pp. 33 – 34. (Rus)

101. Kozlov G.V., Temiraev K.B., MalamatovA.Kh., and Shustov G.B., Izv. KBNTs RAS, 1999, No. 2, pp. 95 – 99. (Rus)

102. Mashukov N.I., Serdyuk V.D., Belousov V.N., Kozlov G.V., Ovcharenko E.N., and Gladyshev G.P., *Izv. AN SSSR, Ser. Khim.*, 1990, No. 8, pp. 1815 – 1817. (Rus)

103. Mashukov N.I., Gladyishev G.P., and Kozlov G.V., *Vysokomol. Soedin.*, 1991, vol. **33A**(12), pp. 2538 – 2546. (Rus)

104. Serdyuk V.D., Kozlov G.V., Mashukov N.I., and Mikitaev A.K., *J. Mater. Sci. Techn.*, 1997, vol. **5**(2), pp. 55 - 60.

105. Krigbaum W.R., Roe R.-J., and Smith K.J., *Polymer*, 1964, vol. **5**(3), pp. 533 - 542.

106. Kozlov G.V. and Zaikov G.E., *Izv. KBNTs RAI*, 2003, No. 9, pp. 126 – 131. (Rus)

107. Sanjuan J. and Lorence M.A., *J. Polymer Sci.: Polymer Phys. Ed.*, 1988, vol. **26**(2), pp. 235 - 244.

108. Popli R. and Mandelkern L., J. *Polymer Sci.: Part B: Polymer Phys.*, 1990, vol. **28**(11), pp. 1917 - 1941.

109. Jiang C.-Y., Carrido L., and Mark J.E., *J. Polymer Sci.: Polymer Phys. Ed.*, 1984, vol. **22**(12), pp. 2281 - 2284.

110. Falender J.R., Yeh G.S.Y., and Mark J.E., *J. Chem. Phys.*, 1979, vol. **70**(11), pp. 5324 - 5325.

111. Milagin M.F. and Shishkin N.I., *Vysokomol. Soedin.*, 1988, vol. **30A**(11), pp. 2249 – 2254. (Rus)

112. Ozden S., Hatsukova M.A., Mashukov N.I., and Kozlov G.V., *Int. Polymer Processing*, 1988, vol. **13**(1), pp. 23 - 26.

113. Ozden S., Hatsukova M.A., Mashukov N.I., and Kozlov G.V., *Plast., Rubber and Composites*, 2000, vol. **29**(5), pp. 212 - 215.

114. Mashukov N.I., Vasnetsova O.A., Kozlov G.V., and Kesheva A.B., *Lakokrasochnye Materialy i Ikh Primenenie*, 1990, No. 5, pp. 38 – 41. (Rus)

115. Vunderlich B., *Physics of Macromolecules*, Vol. **3**, 1984, Moscow, Mir, 484 p. (Rus)

116. Afaunov V.V., Kozlov G.V., and Mashukov N.I., *Doklady Adyigeiskoi (Cherkesskoi) Mezhdunarodnoi AN*, 2001, vol. **5**(2), pp. 114 – 119. (Rus)

117. Seguela R. and Rietsch F., *Polymer*, 1986, vol. **27**(5), pp. 703 - 708.

118. Mashukov N.I., Gladyshev G.P., Kozlov G.V., and Mikitaev A.K., In Coll.: *Proc. VI All-Union Coordinat. Conference on Spectroscopy of Polymers*, 1989, Minsk, BGU, p. 81. (Rus)

119. Okada T. and Mandelkern L., *J. Polymer Sci. Part A-2*, 1967, vol. **5**(2), pp. 239 - 244.

120. Wedgewood A.R. and Seferis J.C., *Pure and Appl. Chem.*, 1983, vol. **55**(5), pp. 873 - 892.

121. Malamatov A.Kh., Mashukov N.I., and Kozlov G.V., *Izv. KBNTs RAI*, 1999, No. 3, pp. 65 – 68. (Rus)

122. Mashukov N.I., Kozlov G.V., Mikitaev A.K., and Vodakhov A.B., In Coll.: *Theory and Practice of Catalytic Reactions and Polymer Chemistry*, 1990, Cheboksary, ChGU, pp. 104 – 108. (Rus)

123. Jones T.E., Butler W.F., Ogden T.R., Gottfredson D.M., and Gullikson E.M., *J. Chem. Phys.*, 1988, vol. **88**(5), pp. 3338 - 3348.

124. Yavorsky B.M. and Detlaf A.A., *Reference Book on Physics*, 1974, Moscow, Nauka, 847 p. (Rus)

125. Shogenov V.N. and Kozlov G.V., *Fractal Clusters in Physicochemistry of Polymers*, 2002, Nalchik, Polygrafservis & T, 270 p. (Rus)

126. Kosobudsky I.D., Kashkina L.V., Gubin S.P., Petrakovsky G.A., Piskorsky V.P., and Svirsky N.M., *Vysokomol. Soedin.*, 1985, vol. **27A**(4), pp. 689 – 695. (Rus)

127. Gladyshev G.P., Mashukov N.I., Eltsin S.A., Mikitaev A.K., Vasnetsova O.A., and Ovcharenko E.N., *Physicochemical Properties of Polyethylene Stabilized by "Non-chain" Inhibitors*, (Preprint), 1985, Chernogolovka, OIKhF AS USSR, 1985, 14 p. (Rus)

128. Piskorsky V.P., Lipanov A.M., and Balusov V.A., *Zh. Vsesouzn. Khim. Obshch.*, 1987, vol. **32**(1), pp. 47 – 50. (Rus)

129. Belyi V.A., Snezhkov V.V., Bezrukov S.V., Voronezhtsev Yu.I., Goldade V.A., and Pinchuk L.S., *Doklady AN SSSR*, 1988, vol. **302**(2), pp. 355 – 357.

130. Privalko V.P., Andrianova G.P., Besklubenko Yu.D., Narozhnaya E.P., and Lipatov Yu.S., *Vysokomol. Soedin.*, 1978, vol. **20**(12), pp. 2777 – 2783. (Rus)

131. Malamatov A.Kh. and Kozlov G.V., *Doklady Adyigeiskoi (Cherkesskoi) Mezhdunarodnoi AN*, 1998, vol. **3**(2), pp. 78 – 81. (Rus)

132. Kavassalis T.A. and Noolandi J., *Macromolecules*, 1988, vol. **21**(9), pp. 2869 - 2879.

133. Mashukov N.I., Mikitaev A.K., Gladyishev G.P., Belousov V.N., and Kozlov G.V., *Plastmassy*, 1990, No. 11, pp. 21 – 23. (Rus)

134. Kitieva L.I., *Candidate's Thesis on Chemistry*, 2000, Nalchik, KBGU, 138 p. (Rus)

Chapter 3.

The concept of structural defect within the cluster model

As mentioned in Chapter 1, the cluster model of the amorphous polymer structure allows introduction of principally new interpretation of the structure defect (with the vengeance of this term) for the mentioned state [1, 2]. The essence of this model will be discussed in detail in this Chapter. It is common knowledge [3] that structure of natural solids contains a significant amount of defects. This concept forms the basis for the dislocation theory, widely applied to description of crystalline solids' behavior. Success in this branch has predetermined attempts of many authors (for example, of works [4 – 11]) to apply this concept to amorphous polymers. Hence, the notions used for crystalline lattices are often transposed to the structure of amorphous polymers. Usually, the basis for this transposition is formal resemblance of stress – deformation (σ - ε) curves for crystalline and amorphous solids. It has been suggested [2] that the defect theory of polymer amorphous state must be based not on the formal resemblance of behavior of the mentioned classes of solids, but on the principal difference in their structures.

In relation to the structure of amorphous polymers, for a long time the most ambiguous point [12 - 14] was the presence or the absence of the local (short range) order in which connection points of view of different authors on this problem were significantly different. The presence of the local order can significantly affect determination of the structure defect in amorphous polymers, if in the general case the order-disorder transition or vice versa is taken for the defect. For example, any distortion (discontinuity) of the long range order in crystalline solids represents a defect (dislocation, vacancy, etc.), and a monocrystal with the proper long range order is the ideal defect-free structure. It is known [15] that sufficiently bulky sample of 100% crystalline polymer cannot be obtained, and all characteristics of such hypothetical polymers are determined by the extrapolation method. That is why the Flory "felt" model [16, 17] can be suggested as the ideal defect-free structure for amorphous polymers. This model suggests that amorphous polymers consist of interpenetrating macromolecular coils interpreting the full disorder (chaos). On this basis, for the defect in amorphous polymers a distortion (discontinuity) of full disorder must be accepted, i.e. formation of the local (or long range) order

[2]. It shall be noted also that formal resemblance of the σ - ε curves for crystalline solids and amorphous polymers appears incomplete and behavior of these classes of materials displays principal differences, which will be discussed in detail below.

Turning back to the suggested concept of amorphous polymer structural defect, let us note that a segment participating in the cluster can be considered the linear defect – the analogue of dislocation in crystalline solids. Since in the cluster model the length of such segment is accepted equal the length of statistical segment, l_{st}, and their amount in the specific volume equals frequency of the cluster entanglement network, v_{cl}, the density of linear defects, ρ_d, in the specific volume of the polymer is presented by the following relation [2]:

$$\rho_d = L_{cl} = v_{cl} l_{st}. \tag{3.1}$$

Suggested interpretation allows application of well developed mathematical apparatus of the dislocation theory to description of amorphous polymer properties. This point will be thoroughly discussed below.

With respect to suggested concept of structural defect in amorphous polymers, of special attention are results of wide-angle X-ray analysis which indicate an antibate change of the relative part of clusters (the local order degree), φ_{cl}, and integral intensity (the area occupied by the amorphous halo) I^{in} of total amorphous halo as the function of cross-linking frequency, v_c, for epoxy polymers [18]. If the consideration is based on the analogy with crystalline materials, this result seems unexpected, because φ_{cl} increase should induce consequent I^{in} growth [19]. Simultaneously, this observation is completely explained by the structure defect concept, suggested in this consideration. In this case, the disorder exponent in the system (i.e. the degree of ideal defect-free structure approaching) can be presented by the relative part of the fluctuation free volume in the glassy state, f_g (also refer to Section 2.3) [20, 21].

Figure 3.1 shows the dependence $I^{in}(f_g)$ which is linear and passes through the origin of coordinates. To put it differently, for polymer amorphous state I^{in} is the value characterizing the disorder degree that explains the above-mentioned antibate type of dependencies $\varphi_{cl}(v_c)$ and $I^{in}(v_c)$ and proves the suggested concept of defect, as well. Generally, an analogous interpretation was suggested by Ginnier [22]. He has assumed that for metals I^{in} can be decreased due to formation of pre-crystallite structures in multicomponent melts.

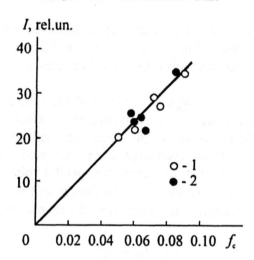

Figure 3.1. The relation between integral intensity, I^{in}, of wide-angle X-ray patterns and relative fluctuation free volume, f_g, for epoxy polymers of amine (1) and anhydride (2) cross-linking type [2]

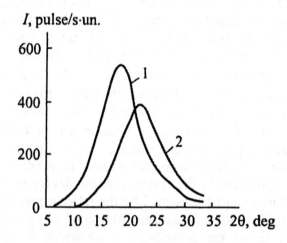

Figure 3.2. Expansion of experimental wide-angle diffraction pattern for PC into components: halo for packless matrix (1) and clusters (2) [18]

Computerized expansion of wide-angle X-ray diffraction patterns into two components of packless matrix and clusters was performed [18]. Figure 3.2 shows an example of such expansion for PC. This procedure allowed separate comparison of characteristics of the amorphous halo appropriate components and parameters of the cluster structure components (Figure 3.3).

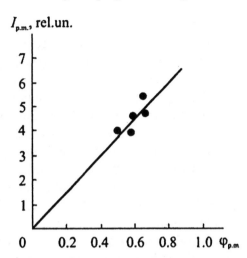

Figure 3.3. Dependence of integral intensity of packless matrix dispersion component, $I_{p.m.}$, on its relative part, $\varphi_{p.m.}$, for epoxy polymers of the amine cross-linking type [18]

Comparison of the results shown in Figures 3.1 and 3.3 with appropriate data on amorphous polymers from the literature indicates that the above-states regularities do not contradict with phenomena, observed by other researchers. For example, the study of annealing temperature (T_{an}) effect on cellulose nitrate structure [23] has detected systematic decrease of amorphous halo intensity and increase of the Bragg interval (d) with temperature T_{an}. It is commonly accepted that annealing of amorphous polymers at temperature below T_g improves molecular packing [24, 25]. As a consequence, this process must be intensified with T_{an} increase. Though different explanation of the effect observed has been suggested [23], experimental data from this work fit the above-suggested interpretation. Decrease of I^{in} with cluster formation was also observed in the work [26].

Generally, it is expected that with temperature increase a solid becomes more disordered. The more so, if similar to the cluster model the order is of thermofluctuational origin, this tendency becomes more obvious. However, studies of inorganic glasses (for example, As_2Se_2 and GeS_2) using X-ray diffraction analysis and small-angle neutron scattering methods indicate that the peak intensity related to the short range order is increased in the amorphous state with temperature [19]. This allowed an assumption that the layered structure embodying the short range order in these glasses becomes more regular with temperature. It should be noted that besides the above-mentioned analogies with the behavior of crystalline solids, no other proofs were presented [19]. The interpretation suggested explains this experimental observation more realistically. Moreover, on the example of different glasses [19], it is indicated that the percentage increase of the peak intensity mentioned is as higher as closer temperature approaches T_g of glass. This observation also correlates with the cluster model (refer to Figures 2.1 and 2.2). Finally, the increase of amorphous halo intensity with temperature both above and below T_g was also observed for PS [27]. Obviously, there are no grounds to suggest intensification of the local order with temperature for this amorphous polymer [28]. That is why it shall be admitted that suggested interpretation of the structural defect gives just a general explanation of experimental results [19, 23, 26 – 28].

An interesting experimental result was obtained for tensile stressed epoxy polymers [29]. As deformation of the epoxy polymer exceeds its strain yield, intensity of the amorphous halo was increased approximately by 17%. This result may also be easily explained in the framework of the interpretation suggested, because as shown below the cluster model suggests that yielding is associated with decomposition of unstable clusters and, consequently, with decrease of ν_{cl} and the local order degree [30, 31]. Concentration of clusters of this type is approximately corresponded to the increase of amorphous halo intensity observed for the epoxy polymer [29, 31].

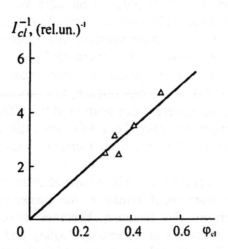

Figure 3.4. Dependence of reverse scattering curve intensity from clusters $\left(I_{cl}^{-1}\right)$ on their relative fracture (φ_{cl}) for epoxy polymers of the amine cross-linking type [18]

Intensity of the halo second component (Figure 3.2, curve 2), associated with clusters (I_{cl}), displays a linear correlation $I_{cl}^{-1}(\varphi_{cl})$ (Figure 3.4). The latter is coordinated with the above-stated ideas on the intensity decrease with ordering increase in amorphous polymers.

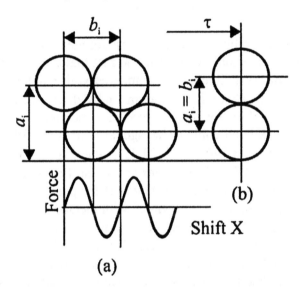

Figure 3.5. Schematic representation of deformation of two sequences of atoms in accordance with the Frenkel model. Positions before (a) and after (b) deformation [3]

Further on, the rightfulness of application of the structural defect concept suggested to polymers wielding will be discussed. As a rule, previously suggested concepts of defects in polymers were primarily used for description of this process or even exclusively for this purpose [4 – 11]. Theoretical shear strength of crystals was primarily calculated by Frenkel. He has based on a simple model of two series of atoms, shifted in relation to one another by the shear stress (Figure 3.5a) [3]. According to this model, critical shear stress τ_0 is expressed as follows [3]:

$$\tau_0 = \frac{G}{2\pi},$$ (3.2)

where G is the shear modulus.

Slightly changed, this model was used in the case of polymer yielding [6], wherefrom the following relation was obtained:

$$\tau_{0,y} = \frac{G}{\pi\sqrt{3}},$$
(3.3)

where $\tau_{0,y}$ is theoretical value of the shear stress for yielding.

Special attention should be paid to the fact that characterizes principally different behavior of crystalline metals compared with polymers. As is known [3, 32], $\tau_{0,y}/\tau_y$ relation (where τ_y is experimentally determined shear stress for yielding) is much higher for metals than for polymers. For five metals possessing the face-centered cubic or hexagonal lattices the following relations were obtained: $\tau_{0,y}/\tau_y = 3,740 - 22,720$ [3], whereas for five polymer this relation equaled 2.9 - 6.3 [6]. As a matter of fact, sufficient closeness of $\tau_{0,y}/\tau_y$ relation values to unit may already be the proof for possibility of realizing the Frenkel mechanism in polymers (in contrast with metals), but it is shown below that for polymers, a small modification of the law of shear stress regular change (τ) used commonly gives $\tau_{0,y}/\tau_y$ values very close to unit [33].

As is shown [34], dislocation analogies are also true for amorphous metals. Essentially, distortion of the composition of atoms (which induces occurrence of elastic stress fields) is considered [34] as a linear defect (dislocation) being practically immovable. Clearly such approach correlates completely with the structural defect concept, suggested in this Chapter. In the framework of this concept, Figure 3.5a may be considered as a cross-section of a cluster (crystallite). As a consequence, the shift of segments in this cluster, according to the Frenkel mechanism, can be taken for the one limiting yielding in polymers. This is proved by the experimental data [35] which show that glassy polymer yielding is realized directly in closely packed zones. Other data indicate [36] that these closely packed zones are clusters. To put it differently, one can state that yielding is associated with clusters (crystallites) stability loss in the shear stresses field [37].

Figure 3.6 shows asymmetrical periodical function [38] that displays dependence of shear stress (τ) on shear deformation (γ_{sh}). As shown before [32], asymmetry of this function and corresponded decrease of the energy barrier height determined by segments of macromolecules at the elementary yielding action is stipulated by formation of fluctuation free volume cavities during deformation (which is the specific feature of polymers [39]). Data in Figure 3.6 indicate that in the initial part of periodical curve from zero to the maximum dependence of τ on displacement x can be simulated by a sine function with the period shorter than in Figure 3.5. In this case, function $\tau(x)$ can be reduced to the following form:

$$\tau = k\sin\left(\frac{6\pi x}{b_i}\right).\qquad(3.4)$$

This expression is fully corresponded to the Frenkel conclusion, except for randomly chosen numerical coefficient in brackets (6 instead of 2).

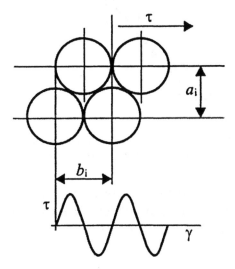

Figure 3.6. Schematic image of shear deformation and corresponded stress – deformation (τ - γ) function [38]

Future calculation of τ_0 by the method, described in ref. [3], and its comparison with experimental values of τ_y indicated their tight correspondence for nine amorphous and amorphous-crystalline polymers (Figure 3.7). This proves the possibility of realization of the above-suggested yielding mechanism at the segment level.

Inconsistency of $\tau_{0,y}$ and τ_y values in metals induced a search for another mechanism of yielding realization. At present, it is commonly accepted that this mechanism is motion of dislocations by sliding planes of the crystal [3]. This implies that interatomic interaction forces, directed transverse to the crystal sliding plane, can be overcome in the case of presence of several local displacements, determined by periodical field of stresses in the lattice. This is strictly different from microscopic shift, during which all bonds are broken simultaneously (the Frenkel model). It seems obvious that with the help of dislocations total shear deformation will occur at applying much lower external stress than for the process including simultaneous break of all atomic bonds by the sliding plane [3].

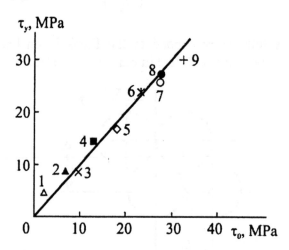

Figure 3.7. Correlation between theoretical (τ_0) and experimental (τ_y) shear stresses for yielding of polytetrafluoroethylene (1), HDPE (2), PP (3), polyamide-6 (4), PVC (5), PHE (6), PC (7), PSP (8) and PAr (9) [33]

Payerls and Nabarro [3] were the first who calculated the shear stress necessary for the dislocation motion, τ_{dm}. They used a sinusoidal approximation and deduced the expression for τ_{dm} as follows:

$$\tau_{dm} = \frac{2G}{1-\nu} e^{-2\pi a_i / b_i (1-\nu)},\quad\quad\quad (3.5)$$

where ν is the Poisson coefficient and parameters a and b are of the same meaning as in Figure 3.6.

By substituting reasonable value of ν, for example, 0.35 [40], and assuming $a_i = b_i$, the following value for τ_{dm} is obtained: $\tau_{dm} = 2\times10^{-4}$ G. Though for metals this value is higher than observed τ_y, it is much closer to them than the stress calculated using simple shear model (the Frenkel model, Figure 3.5).

However, for polymers the situation is opposite: analogous calculation indicates that their τ_{dm} does not exceed 0.2 MPa, which is by two orders of magnitude, approximately, lower than observed τ_y values.

The next point of discussion is the free path length of dislocations, λ_d. As known for metals [3], in which the basic role in elastic deformation is belonged to the mobile dislocations, λ_d is assessed as $\sim 10^4$ Å. For polymers, this parameter is estimated as follows [41]:

$$\lambda_d = \frac{\varepsilon_y}{b\rho_d}, \qquad (3.6)$$

where ε_y is the yielding deformation; b is the Burgers vector; ρ_d is the density of linear defects, determined from the equation (3.1).

Parameter ε_y is assessed as ~ 0.10 [42] and the Burgers vector is $b \approx 3 \times 10^{-10}$ m [9]. Then for different polymers, λ_d assessed by the equation (3.6) is about 2.5 Å. It is the distance passed by a segment at a shift moving it to the position, shown in Figure 3.5b, which can be simply calculated from configuration. Thus this assessment also indicates no reasons for assuming any sufficient free path length of dislocations in polymers rather than transition of a segment (or several segments) of macromolecule from one quasi-equilibrium state to another [43].

It is common knowledge [3, 38] that for crystalline materials the Bale-Hirch relation between shear stress, τ_y, and dislocation density, ρ_d, is fulfilled:

$$\tau_y = \tau_{ini} + \alpha Gb\rho_d^{1/2}, \qquad (3.7)$$

where τ_{ini} is the initial internal stress; α is the efficiency parameter.

Equation (3.7) is also true for amorphous metals [34]. It was used [33] for describing mechanical behavior of polymers on the example of their main classes. For this purpose, data for amorphous glassy PAr [44], amorphous-crystalline HDPE [45] and epoxy polymer networks (EP) of amine and anhydride cross-linking types [30] were used. Different load conditions were used: axial tension of film samples [44], high-speed bending [45], and axial compression [30]. Figure 3.8 shows the relation between $Gb\rho_d^{1/2}$ and experimental τ_y values for mentioned polymers, corresponded to equation (3.7). Clearly they are linear and pass through the origin of coordinates (e.g. $\tau_{ini} = 0$), but α values for linear polymers and polymer networks are different. Thus in the framework of the defect concept suggested, the Baily-Hirch relation is also true for polymers. This indicates that dislocation analogues are correct for any linear defect distorting the ideal structure of the material and creating the field of elastic stresses [33]. From this point of view high defect degree of the polymer structure shall be noted: $\rho_d \approx 10^{10}$ cm^{-2} for crystalline metals [3], $\rho_d \approx 10^9 \div 10^{14}$ cm^{-2} for amorphous metals [34], and $\rho_d \approx 10^{14}$ cm^{-2} for polymers [41].

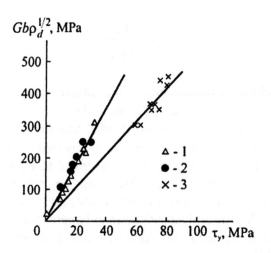

Figure 3.8. Correlation between $Gb\rho_d^{1/2}$ and flow shear stress τ_y by equation (3.7) for PAr (1), HDPE (2), and EP (3) [33]

As a consequence, the results obtained [33] indicate that in contrast with metals, for polymers realization of the Frenkel mechanism during yielding is much more probable rather than the motion of defects (Figure 3.5). This is stipulated by the above-discussed principal (even strictly antipodal) differences in the structure of crystalline metals and polymers [2].

Let us consider the assessment methods for Burgers vector, b. It is suggested that for polymers the value of b must be corresponded to their molecular characteristics. For example, the following formula was deduced [9]:

$$b = \frac{4}{3\pi}\left(r_{ch} + l_m\right), \tag{3.8}$$

where r_{ch} is the distance between neighboring chains; l_m in the monomeric unit length.

One more simple technique for b assessment has been suggested [46]. It is known that there is a linear correlation between the shear modulus G and the glass transition temperature T_g (Figure 3.9). In the framework of the cluster model, theoretical correlation between G and T_g is as follows [46]:

$$\frac{G}{T_g} = \frac{4\pi(1-v)k\ln(1/f_c)}{C_\infty b^2 l_0 \ln(r/r_0)},$$
(3.9)

where k is the Boltzmann constant; r and r_0 are external and internal radii of the dislocation force field, respectively. The value r/r_0 is usually accepted equal 10 [47].

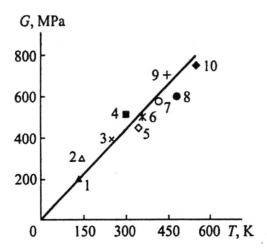

Figure 3.9. Correlation between shear modulus G and glass transition temperature T_g for poly(tetrafluoroethylene) (1), HDPE (2), PP (3), polyamide-6 (4), PETP (5), PVC (6), PC (7) poly(arylate sulfone) (8), PSF (9), and PAr (10). Values of G were measured for 293 K [46]

Substitution of parameters typical of polymers into equation (3.9) ($v = 0.3$ [47], $f_f = 0.07$ [47], $C_\infty = 4.0$ [48], $b = 4$ Å [9], $l_0 = 1.54$ Å [48]) gives $G/T_g = 1.40$ MPa/K, which is close to $G/T_g = 1.50$ MPa/K, obtained from data in Figure

Kozlov G.V. and Zaikov G.E.

3.9. Parameters C_∞ and b in equation (3.9) are the most sensitive to the origin of polymers. The rest characteristics can be accepted practically constant for the majority of polymers. It follows from the condition G/T_g = const and equation (3.9) that Burgers square vector is in inverse proportion to characteristic relation C_∞. If the condition G/T_g = 1.50 MPa/K and the following constants in equation (3.9) are used: v = 0.30; f_f = 0.07; l_0 = 1.54 Å; r/r_0 = 10, then for b assessment the following empirical equation can be obtained [46]:

$$b = \left(\frac{60.7}{C_\infty}\right)^{1/2}, \text{Å}. \tag{3.10}$$

As a consequence, an increase of macromolecule flexibility (C_∞ increase) causes a decrease of the Burgers vector (a decrease of the distance between chains). As C_∞ varies within the range of 2.2 – 10.0 [48], calculation by equation (3.10) gives b = 2.3 – 4.9 Å which is in good correspondence with the known data on polymers [8, 9]. Table 3.1 shows comparison of b values estimated by the mentioned method and the ones obtained from the literature for five polymers. Clearly good conformity is observed again [46].

Table 3.1

Comparison of the Burgers vector b values calculated by equation (3.10) and obtained from the literature [49]

Polymer	b, Å		
	Reference [8]	Reference [9]*	Equation (3.10)**
Polystyrene	4.5	3.1 (2.8)	2.46
Poly(methyl methacrylate)	4.3	2.9 (2.7)	2.94
Polyvinylchloride	5.8	2.3 (2.5)	3.0
Poly(ethylene terephthalate)	6.2	-	3.80
Polycarbonate	5.6	-	5.03

* Values in brackets are experimental [9].

** Calculation of b by equation (3.10) applied C_∞ values, shown in ref. [48].

The Poisson coefficient is one of the most important characteristics of polymers, in many ways determining their mechanical and heat-physical properties [50, 51]. Naturally, a tight interconnection of this parameter with the polymer structure is suggested; however, such correlation has yet been obtained on the basis of the above-discussed structural defect concept [52].

As is known [53], one of the laws in the basis of the theory of continuous medium elasticity is experimentally determined Poisson relation postulating the effect of transverse deformations, $\varepsilon_\perp$:

$$\varepsilon_\perp = v\varepsilon_\parallel, \tag{3.11}$$

where v is the Poisson coefficient; $\varepsilon_\parallel$ is the longitudinal deformation.

One of the methods for quantitative estimation of v is calculation of it from experimental values of elasticity modulus, E, and the strain yield, σ_y [51]:

$$\frac{\sigma_y}{E} = \frac{(1-2v)}{6(1+v)}. \tag{3.12}$$

The equation (3.12) represents quite strict correlation, deduced from the polymer state equation [51]. However, it does not indicate the connection between v and the structure. This connection can be obtained from consideration of the local order zone as a deviation from the ideal structure; hence, the segments in it are identified as linear defects. In such an interpretation using mathematical apparatus of the theory of dislocations, one can obtain the following expression [54]:

$$\frac{\sigma_y}{\sqrt{3}} = \frac{Eb\rho_d^{1/2}}{4\pi(1+v)}. \tag{3.13}$$

Combination of equations (2.8), (3.1), (3.10), and (3.13) gives the following expression for determination of v by structural parameters of polymer only [49]:

$$v = 0.5 - 3.22 \times 10^{-10}(l_0 v_{cl})^{1/2}. \tag{3.14}$$

Figure 3.10 shows comparison of the theoretical temperature dependence of v (equation (3.14)) and its values estimated from equation (3.12) for blow and quasi-static tests. Expectedly [55, 56], v increases with T for all three data series. Maximal deviation between theoretical v values and appropriate blow test results does not exceed 20%, and for quasi-static tension it is even lower – below 10% [49].

Obviously the above-shown data indicate that calculation of the Poisson coefficient by structural characteristics of the polymer only gives good correspondence with results obtained by other methods. It should be noted also that equation (3.14), used for theoretical assessment of v, includes structural characteristics only in the absence of any adjustable parameter [49].

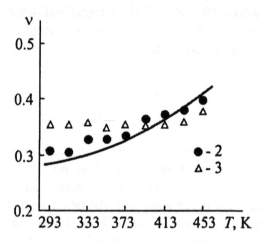

Figure 3.10. Temperature dependence of the Poisson coefficient v for poly(arylate sulfone): 1 – calculation by equation (3.14); 2 and 3 – calculation by equation (3.12) for quasi-static (2) and blow (3) tests [49]

Completing the Chapter, let us discuss an example of "classic" dislocation analogy modification for polymers using the suggested concept of the structural defect. The dislocation model by Crist [57] suggests formation of a spiral dislocation (or a couple of such dislocations) with Burgers vector b in crystalline zones of polymers, and yielding is realized during formation of a critical germinal zone of u^* dimension:

$$u^* = \frac{K_{el}b}{2\pi\tau_y}, \tag{3.15}$$

where K_{el} is an elastic constant; τ_y is the shear strain yield.

In its turn, the zone sized u^* is formed at overcoming the energetic barrier ΔG^* [57]:

$$\Delta G^* = \frac{K_{el} b^2 l_c}{2\pi} \left[\ln\left(\frac{u^*}{r_0}\right) - 1 \right],$$

(3.16)

where l_c is the dislocation length equal the crystallite length; r_0 is the dislocation center radius.

The equation (3.15) allows theoretical assessment of τ_y in the presence of the known parameters u^*, K_{el} and b. Let us consider their determination in relation to the structure of amorphous glassy polymers.

As suggested in the simulation [57], the germinal zone sized u^* is formed in a defect-free area of amorphous-crystalline polymer, i.e. in a crystallite. In the framework of the concept suggested for the structural defect and applied to the structure of amorphous glassy polymer, this area is represented by a packless matrix surrounding a local order zone (cluster). The structure of the latter zone is quite similar to the defect-free polymer, postulated by the Flory "felt" model [16, 17]. It has been shown [31] that during the flow process this structural component is subject to mechanical devitrification. In this interpretation, u^* can be determined as follows [58]:

$$u^* = R_{cl} - r_{cl},$$

(3.17)

where R_{cl} is the half distance between centers of neighbor clusters; r_{cl} is the cluster radius.

For PC and PAr, R_{cl} values are shown in the work [59] (they can also be calculated by the formula (2.14)); r_{cl} can be determined from the relation as follows [60]:

$$r_{cl} = \left(\frac{n_{cl} S}{\pi \eta}\right)^{1/2},$$

(3.18)

where n_{cl} is the number of segments in the cluster: S is the cross-section of the macromolecule; η is the packing factor equal 0.868 (for close packing) [60].

For K the shear modulus, G, must be accepted; b value is determined from the equation (3.10). Similar to the previous, n_{cl} equals $1/2F$. Thus as follows from the above-said, the basic role in τ_y or σ_y determination (because $\sigma_y = \sqrt{3}\ \tau_y$) is played by structural (R_{cl} and F) and molecular (S and C_∞) characteristics of the polymer.

Figure 3.11 shows comparison of temperature dependencies of σ_y for PC and PAr, experimental and calculated by the method described. Clearly good conformity between theory and experiment was obtained. Note that adjustable parameters were not also used for σ_y calculation.

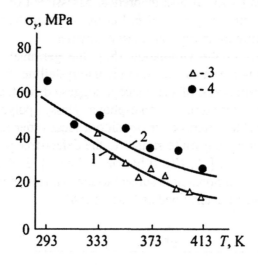

Figure 3.11. Temperature dependencies of σ_y: (1, 2) experimental and (3, 4) calculated by equation (3.15) for PC (1, 3) and PAr (2, 4) [58]

Figure 3.12 presents comparison of $\sigma_y(T)$ dependencies for HDPE, experimental and calculated by the equation (3.10). In this case, no conformity similar to the one observed between PC and PAr (Figure 3.11) was obtained. This is simply explained by the application of HDPE structure only to calculation of its σ_y. That is why σ_y values are much lower than the experimental ones. Nevertheless, of the extreme importance is that for amorphous-crystalline polymers (polyethylenes, in particular), σ_y is fully determined by the contribution of crystalline zones [57, 61 – 63]. The latter can be calculated with regard to technique, described in the work [54], and the totality of contributions of both structural components of HDPE into the strain yield is also presented in Figure 3.12 as the function of test temperature. In the latter case, good conformity between theory and experiment is obtained [58].

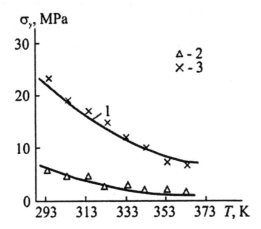

Figure 3.12. Temperature dependencies for HDPE: (1) experimental and (2, 3) theoretical; 2 – calculated by equation (3.15), 3 represents the totality of amorphous and crystalline phases contributions into σ_y [58]

Thus following the interpretation of the ideal structure and defect for the polymer amorphous state, presented in this Chapter, good correspondence of theory to experimental results for both amorphous glassy and amorphous-crystalline polymers was obtained [58]. In this case, molecular and structural characteristics of polymers were used as the initial data. Hence, by virtue of the mentioned reciprocal of ideal structures for crystalline solids and polymer amorphous state, the yielding mechanisms for them is generation and motion of linear defects in the long range order zones and annihilation of the same defects in disordered zones, respectively.

Further on, let us discuss the assessment of the energetic barrier, ΔG^*, in the framework of the cluster model. As shown [47], annihilation (transition into the packless matrix) of a segment participating in the cluster requires energy U_{cl}:

$$U_{cl} = \frac{Gb^2 l_0 C_\infty}{4\pi(1-v)} \ln\left(\frac{r}{r_0}\right). \qquad (3.19)$$

Figure 3.13 shows comparison of ΔG^* values, calculated from the equation (3.16), where for l_s the length of statistical segment (l_{st}) is taken, and U_{cl} ones, calculated by the equation (3.19), for PC and PAr. These data indicate good quantitative conformity of ΔG^* и U_{cl} parameters. This proves the conclusion, made in the framework of the model [2] that energy U_{cl}, required for segment separation from the cluster (e.g. annihilation of a linear defect), represents the energetic barrier for the flow process of amorphous glassy polymers.

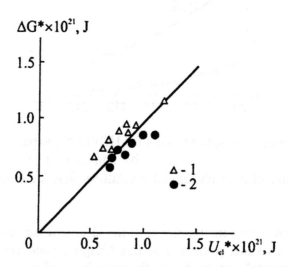

Figure 3.13. Correlation between critical energy of the germinal zone formation under yielding, ΔG^*, and linear defect annihilation energy, U_{cl}, for PC (1) and PAr (2) [58]

The interpretation suggested in this Chapter is in good conformity with other models, developed for description of polymer yielding. The equation for assessment of critical radius of the dislocation loop, R_l, is shown below [6]:

$$R_l = \frac{Gb}{4\pi\tau_y}\left[\ln\left(\frac{2R_l}{b\sqrt{3}}\right)+2\right]. \tag{3.20}$$

Calculations performed by equation (3.20) give R_l equal 38.7 Å for PC and 26 Å for PAr (at $T = 293$ K). These values are in good conformity with appropriate R_{cl} values, obtained in the framework of the cluster model (31.1 Å and 25 Å, respectively [59]).

In the work [64] the flow process in amorphous glassy PC is associated with generation of quasi-point defects with increased molecular mobility. Obviously in the framework of the current model, mechanically devitrified germinal zone sized u^* represents the analogue for the mentioned defect.

Thus the results obtained [58] have proved once again correctness of the cluster model of the amorphous polymer state structure, interpretation of linear defect in this structure and the yield mechanism applied to description of amorphous glassy and amorphous-crystalline polymer yielding. Of special attention is the existence of significant (about 25%) contribution of the amorphous phase in amorphous-crystalline polymers into their strain yield, σ_y [58].

REFERENCES

1. Kozlov G.V., Beloshenko V.A., and Varyukhin V.N., *Applied Mechanics and Technical Physics*, 1996, vol. 37(3), pp. 115 – 119. (Rus)
2. Kozlov G.V., Belousov V.N., Serdyuk V.D., and Kuznetsov E.N., *High Pressure Physics and Technology*, 1995, vol. 5(3), pp. 59 – 64. (Rus)
3. Honeycombe R.W.K., *The Plastic Deformation of Metals*, London, Edward Arnold Publishers Ltd., 1968, 497 p.
4. Argon A.S., *Phil. Mag.*, 1973, vol. 28(4), pp. 839 - 865.
5. Argon A.S., *J. Macromol. Sci.-Phys.*, 1973, vol. B8(3-4), pp. 573 - 596.
6. Bowden P.B. and Raha S., *Phil. Mag.*, 1974, vol. 29(1), pp. 148 - 165.
7. Escaig B., *Ann. Phys.*, 1978, vol. 3(2), pp. 207 - 220.
8. Pechhold W.R. and Stoll B., *Polymer Bull.*, 1982, vol. 7(4), pp. 413 - 416.
9. Sinani A.B. and Stepanov V.A., *Mechanics of Composite Materials*, 1981, vol. 17(1), pp. 109 – 115. (Rus)
10. Oleinik E.F., Rudnev S.N., Salamatina O.B., Nazarenko S.I., and Grigoryan G.A., *Doklady AN SSSR*, 1986, vol. 286(1), pp. 135 – 138. (Rus)
11. Melot D., Escaig B., Lefebvre J.M., Eustache R.P., and Laupretre F., *J. Polymer Sci.: Part B: Polymer Phys.*, 1994, vol. 32(11), pp. 1805 - 1811.
12. Boyer R.F., *J. Macromol. Sci.-Phys.*, 1976, vol. B12(12), pp. 253 – 301.
13. Fischer E.W. and Dettenmaier M., *J. Non-Cryst. Solids*, 1978, vol. 31(1-2), pp. 181 - 205.
14. Wendorff J.H., *Polymer*, 1982, vol. 23(4), pp. 543 - 557.

15. Nikol'sky V.G., Plate I.V., Fazlyiev F.A., Fedorova E.A., Filippov V.V., and Yudaeva L.V., *Vysokomol. Soedin.*, 1983, vol. A25(11), pp. 2366 – 2371. (Rus)
16. Flory P.J., *Pure and Appl. Chem.*, 1984, vol. 56(3), pp. 305 - 312.
17. Flory P.J., *Brit. Polymer J.*, 1976, vol. 8(1), pp. 1 - 10.
18. Kozlov G.V., Kuznetsov E.N., Beloshenko V.A., and Lipatov Yu.S., *Doklady NAN Ukrainy*, 1995, No. 11, pp. 102 – 104. (Rus)
19. Lin C., Busse L.E., Nagel S.R., and Faber J., *Phys. Rev. V.*, 1984, vol. 29(9), pp. 5060 - 5062.
20. Kozlov G.V., Beloshenko V.A., and Lipskaya, *Ukrainsky Fizichesky Zhurnal*, 1996, vo. 41(2), pp. 222 – 225. (Rus)
21. Kozlov G.V. and Zaikov G.E., *Izv. KBNTs RAI*, 2003, No. 9, pp. 132 – 140. (Rus)
22. Ginier A., *X-Ray Diffraction Analysis of Crystals*, Moscow, Fizmatgiz, 1961, 604 p. (Rus)
23. Majid C.A., Ahmad A.U., Akber R.A., and Khan H.A., *Radiat. Phys. Chem.*, 1980, vol. 16(5), pp. 379 - 383.
24. Petrie S.E.B., *J. Macromol. Sci.-Phys.*, 1976, vol. B12(2), pp. 225 - 247.
25. Kozlov G.V., Sanditov D.S., and Lipatov Yu.S., In Coll.: *Fractals and Local Order in Polymeric Materials*, Ed. G.V. Kozlov and G.E. Zaikov, Huntington, New York, Nova Science Publishers, 2001, pp. 65 - 82.
26. Ball R.C., *Physica D.*, 1989, vol. 38(1), pp. 13 - 15.
27. Song H.-H. and Roe R.-J., *Macromolecules*, 1987, vol. 20(11), pp. 2723 - 2732.
28. Murthy N.S., Minor H., Bednarczyk C., and Krimm S., *Macromolecules*, 1993, vol. 26(7), pp. 1712 - 1721.
29. Kumar S. and Adams W.W., *Polymer*, 1987, vol. 28(8), pp. 1497 - 1504.
30. Beloshenko V.A. and Kozlov G.V., *Mechanics of Composite Materials*, 1994, vol. 30(4), pp. 451 – 454. (Rus)
31. Kozlov G.V., Beloshenko V.A., Gazaev M.A., and Novikov V.U., *Mechanics of Composite Materials*, 1996, vol. 32(2), pp. 270 – 278. (Rus)
32. Kozlov G.V., Shogenov V.N., and Mikitaev A.K., *Doklady AN SSSR*, 1988, vol. 298(1), pp. 142 – 144. (Rus)
33. Kozlov G.V., Afaunova Z.I., and Zaikov G.E., 'Theoretical assessment of polymer yielding', *Electron J. "Studied in Russia"*, 2002, vol. 98, pp. 1071 – 1080; http: //zhurnal.ape.relarn.ru/articles/2002/098.pdf//. (Rus)
34. Liu R.S. and Li J.Y., *Mater. Sci. Eng.*, 1989, vol. A114(1), pp. 127 - 132.
35. Aleksanyan G.G., Berlin Al.Al., Gol'dansky A.V., Grineva N.S., Onishchuk V.A., Shantarovich V.P., and Safonov G.P., *Khimicheskaya Fizika*, 1986,

vol. **5**(9), pp. 1225 – 1234. (Rus)

36. Balankin A.S., Bugrimov A.L., Kozlov G.V., Mikitaev A.K., and Sanditov D.S., *Doklady Akademii Nauk*, 1992, vol. **236**(3), pp. 463 – 466. (Rus)
37. Kozlov G.V., and Novikov V.U., *Uspekhi Fizicheskikh Nauk*, 2001, vol. **171**(7) pp. 717 – 764. (Rus)
38. McClintock F.A. and Argon A.S., *Mechanical Behavior of Materials*, Addison-Wesley Publishing Company, Inc., Reading, Massachusetts, 1966, 514 p.
39. Wu S., *J. Appl. Polymer Sci.*, 1992, vol. **46**(4), pp. 619 - 624.
40. Sanditov D.S. and Bartenev G.M., *Physical Properties of Irregular Structures*, 1982, Novosibirsk, Nauka, 256 p. (Rus)
41. Milman L.D. and Kozlov G.V., In Coll.: *Polycondensation Processes and Polymers*, Ed. A.K. Mikitaev, 1986, Nalchik, KBGU, pp. 130 – 141. (Rus)
42. Shogenov V.N., Kozlov G.V., and Mikitaev A.K., *Vysokomol. Soedin.*, 1989, vol. **A31**(8), pp. 1766 – 1770. (Rus)
43. Peschanskaya N.N., Bernstein V.A., and Stepanov A.A., *Fizika Tverdogo Tela*, 1978, vol. **20**(11), pp. 3371 – 3374. (Rus)
44. Shogenov V.N., Belousov V.N., Potapov V.V., Kozlov G.V., and Prut E.V., *Vysokomol. Soedin.*, 1991, vol. **A33**(1), pp. 155 – 160. (Rus)
45. Mashukov N.I., Gladyishev G.P., and Kozlov G.V., *Vysokomol. Soedin.*, 1991, vol. **A33**(12), pp. 2538 – 2546. (Rus)
46. Sanditov D.S. and Kozlov G.V., *Fizika I Khimia Stekla*, 1993, vol. **19**(4), pp. 593 – 601. (Rus)
47. Kozlov G.V. and Sanditov D.S., *Anharmonic Effects and Physicomechanical Properties of Polymers*, 1994, Novosibirsk, Nauka, 261 p. (Rus)
48. Aharoni S.M., *Macromolecules*, 1983, vol. **16**(9), pp. 1722 - 1728.
49. Kozlov G.V., Belousov V.N., Serdyuk V.D., Mikitaev A.K., and Mashukov N.I., *Doklady Adyigeiskoi (Cherkesskoi) Mezhdunarodnoi AN*, 1997, vol. **2**(2), pp. 88 – 93. (Rus)
50. Balankin A.S., *Synergism of Deformable Body*, 1991, Moscow, MO SSSR, 404 p. (Rus)
51. Novikov V.U. and Kozlov G.V., *Analysis of Polymer Degradation in the Framework of the Fractal Concept*, 2001, Moscow, Izd. MGOU, 135 p. (Rus)
52. Kozlov G.V., Belousov V.N., Sanditov D.S., and Mikitaev A.K., *Izv. VUZov, Severo-Kavkazskii Region. Estestvennye Nauki*, 1994, No. 1-2(86), pp. 54 – 58. (Rus)
53. Balankin A.S., *Doklady AN SSSR*, 1991, vol. **319**(5), pp. 1098 – 1101. (Rus)

54. Belousov V.N., Kozlov G.V., Mashukov N.I., and Lipatov Yu.S., *Doklady RAN*, 1993, vol. **328**(6), pp. 706 – 708. (Rus)
55. Filyanov E.M., *Vysokomol. Soedin.*, 1987, vol. **A29**(5), pp. 975 – 981. (Rus)
56. Hohg S.-D., Chung S.Y., Fedors R.F., and Moacanin J., *J. Polymer Sci.: Polymer Phys. Ed.*, 1983, vol. **21**(9), pp. 1647 - 1660.
57. Crist B., *Polymer Commun.*, 1989, vol. **30**(3), pp. 69 - 71.
58. Kozlov G.V., Zaikov G.E., Burmistr M.V., and Korenyako V.A., *Voprosy Khimii i Khimicheskoi Tekhnologii*, 2002, No. 6, pp. 81 – 84. (Rus)
59. Kozlov G.V., Belousov V.N., and Mikitaev A.K., *High Pressure Physics and Technology*, 1998, vol. **8**(1), pp. 64 – 70. (Rus)
60. Kozlov G.V., Shogenov V.N., and Mikitaev A.K., *Inzhenerno-Fizicheskii Zhurnal*, 1988, vol. **71**(6), pp. 1012 – 1015. (Rus)
61. Popli R. and Mandelkern L., *J. Polymer Sci.: Part B: Polymer Phys.*, 1987, vol. **25**(3), pp. 441 - 483.
62. Bernstein V.A., Sirota A.G., Egorova L.M., and Egorova V.M., *Vysokomol. Soedin.*, 1989, vol. **A31**(4), pp. 776 – 779. (Rus)
63. Kennedy M.A., Peacock A.J., and Mandelkern L., *Macromolecules*, 1994, vol. **27**(19), pp. 5297 - 5310.
64. G'Sell C., Bari H.E., Perez J., Cavaille J.Y., and Johari G.P., *Mater. Sci. Eng.*, 1989, vol. **A110**(2), pp. 223 - 229.

Chapter 4.

Experimental proof of the cluster model

As indicated in Chapter 1, existence of the local order in the polymer amorphous state directly or indirectly proves practically all modern experimental techniques. In the current Chapter the authors will briefly consider the experimental observations which prove the cluster model, i.e. type of packing, existence range, etc.

4.1. X-RAY DIFFRACTION ANALYSIS

It has been assumed for a long time that techniques applying wide-angle X-ray diffraction give short information about the structure of amorphous polymers, and that interpretation of experimental results obtained is very complicated and ambiguous [1 – 3]. Nevertheless, the study of the totality of polymers with regularly variable structure can give important data on it. In particular, for a group of polymers, comparison of data on the position of their amorphous halo peak with regard to the change of their structures (size of side groups, macromolecule cross-section) has induced a conclusion [1 – 3] about existence of the local order in amorphous polymers. These observations were not studied in detail, which is probably caused by some circumstances and, first of all, by the absence of an appropriate model of amorphous polymer structure considering the presence of the local order in them. Moreover, in accordance with the authors' point of view, in the works [1 – 3] an insufficient attention was paid to analysis of behavior of the amorphous halo intensity, which is also the sensitive structural parameter of the material. The more so, it has been concluded [4] that amorphous halo intensity gives an approximate measure of the local order in the structure.

Recently, a series of works was performed [4 – 9], in which amorphous halos of amorphous and amorphous-crystalline polymers were studied in much more details and, which is most important, quantitatively. Conclusions made in the mentioned works have been already compared with the results, obtained by the authors in Chapter 3 of the present monograph. These comparisons will be continued below, but one important qualitative difference in interpretations of wide-angle X-ray diffraction data, displayed in the works [4 – 9] and discussed below, should be noted. The difference concludes in the absence of any quantitative structural model of the polymer amorphous state in these works. Essentially, [4 – 9] this model was composed on the basis on the data, obtained by analysis of the wide-angle X-ray diffraction. Considered in the previous Chapter were some aspects of X-ray diffraction data applied to structural defects. More general interpretation of this experimental technique applied to the entire cluster model will be given below.

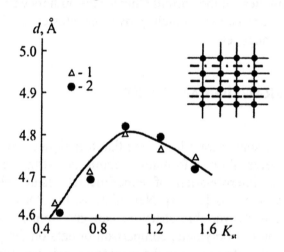

Figure 4.1. Dependence of the Bragg interval d on curing agent : oligomer ratio K_{st} for EP-1 (1) and EP-2 (2). Insertion presents scheme of the polymer network structure [10]

For epoxy polymers, derived from ED-22 resin of amine (EP-1) and anhydride (EP-2) cross-linking type, dependence of the Bragg interval d on the curing agent : oligomer ratio, K_{st}, possesses a maximum near the stoichiometric ratio $K_{st} = 1.0$ (Figure 4.1). Chemical cross-link network frequency, ν_{nw}, varies similarly (Table 4.1). Miller and Boyer [2] have suggested the following correlation between d and effective diameter of macromolecule, D (the distance between centers of macromolecules), simulated by a cylinder:

$$D = d^{1.22}. \tag{4.1}$$

Table 4.1

Structural characteristics of epoxy polymers [10]

Epoxy polymer	K_{st}	$v_{nw} \times 10^{-26}$, m^{-3}	T_g, K	$v_{cl} \times 10^{-27}$, m^{-3}
	0.50	2	326	1.3
	0.75	6	366	1.9
EP-1	1.0	17	423	2.7
	1.25	12	405	3.5
	1.50	8	390	2.0
	0.50	3	333	1.5
	0.75	12	389	2.4
EP-2	1.0	19	423	2.9
	1.25	12	408	2.9
	1.50	8	393	2.0

In accordance with the equation (4.1) the extreme value of $d(K_{st})$ function indicates that the macromolecule effective diameter increases with the cross-linking frequency. This fact is simply understood with the help of simplified planar scheme of the polymer network structure (insertion in Figure 4.1). If for macromolecule a configuration located between two sectional planes depicted by dashed lines is taken, it will appear similar to the case of linear polymer shaped as a backbone with side groups [11]. It is common knowledge [2] that the latter inhibit dense packing of chains expanding distance between them and, as a consequence, D [10]. Parameter v_{nw} is increased analogously to the number of side groups, wherefrom, symbate change of D and v_{nw} follows [10].

For linear polymers, it was also suggested to approximate glass transition temperature T_g by the following empirical correlation:

$$T_g = 129 \left(\frac{S}{C_\infty} \right)^{1/2}, \text{K}, \quad (4.2)$$

where S is the macromolecule cross-section, Å^2; C_∞ is the characteristic relation. The equation (4.2) can also be applied to EP under consideration under the assumption that C_∞ varies with K_{st} and, as a consequence, with v_{nw}. Figure 4.2 shows results of C_∞ calculations, performed by the above-mentioned method. Clearly C_∞ is decreased monotonously with the cross-linking frequency increase, and compactness of the macromolecular coil increases [13]. According to [2], collectionwise with simultaneous increase of S, this circumstance must promote an increase of the local order parameter.

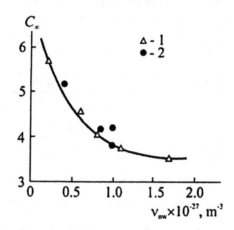

Figure 4.2. Dependence of the characteristic relation C_∞ on the chemical cross-link network frequency v_{nw} for EP-1 (1) and EP-2 (2) [10]

Let us check concordance between this conclusion and ideas of the cluster model [14]. The calculation data on macromolecular entanglement cluster network frequency, v_{cl}, and the relative part of clusters, φ_{cl}, confirm the above statement about the local order increase with v_{nw} (Table 4.1 and Figure 4.3). Contrary to assumptions [2], where the change of d is directly associated with the local order change, the authors assume that d is just a factor characterizing D and C_∞, i.e. chain rigidity. Simultaneously, the latter factor determines φ_{cl} variation [10]. Direct interrelation between structural and X-ray diffraction characteristics is shown in Figures 3.1, 3.3 and 3.4.

Usually, wide-angle X-ray diffraction halos of amorphous polymers deviate from the ideal shape (asymmetry, unclear maximum, etc.) that allows a supposition about superposition of several simpler dispersions [8, 9]. Figure 4.4 shows a wide-angle diffraction pattern typical for EP-1, corresponded to $K_{st} =$ 1.0 [15]. By studying the shape of amorphous halos for EP-1 and EP-2 epoxy polymers, it has been detected that the best description is obtained under the assumption of two-component structure of the halo (Figure 3.2). Using ideas of the cluster model [14], one can associate a halo component with the maximum disposed at lower θ (to which broader Bragg interval is corresponded, Figure

3.2, curve 1) with a packless matrix, and the one at higher θ (Figure 3.2, curve 2) with clusters [15].

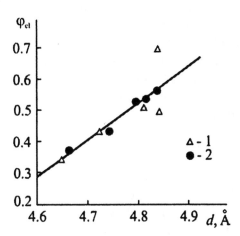

Figure 4.3. Dependence of the relative fracture of clusters, φ_{cl}, on the Bragg interval, d, for EP-1 (1) and EP-2 (2) [10]

The integral intensity of both components and the total halo as well displays an extreme variation with K_{st} possessing a maximum at $K_{st} = 1.0$. Comparison of the dispersion integral intensity from packless matrix ($I_{p.m.}$) with $\varphi_{p.m.}$ indicates linearity of $I_{p.m.}(\varphi_{p.m.})$ dependence and its passing through the origin of coordinates. This gives supplementary grounds for associating the dispersion curve 1 (Figure 3.2, curve 1) for greater d with the packless matrix [10, 15].

X-ray diffraction analysis of more complicated cross-linked systems was performed [16] on the example of EP-2 epoxy polymer of the anhydride cross-linking type, modified by adamantane carboxylic acids. It is common knowledge that injection of adamantane fragments in EPs significantly affects their characteristics. The action of such carcass structures is considered on the example of EP-2 [17], and interpretation of experimental results within the framework of the cluster model allowed supposition of the existence of two types of clusters in EP-2 modified by adamantane carboxylic acids: stable ones formed by segments and unstable ones formed due to interaction between the adamantane fragments. The conformity degree between these ideas and the real structure of modified EPs was also determined [16]. Two series of EPs modified by adamantane mono- (EP-3) and dicarboxylic (EP-4) acids were studied.

The studies of the halo shape observed for EP-3 and EP-4 indicated their best description in the presence of ??? halos, whereas for EP-2 with two halos only. Occurrence of the amorphous halo third component for EP with carcass fragments indicates that the structural state of EP-3 and EP-4 is more complicated (inhomogeneous) compared with unmodified EP-2 (Figure 4.5). Within the framework of the cluster model, occurrence of three halos may be induced by the presence of three structural components: packless matrix and two types of clusters. The latter are formed by collinear segments of EP network and adamantane fragments.

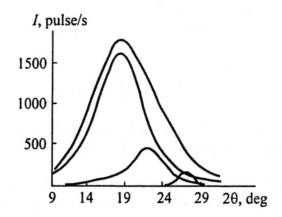

Figure 4.5. Experimental X-ray diffraction pattern (1) and its factorization for EP-3 [16]

Table 4.2 presents values of the Bragg interval, d, calculated from position of the maximum of experimentally observed halo and its components, separated after computerized treatment of diffraction patterns (d_1, d_2, d_3) and corresponded amplitude (I^a) and integral (I^i) intensities. A tendency of the Bragg interval growth with modifier concentration C was observed. As a consequence, introduction of carcass fragments induced loosening of EP structure.

As before, the halo component with the greatest corresponded Bragg interval d_1 should be associated with the packless matrix. In this case, d_2 and d_3 characterize the network of physical entanglements. Because d_2 values for EP-2, EP-3 and EP-4 are quite close (especially at low concentrations of the modifier), this parameter must be corresponded to entanglements (clusters), formed by EP segments. Thus d_3 is corresponded to the network of physical entanglements, formed by interaction of adamantane fragments.

Table 4.2
The effect of modifier concentration on X-ray diffraction characteristics of EP [16]

Epoxy polymer	C, %	d	d_1	d_2	d_3	I_l	I_1^a	I_2^a	I_3^a	I_1^i	I_2^i	I_3^i
		Å				pulse/s $\times 10^{-3}$				relative units		
EP-2	0	4.76	4.84	3.86	-	1.65	1.22	0.67	-	7.7	6.8	-
EP-3	1.0	5.0	5.07	3.86	3.03	1.86	1.53	0.96	0.14	10.2	7.9	0.3
	2.5	4.66	4.55	-	3.18	1.91	1.84	-	0.41	15.6	-	2.1
	5.0	5.05	5.04	4.21	3.52	2.12	1.75	0.79	0.57	11.5	3.9	3.3
	10.0	4.93	4.76	-	3.31	1.98	1.99	-	0.42	8.3	-	0.8
	15.0	4.98	5.15	4.27	3.34	1.92	2.12	1.45	0.44	4.3	3.0	0.6
EP-4	1.4	4.78	4.89	4.05	3.29	1.80	1.63	0.47	0.17	14.6	2.6	0.4
	2.9	4.83	4.91	3.96	3.09	1.70	1.69	0.47	0.25	11.6	1.4	0.5
	4.3	4.98	4.83	-	3.16	2.37	2.21	-	0.75	18.1	-	4.4
	9.2	5.19	5.12	3.95	-	1.97	1.88	0.47	-	13.5	3.0	-
	13.2	5.28	5.36	4.54	3.45	2.06	1.43	0.11	0.25	7.6	6.6	1.3

First of all, structural changes, which may be obtained from the analysis of parameters of EP general amorphous halo, should be considered. However, note first that addition of adamantane groups makes the network of chemical cross-links less frequent (Figure 4.6) that obviously happens due to steric obstacles, formed by them [17]. Intensity and rate of v_{nw} decrease are independent of the modifier type and the method of its injection, and at concentration $C \approx 15\%$, v_{nw} apparently reaches the near-border values. Thus $0 \leq C \leq 15\%$ range is of the greatest interest for the study.

Figure 4.7 shows $S(v_{nw})$ dependencies calculated by the equation (4.1) for EP-3 and EP-4 systems. In contrast with EP-2, they display S decrease with v_{nw} growth. For EP-3 and EP-4, v_{nw} decrease is accompanied by increase of concentration of side groups formed by adamantane fragments (refer to Figure 4.6), which displays a high influence on S inducing its significant increase.

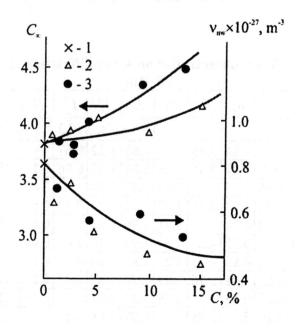

Figure 4.6. Dependencies of chemical cross-link network frequency ν_{nw} and characteristic relation C_∞ on the modifier concentration C for EP-2 (1), EP-3 (2) and EP-4 (3) [16]

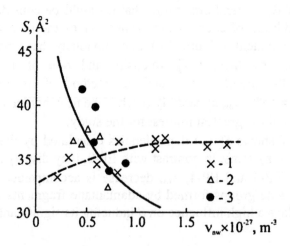

Figure 4.7. Dependence of macromolecular cross-section S on chemical cross-link network frequency ν_{nw} for EP-2 (1), EP-3 (2) and EP-4 (3) [16]

Dependencies $C_\infty(C)$ for modified EPs, shown in Figure 4.6, indicate more intensive C_∞ increase for EP-4 rather than for EP-3 due to differences in chemical structures of these systems. In the case of EP-3, adamantane fragment is linked by two bonds to the EP network or partly forms a card polymer structure. For EP-4, card polymer structure with a single bond at the adamantane fragment is the most probable alternative. Thus for macromolecular coil, card elements of EP-4 chain structure are more effective as steric hindrances. They inhibit the macromolecular coil degree similar to EP-2 or EP-3 [16].

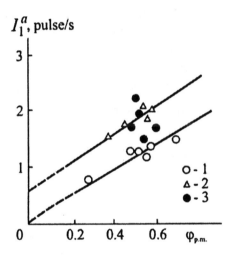

Figure 4.8. Dependencies of amplitude intensity I_1^a on relative fracture of the packless matrix $\varphi_{p.m.}$ for EP-2 (1), EP-3 (2) and EP-4 (3) [16]

Let us pass to the study of parameters of amorphous halo components (Table 4.2). Figure 4.8 shows dependencies of amplitude intensity of the halo first component, I_1^a, on the packless matrix fracture, $\varphi_{p.m.}$. For all EPs, I_1^a increases with $\varphi_{p.m.}$, i.e. the structure disorder degree similar to EP-1 in Figure 3.3 is observed. However, in contrast with $I_1^a(\varphi_{p.m.})$ dependencies for unmodified EP-1 and EP-2, analogous dependence for EP-3 and EP-4 at $\varphi_{p.m.} = 0$ is extrapolated to a non-zero value I_1^a. Obviously the difference is associated with the presence of adamantane fragments in the packless matrix of EP-3 and EP-4, which form the third component of the amorphous halo (Table 4.2).

Kozlov G.V. and Zaikov G.E.

Clearly for EP-3 and EP-4, extrapolated value I_1^a at $\varphi_{p.m.} = 0$ equals, approximately, 0.6×10^3 pulse/s, and the amplitude intensity of the third component, I_3^a, varies within the range of $(0.14 - 0.75) \times 10^3$ pulse/s (Table 4.2), i.e. these values are quite close.

Let us now consider the interconnection between parameters of the amorphous halo second component (I_2^a), d_2 and the local order zones. As mentioned above, the presentation of segments composing clusters in the form of linear defects assumes that φ_{cl} increase must induce I_2^a decrease (refer to Figure 3.4). Actually, such tendency is observed (Figure 4.9), data for unmodified EP-2 also fitting the general line. The modifier concentration increase induces d_2 growth (Table 4.2). Cluster formation from segments at greater distance between their axial lines means that the local order zones are formed easier with C increase. That is why a correlation between parameters I_2^a and d_2 should be expected that was observed in the work [16].

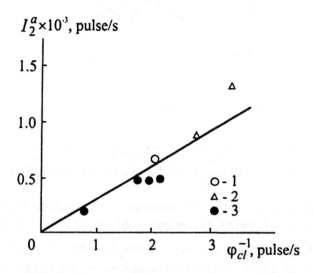

Figure 4.9. Dependence of the amplitude intensity I_2^a on reverse relative part of clusters φ_{cl} for EP-2 (1), EP-3 (2) and EP-4 (3) [16]

For modified and unmodified EP, comparison of d and d_2 dependencies on v_{nw} displays a significant difference in their behavior. As the function of v_{nw}, parameter d_2 values fit a single curve for all studied EP [16] (Figure 4.10). At the same time, in the case of unmodified EP, d is weakly increased with v_{nw} and intensively decreased for modified EP (refer to Figure 4.7). Probably, though changing the chain structure, adamantane fragments do not directly present in the local order zones, formed by segments of epoxy polymeric network.

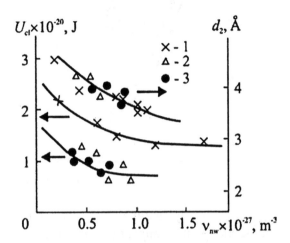

Figure 4.10. Dependencies of the Bragg interval d_2 and association energy of a couple of segments in the cluster U_{cl} on the chemical cross-link network frequency v_{nw} for EP-2 (1), EP-3 (2) and EP-4 (3) [16]

Energy for coupling collinear segments to a cluster, U_{cl}, can be calculated by equation (3.19). Dependencies $U_{cl}(v_{nw})$ for studied EP [16] were found symbate (Figure 4.10); however, absolute values of U_{cl} for modified epoxy polymers are much lower. This means that formation of the local order zones in EP-3 and EP-4 proceeds much easier. Apparently, strong interactions created by adamantane fragments promote for collinear disposition of EP segments that simplifies formation of the fluctuation cluster network [7].

Dependencies of the entanglement cluster network frequency, v_{cl}, on U_{cl} fit a single curve for EP-2, EP-3 and EP-4, whereas dependence $I_2^a(U_{cl})$ is linear and passes through the origin of coordinates (Figure 4.11). As a consequence, intensity of the amorphous halo second component characterizes the energetic state of clusters: the higher U_{cl} is, the greater I_2^a is. Increasing U_{cl} reduces probability of the local order formation (Figure 4.11) that explains I_2^a decrease with φ_{cl} growth (refer to Figures 3.4 and 4.9).

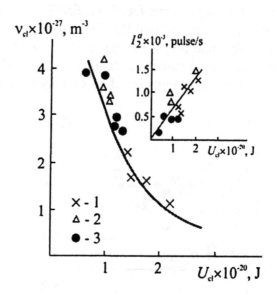

Figure 4.11. Dependence of entanglement cluster network frequency, ν_{cl}, on the association energy of a couple of segments in a cluster, U_{cl}, for EP-2 (1), EP-3 (2) and EP-4 (3). Insertion shows dependence of amplitude intensity I_2^a on U_{cl} [16]

Figure 4.12 shows dependencies of amplitude intensity and Bragg interval of the amorphous halo third component corresponded to the entanglement network of adamantane fragments on the modifier concentration C. The parameters mentioned vary extremely. Dependence $d_3(C)$ indicates that in the area of small C modifier concentration increase and corresponded ν_{nw} decrease (refer to Figure 4.6) significantly simplify formation of a network from the adamantane fragments: it can be formed at more and much more higher d_3. However, at $C > 5\%$ parameter d_3 increase is stopped and a slight decrease testifying about the reverse effect is observed [16].

Calculation of average I_2^a and I_3^a values for EP-3 and EP-4 displays two-fold exceeding of the former over the latter (Table 4.2). Figure 4.11 shows that higher difference will be observed in U_{cl} of the mentioned types of networks. A decrease of U_{cl} for a network formed due to strong interactions of adamantane fragments means the possibility of easier decomposition of clusters composing it. The highest stability of the latter is reached at optimal modifier concentration equal $C \approx 5\%$. As $C < 5\%$, the lack of adamantane fragments impedes formation of stable clusters, and at $C > 5\%$ their excess creates steric hindrances to the mentioned process.

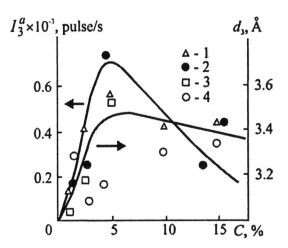

Figure 4.12. Dependence of amplitude intensity I_3^a (1, 2) and the Bragg interval d_3 (3, 4) on modifier concentration C for EP-3 (1, 3) and EP-4 (2, 4) [16]

Basing on the equation (3.19) one may assume that lower values of U_{cl} for clusters from adamantane fragments are caused by their shorter length, compared with EP clusters (the length of the latter equals l_{st}). This means that the length l_a of adamantane fragments in the cluster is K times smaller than l_{st}, where K is the ratio of appropriate energies U_{cl}. It can be simply calculated from the data in Figure 4.11. Then taking the mean value of K and calculating the relative part of clusters of adamantane fragments, φ_A, by the known technique [18], the dependence $I_3^a\left(\varphi_A^{-1}\right)$ for EP-3 and EP-4 is obtained. The results are shown in Figure 4.13, where the dashed line shows the run of $I_3^a\left(\varphi_A^{-1}\right)$ dependence, plotted by the data from Figure 4.9.

Note also that analogous dependence of "backlashes" [19], presented in coordinated of the Figure 4.13, will be shaped as a straight line parallel to the abscissa axis, because frequency of backlashes is constant [14, 18]. As a consequence, by energetic parameters, the network of macromolecular entanglements composed from adamantane fragments is the intermediate stage between EP cluster network and the "backlash" one. Two points corresponded to EP-3 ($C = 5\%$) and EP-4 ($C = 4.3\%$) fit the line 1 in Figure 4.13. As mentioned above, these epoxy polymers possess the most stable clusters of adamantane fragments, characteristics of which are close to EP clusters. The rest points fit the straight line 2 by slope more close to the abscissa axis, i.e. clusters in these epoxy polymers, composed by adamantane fragments, are rather incomplete local order zones approaching the properties of macromolecular backlashes [19].

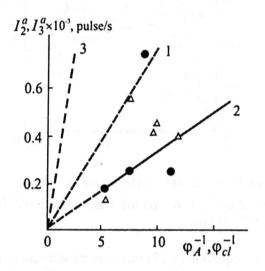

Figure 4.13. Dependence of amplitude intensities I_3^a (1, 2) and I_2^a (3) on reverse relative fractures of clusters from adamantane fragments φ_A (1, 2) and segments φ_{cl} (3). Refer to Figure 4.12 for the legend [16]

Thus results discussed in this Section indicate that application of the wide-angle X-ray diffraction data involving ideas of the cluster model allow quite thorough analysis of structural organization of complex polymeric systems, for example, epoxy polymers modified by carcass fragments [10, 11, 15, 16].

4.2. IR-SPECTROSCOPY

IR-spectroscopy is the well-known and successful method for studying the structure of polymeric materials [20, 21]. Section 2.6 presented the detailed analysis of IR-spectroscopy data within the framework of a model [20 – 24] for HDPE + Z composites. However, this analysis is of the qualitative type, whereas application of the cluster model allows obtaining of quantitative relations of structural parameters and characteristics of IR-spectra of these composites [25]. Further on, such relations can be used for direct assessment of the structure of crystallized polymer non-crystalline zones by IR-spectroscopy methods.

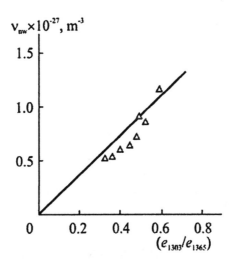

Figure 4.14. Dependence of entanglement cluster network frequency, v_{cl}, on relation of extinction coefficients $e_{1,303}/e_{1,368} = B$ for HDPE + Z composites [25]

To add the data discussed in Section 2.6, it should be noted that Wedgewood and Seferis [21] have also suggested that the change in relations of extinction coefficients $e_{1,303}/e_{1,352}$ and $e_{1,303}/e_{1,368}$ can be direct consequence of intensification of differences between structures of amorphous and interphase zones of crystallized polymers. Thus the decrease of mentioned relations, B, is stipulated by increase of macromolecular unit concentration in stretched *trans*-conformations and, as a consequence, by increase of the regular order in the interphase zones. Quantitatively, the mentioned degree can be expressed via v_{cl} [25]. Therefore, there are grounds for expecting a correlation between v_{cl} and

B^{-1} (Figure 2.38). Figure 4.14 shows the dependence $v_{cl}(B^{-1})$, composed by IR-spectroscopy data, that proves this conclusion and allows the local order assessment for non-crystalline zones of HDPE + Z composites.

   If the interpretation suggested is correct, it should be expected that decrease of $e_{1,303}/e_{1,352}$ and $e_{1,303}/e_{1,368}$ relations is associated with increase of the relative part of interphase zones, v_{ip} [26]. Figure 4.15 shows dependence of v_{ip} on reverse $e_{1,303}/e_{1,352}$ relation. Quite good linear correlation between these parameters proves the interpretation suggested [25].

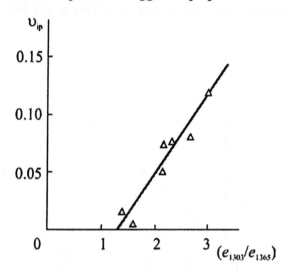

Figure 4.15. Dependence of interphase zones relative part, v_{ip}, on the reverse relation of extinction coefficients, $e_{1,303}/e_{1,352}$, for HDPE + Z composites [25]

   Figure 4.16 shows $v_{cl}(v_{ip})$ correlation. It should be noted that for HDPE + Z composite v_{cl} increase (the ordering degree) with v_{ip} is observed. As this dependence is extrapolated to $v_{ip} = 0$, $v_{cl} = 0$ is obtained. Thus for HDPE + Z composites with $v_{ip} \neq 0$, all clusters are concentrated in the interphase zones. However, the initial linear HDPE possesses $v_{cl} \neq 0$ at $v_{ip} = 0$. This means that the absence of interphase zones is not equal to the absence of the local order in non-crystalline zones of polyethylenes; therefore, in this case clusters are concentrated in the amorphous phase [25].

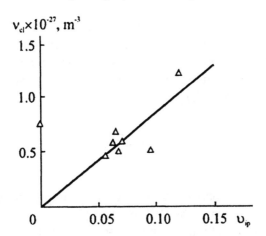

Figure 4.16. Correlation between entanglement cluster network frequency, ν_{cl}, and the relative part of interphase zones, υ_{ip}, for HDPE + Z composites [25]

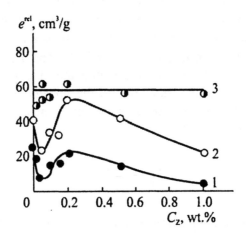

Figure 4.17. Dependencies of relative extinction coefficient, e^{rel}, on modifier Z concentration, C_Z, in HDPE + Z composites for IR-bands at: 1,303 cm^{-1} (1), 1,352 cm^{-1} (2) and 1,078 cm^{-1} (3) [23]

Three studied IR-bands (1,030, 1,352 and 1,368 cm^{-1}) are induced by oscillations of CH$_2$-groups in a specific sequence which includes several groups, some of which are present in the *gauche*-conformation [20, 21, 27]. For CH$_2$-groups in *trans*- and *gauche*-conformations located in non-crystalline zones, the band at 1,078 cm^{-1} is related to the stretching mode in C-C bond [27]. Changes of macromolecular unit conformations induced by introduction of Z modifier into HDPE can be assessed using the relative extinction coefficients, e^{rel}, which corresponds to the specific mass of non-crystalline component. The value of e^{rel} can be calculated as follows [27]:

$$e^{rel} = \frac{e}{\upsilon_{nc}},$$ (4.3)

where υ_{nc} is the relative part of non-crystalline component, determined using crystallinity degrees, estimated by DCS method [24].

Dependencies of e^{rel} for 1,078, 1,303 and 1,352 cm^{-1} IR-absorption bands were also calculated [23]. For the bands at 1,303 and 1,352 cm^{-1} at Z concentration of 0.05 wt.%, a minimum of e^{rel} is observed, whereas for the band at 1,078 cm^{-1} e^{rel} is independent of Z concentration within the limits of experimental error (Figure 4.17). For three above-mentioned IR-bands, these dependencies of e^{rel} characterize extreme decrease of the number of CH$_2$-groups in *gauche*-conformation with simultaneous increase of their number in *trans*-conformations. Since all IR-bands used in the analysis are corresponded to non-crystalline zones, one may conclude that for 1,303 and 1,352 cm^{-1} bands at Z concentration of 0.05 wt.%, e^{rel} decrease indicates the increase of tie chain frequency in stretched conformations due to the presence of loops [27]. In the framework of the cluster model, frequency of tie chains, V_{tr}, can be calculated by the following equation [28]:

$$V_{tr} = \frac{2v_{cl}}{F}.$$ (4.4)

Figure 4.18 shows dependence of V_{tr} on reverse e^{rel} of the band at 1,352 cm^{-1} for HDPE + Z composites. Good correlation proving the above assumption was again obtained.

Thus, combined application of the cluster model of the polymer amorphous state structure with IR-spectroscopy data allows obtaining of quantitative information about structure of non-crystalline zones in HDPE + Z composites. Figures 4.14, 4.15 and 4.18 present graphs, which can be used for calibration for this purpose [25].

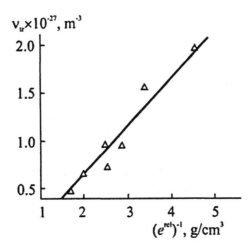

Figure 4.18. Dependence of tie chain frequency, v_{tr}, on reverse relative extinction coefficient, e^{rel}, of IR-band at 1,352 cm^{-1} for HDPE + Z composites [25]

4.3. LOW-FREQUENCY RAMAN SCATTERING

One of the properties of interest, general for all glassy-like systems and absent for the appropriate crystals, is the presence of excessive low-energy density of oscillating states in the acoustic zone of the spectrum. For low energies the spectrum area with energies in the range of 2 – 10 MeV (≤ 100 cm^{-1}) are accepted [29]. For crystals in the mentioned frequency range, the Debye model of oscillatory actuations is usually applied well. As for amorphous bodies, the situation is different. Besides acoustic oscillations, additional quasi-local oscillatory actuations are present in the low-energy area of the spectrum of these materials. These actuations induce anomalous low-temperature properties of amorphous bodies. Being quite different from the properties of crystals, these low-temperature properties of amorphous bodies, nevertheless, are universal, which is expressed in independence of particular chemical composition and type of the short range order of glasses.

At the present time, data on density of oscillatory states in the low-frequency zone are displayed for a series of metallic glasses, amorphous and amorphous-crystalline polymers, inorganic glasses, and amorphous silicon and germanium [30 – 35].

To the point of view of Malinovsky, Novikov and Sokolov [30], there are grounds for associating excessive low-energy density of oscillatory states in glasses with the presence of typical length, which is the mean (intermediate) order radius of the nanometer size, in them. In the model presented by these authors, low-energy oscillatory actuations responsible for excessive density of oscillatory states are assumed localized on nano-inhomogeneities of the structure.

These observations should be discussed in more detail and compared with characteristics of the cluster model. A model based on the results of neutron and low-frequency Raman scattering was presented [36], in which discrete structure of glasses was suggested. The "boson" peak in the Raman scattering is associated with the excessive density of oscillatory states. As assumed, these two interrelated features represent the result of oscillations, localized in structural units (blobs), which form the glass structure. The formula determining the blob size, $2a$, was deduced as follows [36]:

$$\omega_0 = S_f\left(\frac{\upsilon}{2a}\right), \tag{4.5}$$

where ω_0 is the "boson" peak frequency; S_f is the blob shape factor (for spherical particles S_f is close to 0.8; for linear objects $S_f = 0.5$); υ is the sonic speed in the blob.

For inorganic As_2S_3 and SiO_2 glasses, the blob size $2a$ was estimated as 21 Å and 22 Å, respectively. These values are in good concordance with the size of structural units approximately equal 20 Å, estimated from width of the first sharp diffraction peak for As_2S_3 [36].

An interesting interpretation of inelastic scattering of neutrons was suggested by Novikov [37]. He has assumed that it is stipulated by the presence of oscillations localized on disclination defects, which are added to usual Debye oscillations. He has shown that the number of "defect" atoms which form disclinations equals $10 - 15\%$ of the total number of atoms. Of special attention is correspondence of this interpretation and the concept of structural defect in the cluster model (refer to Chapter 3).

Some ambiguity of the model presented is also accepted [36]. The main and the most frequently put forward objection is the following: the suggested cluster structure of glasses was not yet directly and clearly displayed, for example, by data of electron microscopy of small-angle neutron and X-ray diffraction analysis. Nevertheless, it is possible that the discrete structure observed in oscillation dynamics does not display well-resolved fluctuation density. The reason for existence of blobs is in weaker bonds between blobs than between atoms in them. Moreover, blobs display not a simple shape, but consist of more or less branched chains [36].

The structure of amorphous polymers (PETP and PMA) and tension effect on it were studied with the help of low-frequency Raman scattering [38]. Structural characteristics of these and some other polymers were calculated by equation (4.5), i.e. it is assumed that similar to [36] the structure of these polymers consists of blobs. As observed, the blob size, $2a$, does not significantly differ from the distance between macromolecular entanglements (backlashes) in appropriate polymeric melts, determined by the plateau modulus [39, 40]. The distance between entanglements, ζ, is given by the following relation [40]:

$$\zeta = N_l^{1/2} C_\infty^{1/2} l_c,$$ (4.6)

where N_l is the average number of bonds l_c long between entanglements.

Quite close conformity between the blob size $2a$ and the distance ζ was obtained [36]. On this basis a conclusion was made that macromolecular "backlash"-specific memory is preserved at the melt – solid polymer transition. In this connection, two remarks should be made. Firstly, as follows from equations (2.2) – (2.4), for unspecified polymer, distances between "backlashes" and clusters are approximately equal. This means correctness of approximate equality of $2a$ and ζ for the entanglement cluster network also. Secondly, $2a$ value for a series of polymers, calculated in the work [38], varies within the range of 40 – 70 Å which conforms to the cluster structure dimensions [41].

Stretching of a polymer causes increase of the longitudinal blob size with stretch [38]. The same effect, namely, R_{cl} increase, is observed in the framework of the cluster model [42].

It is common knowledge that the presence of the first sharp diffraction peak of X-ray and neutron structural factor and the boson peak are frequently assigned to structural correlations of an intermediate scale. However, dimensions of structural units obtained from these experimental data are different, though linear correlation between them (R_c and $2a$, respectively) is observed [43]. It is assumed that $4a$ equals $(3 - 5)R_c$. Note also that structural parameters R_c and $2a$ are identified by different techniques. The value R_c is determined by the width of the first sharp diffraction peak in accordance with the Scherrer formula. It is common knowledge that for crystalline materials this formula gives the size of regular zones (microcrystallites). The value $2a$ is calculated by the boson peak position (equation (4.5)) and characterizes the structure periodicity [38]. Making comparison with the cluster model, it should be indicated that for the latter R_c is analogous to radius r_{cp} of closely packed local order zone (cluster), which represents the amorphous analogue of microcrystallite with stretched chains [14]. Therefore, $2a$ value corresponds to the distance between clusters, R_{cl}. Values of r_{cp} and R_{cl} can be calculated by equations (3.18) and (2.14), respectively. Figure 4.19 shows the interrelation between R_{cl} and r_{cp} for five amorphous and five amorphous-crystalline polymers. Clearly for both groups of polymers linear approximations possessing different slopes are obtained. According to the data present [43], for inorganic materials this slope varies from 0.40 to 0.66. The mentioned limits correlate well with the slopes determined from data in Figure 4.19: 0.31 and 0.70, respectively. Different slopes of linear approximations in Figure 4.19 are obviously stipulated by different mechanisms of the cluster formation for the mentioned polymer classes (refer to Section 2.5).

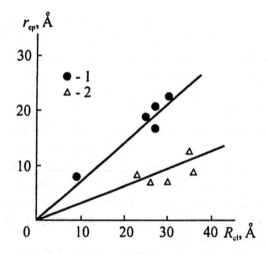

Figure 4.19. Correlation between the radius r_{cp} of closely packed zone and the distance between clusters R_{cl} for amorphous (1) and amorphous-crystalline (2) polymers [41]

Finally, structural mechanism of elastic deformation for PC was studied by the low-frequency Raman spectroscopy method [45]. It has induced a conclusion that that invariability of the boson peak position at low test temperatures assumes invariability of zones with higher cohesion during elastic deformation. As a consequence, this deformation is realized in the zones with lower cohesion, where molecular reconstruction proceeds. The zones with higher cohesion are oriented parallel to the deformation axis. Since in the framework of the cluster model the zones with higher cohesion are identified with clusters and the ones with lower cohesion with packless matrix, the model suggested [45] completely correlates with description of the structural mechanism of elastic deformation in the framework of the cluster model [46].

For PC, at temperatures close to T_g elastic deformation is also realized by partial orientation of macromolecular rovings in the zones with lower cohesion and general deformation of the zones with higher cohesion [45]. This does also correlate with interpretation of the cluster model, because at temperatures mentioned (above T_g') the packless matrix is thermally devitrified [44, 46].

Thus the data applying ideas of the discrete structure of amorphous materials [30, 34, 36 - 38, 43] completely correlate with the cluster model statements both qualitatively and quantitatively.

REFERENCES

1. Miller R.L., Boyer R.F., and Heijboer J., *J. Polymer Sci.: Polymer Phys. Ed.*, 1984, vol. **22**(12), pp. 2021 - 2041.
2. Miller R.L. and Boyer R.F., *J. Polymer Sci.: Polymer Phys. Ed.*, 1984, vol. **22**(12), pp. 2043 - 2050.
3. Miller R.L., *Proc. 17th Int. Symp. "Order Amorphous State"*, Midland, Mich., August 18-21, 1985, New York – London, 1987, pp. 33 - 49.
4. Song H.-H. and Roe R.J., *Macromolecules*, 1987, vol. **20**(11), pp. 2723 - 2732.
5. Lin C., Busse L.E., Nagel S.R., and Faber J., *Phys. Rev. B.*, 1984, vol. **29**(9), pp. 5060 - 5062.
6. Murthy N.S., Minor H., Bedna'rczyk C., and Krimm S., *Macromolecules*, 1993, vol. **26**(7), pp. 1712 - 1721.
7. Kumar S. and Adams W.W., *Polymer*, 1987, vol. **28**(8), pp. 1497 - 1504.
8. Murthy N.S. and Minor H., *Polymer*, 1990, vol. **31**(6), pp. 996 - 1002.
9. Murthy N.S., Correale S.T., and Minor H., *Macromolecules*, 1991, vol. **24**(5), pp. 1185 - 1189.
10. Kozlov G.V., Beloshenko V.A., Kuznetsov E.N., and Lipatov Yu.S., *Ukrainski Khimicheski Zhurnal*, 1997, vol. **63**(7), pp. 52 – 61. (Rus)

11. Kozlov G.V., Beloshenko V.A., Kuznetsov E.N., and Lipatov Yu.S., *Doklady NAN Ukrainy*, 1994, No. 12, pp. 126 – 128. (Rus)
12. Shogenov V.N., Beloshenko V.A., Kozlov G.V., and Varyukhin V.N., *High Pressure Physics and Technology*, 1999, vol. 9(3), pp. 30 – 36. (Rus)
13. Aharoni S.M., *Macromolecules*, 1983, vol. 16(9), pp. 1722 - 1728.
14. Kozlov G.V. and Novikov V.U., *Uspekhi Fizicheskikh Nauk*, 2001, vol. 171(7), pp. 717 – 764. (Rus)
15. Kozlov G.V., Kuznetsov E.N., Beloshenko V.A., and Lipatov Yu.S., *Doklady NAN Ukrainy*, 1995, No. 11, pp. 102 – 104. (Rus)
16. Beloshenko V.A., Kozlov G.V., Kuznetsov E.N., Stroganov I.V., and Stroganov V.F., *High Pressure Physics and Technology*, 1997, vol. 7(3), pp. 62 – 71. (Rus)
17. Beloshenko V.A., Kozlov G.V., Stroganov I.V., and Stroganov V.F., *High Pressure Physics and Technology*, 1994, vol. 4(3-4), pp. 113 – 117. (Rus)
18. Sanditov D.S., Kozlov G.V., Belousov V.N., and Lipatov Yu.S., *Ukrainian Polymer J.*, 1992, vol. 1(3-4), pp. 241 - 258.
19. Wu S., *J. Polymer Sci.: Part B: Polymer Phys.*, 1989, vol. 27(4), pp. 723 - 741.
20. Okada T. and Mandelkern L., *J. Polymer Sci., A-2*, 1967, vol. 5(2), pp. 239 - 242.
21. Wedgewood A.R. and Seferis J.C., *Pure and Appl. Chem.*, 1983, vol. 55(5), pp. 873 - 892.
22. Mashukov N.I., Gladyishev G.P., Kozlov G.V., and Mikitaev A.K., *Thes. Rep. VI All-Union Coord. Meeting of Polymer Spectroscopy*, 1989, Minsk, p. 81. (Rus)
23. Mashukov N.I., Serdyuk V.D., Kozlov G.V., Ovcharenko E.N., Gladyishev G.P., and Vodakhov A.B., *Stabilization and Modification of Polyethylene by Oxygen Acceptors* (Preprint), 1990, Moscow, IchF AS USSR, 64 p. (Rus)
24. Mashukov N.I., Gladyishev G.P., and Kozlov G.V., *Vysokomol. Soedin.*, 1991, vol. A33(12), pp. 2538 – 2546. (Rus)
25. Malamatov A.Kh., Mashukov N.I., and Kozlov G.V., *Izv. KBNTs RAN*, 1999, No. 3, pp. 65 – 68. (Rus)
26. Mashukov N.I., Serdyuk V.D., Kozlov G.V., and Khatsukova M.A., *Interphase Zone Structure and Blow Viscosity of High Density Polyethylenes Modified by High-Dispersion Fe/FeO Mixture*, Manuscript deposited to VINITI RAS, Moscow, June 20, 1994, No. 1538-V94. (Rus)
27. Glenz W. and Peterlin A., *J. Macromol. Sci. Phys.*, 1970, vol. B4(3), pp. 473 - 490.
28. Flory P.J., *Polymer J.*, 1985, vol. 17(1), pp. 1 - 12.

29. Kozlov G.V. and Sanditov D.S., *Anharmonic Effects and Physicomechanical Properties of Polymers*, 1994, Novosibirsk, Nauka, 261 p. (Rus)

30. Malinovsky V.K., Novikov V.N., and Sokolov A.P., *Uspekhi Fizicheskikh Nauk*, 1993, vol. **163**(5), pp. 119 – 124. (Rus)

31. Bagryansky V.A., Malinovsky V.K., Novikov V.N., Pushchaeva L.M., and Sokolov A.P., *Fizika Tverdogo Tela*, 1988, vol. **30**(8), pp. 2360 – 2366. (Rus)

32. Zemlyanov M.G., Malinovsky V.K., Novikov V.N., Parshin P.P., and Sokolov A.P., *Zhurnal Eksperimental'noi I Teoreticheskoi Fiziki*, 1992, vol. **101**(1), pp. 284 – 293. (Rus)

33. Klinger M.I., *Fizika I Khimia Stekla*, 1989, vol. **15**(3), pp. 377 – 396. (Rus)

34. Karpov V.G. and Parshin D.A., *Pis'ma v ZhETF*, 1983, vol. **38**(11), pp. 536 – 539. (Rus)

35. Dyadyina G.A., Karpov V.G., Soloviev V.N., and Khrisanov V.A., *Fizika Tverdogo Tela*, 1989, vol. **31**(4), pp. 148 – 154. (Rus)

36. Duval E., Boukenter A., and Achibat T., *J. Phys.: Condens. Matter*, 1990, vol. **2**(11), pp. 10227 - 10234.

37. Novikov V.N. In Coll.: *Advanced Solid State Chemistry*, Amsterdam, Elsevier, 1989, p. 349.

38. Achibat T., Boukenter A., Duval E., Lorentz G., and Etienne S., *J. Chem. Phys.*, 1991, vol. **95**(4), pp. 2949 - 2954.

39. Graessley W.W. and Edwards S.F., *Polymer*, 1981, vol. **22**(10), pp. 1329 - 1334.

40. Kavassalis T.A. and Noolandi J., *Macromolecules*, 1988, vol. **21**(9), pp. 2869 - 2879.

41. Kozlov G.V., Belousov V.N., and Mikitaev A.K., *High Pressure Physics and Technology*, 1998, vol. **8**(1), pp. 101 – 107. (Rus)

42. Aloev V.Z., Kozlov G.V., and Beloshenko V.A., *Izvestiya VUZov, Severo-Kavkazsky Region, Estestvennye Nauki*, 2001, No. 1(113), pp. 53 – 56. (Rus)

43. Sokolov A.P., Kisliuk A., Soltwisch M., and Quitmann D., *Phys. Rev. Lett.*, 1992, vol. **69**(10), pp. 1540 - 1543.

44. Belousov V.N., Kozlov G.V., Mikitaev A.K., and Lipatov Yu.S., *Doklady AN SSSR*, 1990, vol. **313**(3), pp. 630 – 633. (Rus)

45. Mermet A., Duval E., Etienne S., and G'Sell C., *Polymer*, 1996, vol. **37**(4), pp. 615 - 623.

46. Kozlov G.V., Beloshenko V.A., Gazaev M.A., and Novikov V.U., *Mekhanika Kompozitnyikh Materialov*, 1996, vol. **32**(2), pp. 270 – 278. (Rus)

Chapter 5.

Interrelation between the cluster model and modern physical concepts

5.1. FRACTALITY OF THE POLYMER STRUCTURE AND THE CLUSTER MODEL

During recent 20 years methods of fractal analysis were widely spread in theoretical physics [1] and general material science [2], in physicochemistry of polymers [3 – 8], in particular. This tendency is explained by a broad spreading of fractal objects in the nature.

There are two basic factors determining interrelation between the local order and fractal origin of the polymer solid phase: thermodynamic non-equilibrium and dimensional periodicity of the structure. The connection between the local order and thermodynamic non-equilibrium was discussed above (Section 2.2). A simple relation between the non-equilibrium index, which is the Gibbs function variation, $\Delta \widetilde{G}^{im}$, at polymer self-assembly (the cluster structure formation), and a relative part of clusters, φ_{cl}, as follows:

$$\Delta \widetilde{G}^{im} \sim \varphi_{cl}. \tag{5.1}$$

Figure 5.1 shows graphical presentation of this relation for amorphous glassy PC and PAr polymers. Since at $T = T_g(T_m)$ $\Delta \widetilde{G}^{im} = 0$ [10, 11], the relation (5.1) indicates necessary complete decomposition of the cluster system ($\varphi_{cl} = 0$) at these temperatures or transition to thermodynamically equilibrium structure.

As for interrelation between fractality and thermodynamic non-equilibrium, Hornbogen [12] has indicated that exactly thermodynamically non-equilibrium processes form fractal structures. The classic example of this regularity fulfillment is represented by fracture surfaces of solids: many experimental researches have indicated their fractality independently of thermodynamic conditions of the solid itself [12]. This is stipulated by thermodynamic non-equilibrium of the degradation process [13]. This very fact does also determine experimentally proved fractality of the polymer structure [14 – 16]. Similar to every real (physical) fractal, fractal properties of the polymer structure are limited by definite linear scales. For example, in works [14, 17] these scales are limited by several angstroms (the lower limit) to several ten angstroms (the upper limit). The lower limit is associated with the finite size of structural elements and the upper limit with irregular approaching the limit of the fractal dimensionality of the structure, d_f [18]. The above-mentioned scales are in good keeping with the border dimensions of the cluster structure: the lower with the static segment length, l_{st}, and the upper with the distance between clusters, R_{cl} [19].

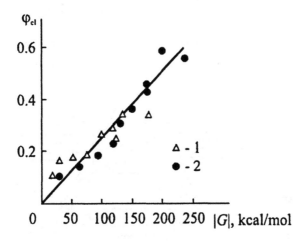

Figure 5.1. Dependence of the relative cluster part, φ_{cl}, on the absolute value of specific Gibbs function for non-equilibrium phase transition $\left|\Delta G^{im}\right|$ for PC (1) and PAr (2) [8]

In these scales, fractality of the polymer structure assumes dependence of density, ρ, on scale, L, of the following shape (Figure 5.2) [1]:

$$\rho \sim L^{d_f - d}, \tag{5.2}$$

where d is dimensionality of the Euclidean space, to which the fractal is inserted.

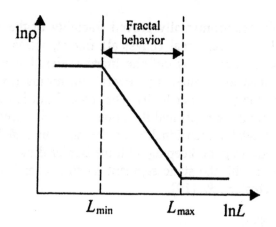

Figure 5.2. Dependence of density ρ on linear scale L for a solid in double logarithmic coordinates. Fractal behavior is observed in the L_{min}-L_{max} range [18]

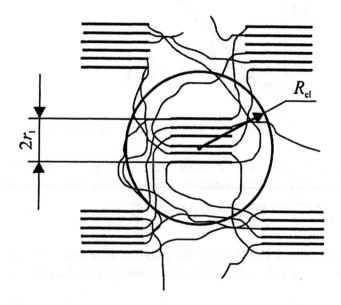

Figure 5.3. Schematic presentation of the cluster structure of the polymer amorphous state [5]

Figure 5.3 presents schematic image of the cluster structure of the polymer amorphous state. Clearly within the borders of the above-mentioned dimensional periodicity data in Figures 5.2 and 5.3 correspond to one another, i.e. the cluster model suggests ρ decrease with removing from the cluster center. At the same time, the Flory "felt" model [20] does not fit this criterion (for it ρ = const). Since fractality of the polymer structure was proved in experiments [14 – 16], it is obvious that the cluster model conforms to the structure of real solid polymers, whereas the "felt" model does not correspond to it. Obviously the reverse dependence for final ρ is also true: ρ variation in definite limits necessarily suggests structure periodicity.

However, one should remember that fractal analysis gives just general mathematical description of polymers, i.e. it does not identify structural units (elements), from which any real polymer is formed. As indicated in Chapter 2, physical description of thermodynamically non-equilibrium polymer structure in the framework of the local order ideas gives the cluster model of the polymer amorphous state structure that quantitatively identify its elements. Since these models consider the polymer structure somewhat from two sides, they are excellent completing one another [21].

It is common knowledge [22] that structures displaying fractal behavior on small length scale and being homogeneous on large length scale are named homogeneous fractals. These fractals are percolation clusters at the percolation threshold [1]. As shown below, the cluster structure represents the percolation system and, due to the above-said, the homogeneous fractal. To put it differently, the presence of the local order in the condensed phase of polymers determines fractality of their structures.

Fractal dimensionality, d_f, of the percolation system can be expressed by the equation as follows [22]:

$$d_f = d - \frac{\beta_p}{v_p},$$ (5.3)

where β_p and v_p are critical indices in the percolation theory.

As a consequence, the condition $\beta_p \neq 0$ that follows from Table 5.1 does also determine fractality of the polymer structure. Obviously this condition and, consequently, fractality of the polymer structure determined temperature dependence of φ_{cl}, e.g. thermofluctuation origin of the local order. To put it differently, the structure fractality and the local order "frozen" below T_g (φ_{cl} = const) represent mutually exclusive notions.

Table 5.1

Percolation cluster characteristics [5]

Parameter	Experimentally determined indices		Theoretical "configuration" indices [22]
	EP-1	EP-2	
β_p	0.54	0.58	0.40
v_p	1.20	1.15	0.88
β_p/v_p	0.45	0.504	0.455
d_f	2.55	2.50	2.545

The above-considered statements allow binding the fractal dimension, d_f, and local order characteristics (φ_{cl} or v_{cl}) by an analytical relation. The following equation binds d_f value to the Poisson coefficient v [18]:

$$d_f = (d-1)(1+v); \qquad (5.4)$$

in its turn, v value is associated with v_{cl} by the relation (3.14). Combination of equations (2.11), (3.14) and (5.4) allows obtaining of the interconnection between d_f and φ_{cl} or v_{cl} as follows:

$$d_f = 1.5(d-1) - 3.22(d-1)\times 10^{-10}\left(\frac{\varphi_{cl}}{SC_\infty}\right)^{1/2}, \qquad (5.5)$$

or

$$d_f = 1.5(d-1) - 3.22(d-1)\times 10^{-10}(l_0 v_{cl})^{1/2}. \qquad (5.6)$$

These expressions, most frequently used for the case $d = 3$, are reduced to the simple form as follows [4]:

$$d_f = 3 - 6.44\times 10^{-10}\left(\frac{\varphi_{cl}}{SC_\infty}\right)^{1/2}, \qquad (5.7)$$

or

$$d_f = 3 - 6.44\times 10^{-10}(l_0 v_{cl})^{1/2}. \qquad (5.8)$$

Turning back to correlation (5.1), note that its substitution into equation (5.5) or (5.7) allows obtaining of the following expression (for $d = 3$) [9]:

$$\Delta \widetilde{G}^{im} \sim C_\infty S\big(3-d_f\big)^2 .\tag{5.9}$$

The Gibbs specific function of supermolecular structure formation, $\Delta \widetilde{G}^{im}$, is given for non-equilibrium "overcooled liquid – solid" phase transition [11]. As follows from the relation (5.9), $\Delta \widetilde{G}^{im} = 0$ at $d_f = 3$, e.g. at $d_f = d$ and transition to the Euclidean behavior. To put it differently, fractal structures are induced by non-equilibrium processes only, as it has been assumed before [12].

One more noteworthy analogy can be traced between glassy polymers and homogeneous fractals. It is known [23] that the mentioned polymers do not fit the Prigozhin-Defye criterion that means that to be described they require more than one order parameter. Concerning Euclidean objects (spaces), they possess translational symmetry and, consequently, require one parameter to be described, namely, the Euclidean dimensionality d [24]. Fractal objects (spaces) possess dilatation symmetry and, therefore, require three dimensionalities for characterization: the Hausdorff d_f, spectral d_s, and d. Even if in the latter case d is assumed constant and equal 3, correct description of the fractal requires, at least, two dimensionalities: d_f and d_s. Even this qualitative comparison proves fractality of the glassy polymer structure.

In the end of Section 2.1 general ideas for the benefit of the local order existence in polymers were considered based on the Ramsey theorem [18, 21]. By analogy, it can be proved that any structure consisting of N elements ($N > B_N(i)$) represents an aggregate of a finite number $k \leq j$ of self-similar structures enclosed in one another, the Hausdorff dimensionality of which, in general, can be different. This means that any structure consisting of quite large number of elements, independently of its physical origin, represents a multifractal (fractal, in particular) and is characterized by a spectrum of Regnier dimensionalities d_q, $q = 0, 1, 2, \ldots$ [18]. It has been shown [25] that the tendency of condensed systems to self-organization in the scale-invariant multifractal forms originate from the fundamental thermodynamic principles of open systems, and d_q is determined by the competition of short- and long-range interatomic correlations, which determine volumetric compressibility and shear rigidity of solids, respectively.

As mentioned above, the cluster model and the fractal analysis are complement approaches and their combined application discloses new possibilities for simulation of the polymer structure. This point is considered on the example of forecasting parameters of polymer network cluster structure [26]. It is known [27 – 29] that computer simulation of the processes has induced a conclusion that in the gel-formation point cross-linked backbones of these polymers represent fractal structures, cross-linking frequency increase inducing an increase of the fractal dimensionality of such backbone. However, from positions of practice the cross-linking frequency is not suitable enough, though at the present time polymer networks are preferably characterized using the value v_{nw} [30]. Nevertheless, it has been shown [31] that at constant v_{nw} for one and the same polymer network, different properties (for example, different T_g) can be obtained. This is explained by the change of supermolecular (cluster) structure of polymer network during physical aging, namely, by increase of the local order degree [31]. In its turn, the local order degree determines d_f value (equations (5.5) – (5.8)). Thus there occurs a possibility of forecasting structural characteristics (and as a consequence properties) of polymer networks in the glassy state basing on chemical cross-linking degree in the gel formation point. One of possible alternatives of solving this problem was discussed [32] on the example of typical representatives of the polymer network class – epoxy polymers of amine (EP-1) and anhydride (EP-2) cross-linking type, as well as their homologues solidified under pressure of 200 MPa (EP-1-200 and EP-2-200).

The relation between v_{nw} and d_f of studied EPs was deduced on the basis of the model suggested in the work [33]. The essence of this model is as follows. Competing processes of crystal growth and nucleation were described in the framework of two kinetic equations [33]:

$$dR/dt = k_1 c^m, \qquad (5.10)$$
$$dN/dt = k_2 c^p, \qquad (5.11)$$

where R is the crystal dimension; t is the current time; k_1 and k_2 are equilibrium constants; N is the number of crystals; c is the concentration of molecules at the surface; m and p are indices.

The application of equations (5.10) and (5.11) to description of the local order zones (clusters) nucleation and growth in EP is based on the fact that clusters in itself represent analogues of crystallites with extended chains (CEC) [34]. The characteristic size of cluster is accepted equal the length of segments in it, i.e. the statistic segment length l_{st}. Similar to [33], $m = 1$, because cluster growth represents addition of a single segment with the length l_{st} [35]; therefore, $p > 1$, because cluster nucleation requires occurrence of two or more segments in the nucleation point. However, application of the equation (5.10) to a single separate EP requires sufficient modification. Obviously, if $R = l_{st}$ is accepted, the equation (5.10) becomes meaningless, because for every studied EP $l_{st} =$

const and $dl_{st}/dt = 0$. The latter means that clusters has not the ability to grow. That is why, denoting the number of segments in a single cluster by n_{seg}, the equation (5.10) will be presented as follows [32]:

$$\frac{dn_{seg}}{dt} = k_1 c. \tag{5.12}$$

The border value of n_{seg} is limited by nucleation of new clusters of the thermofluctuation origin. That is why [33]:

$$l_{st} = \frac{\left(dn_{seg}/dt\right)}{\left(dN/dt\right)}. \tag{5.13}$$

Obviously the part of segments of macromolecules, c, capable of forming clusters must be decreased with time during the cluster formation and, taking into account the condition $l_{st} = $ const for every EP, n_{seg} must also be decreased. This fact explains the presence of unstable (possessing low n_{seg}) clusters in the structure of the polymer amorphous state [36, 37]. This also explains self-similarity of the cluster structure in the scale range of its existence. Note back again the important circumstance that the system of equations (5.10) and (5.11) can be applied to the totality of studied EPs, and the system of equations (5.11) and (5.12) is applicable to any separate EP of the totality [32].

For the totality of studied EPs, one can write down the following equations [33]:

$$N(l_{st}) = \left(dN_{cl}/dt\right) t_{gel}, \tag{5.14}$$

and

$$l_{st} \sim \frac{\left(dl_{st}/dt\right)}{\left(dN_{cl}/dt\right)}, \tag{5.15}$$

where N_{cl} is the number of clusters; t_{gel} is the time interval of the gel formation.

The relation (5.15) forecasts C_∞ decrease with cross-linking frequency, v_{nw}, increase and consequent l_{st} decline. This intensifies nucleation of new clusters, i.e. N_{cl} increase. Assuming $R = l_{st}$, $m = 1$ and dividing equation (5.10) by (5.11), the following correlation is obtained:

$$l_{st} \sim \frac{dl_{st}/dt}{dN_{cl}/dt} \sim \frac{c}{c^P} = c^{1-P},$$

(5.16)

because k_1 and k_2 are constants.

The equation (5.14) combined with the expression (5.11) under the conditions t_{gel} = const, k_2 = const and, as before, $N = N_{cl}$, gives the following correlation:

$$N_{cl}(l_{st}) = k_2 t_{gel} c^P,$$

(5.17)

or with regard to $k_2 t_{gel}$ = const:

$$N_{cl}(l_{st}) \sim c^P.$$

(5.18)

Equations (5.10) and (5.11) allow determination of c as follows:

$$c \sim l_{st}^{-1/(p-1)},$$

(5.19)

and

$$c \sim N_{cl}^{1/p}.$$

(5.20)

Making equal the right parts of equations (5.19) and (5.20), it is obtained that [32]:

$$N_{cl}^{1/p} \sim l_{st}^{-1/(p-1)}$$

(5.21)

or

$$N_{cl}(l_{st}) \sim l_{st}^{-p/(p-1)}.$$

(5.22)

It has been indicated [33] that the classic nucleation theory requires $p \leq 2$ and, consequently, the condition occurs as follows:

$$\frac{p}{p-1} \geq 2.$$

(5.23)

Moreover, there is one more restriction that states that the volume of clusters must not exceed total volume of the polymer, wherefrom one more condition follows [33]:

$$\frac{p}{p-1} \leq 3. \tag{5.24}$$

Thus index in the relation (5.22) falls within the same limits as the fractal dimensionality of the object in the three-dimensional space [34]. The accurate interrelation between $p/(p-1)$ and d_f can be obtained from comparison of the expression (5.22) with the ones deduced in the work [39]:

$$N_{cl}(l_{st}) \sim l_{st}^{-d_f}, \tag{5.25}$$

therefore [32]:

$$p/(p-1) = d_f. \tag{5.26}$$

The value N_{cl} can be presented as follows [34]:

$$N_{cl} = \frac{V_{cl}}{n_{seg}} = \frac{2v_{cl}}{F}. \tag{5.27}$$

One more important circumstance shall be mentioned. The comparison of equations (5.11) and (5.25) indicates that the maximal rate of the cluster formation is reached at $p = 2$ or $d_f = 2$. To put it differently, it is assumed that the minimal fractal dimensionality of the polymer structure is corresponded to the maximal local order.

In the framework of the cluster model [34] polymer network can be considered as a superposition of two carcasses – chemical cross-link network and physical entanglement cluster network possessing frequencies v_{nw} and v_{cl}, respectively. As each of these carcasses is accepted as the ideal structure (i.e. the effect of end chains is neglected), one can write down as follows [40]:

$$M_c^{eff} = \frac{M_c f}{2}, \tag{5.28}$$

where M_c^{eff} is the effective molecular mass of a chain part between chemical cross-link points; M_c is the value of the latter; f is the functionality of chemical cross-link point. For the cluster network, similar expression is presented by the equation (2.4).

With regard to equations (2.4), (5.27) and (5.28), the relation (5.25) can be reduced to the following form [32]:

$$\frac{2v_{nw}}{f} \sim l_{st}^{-d_f} .$$

(5.29)

An equation linking parameters C_∞ and d_f for $d = 3$ was deduced [41]. Full deduction and complete form of this equation will be given in the following Section:

$$C_\infty = \left(\frac{d_f}{3(3 - d_f)} + \frac{4}{3} \right).$$

(5.30)

Combined equations (5.29) and (5.30) give the following expression [32]:

$$l_0^{d_f} \left(\frac{d_f}{3(3 - d_f)} + \frac{4}{3} \right)^{d_f} = 3.75 \times 10^{28} \frac{f}{v_{nw}},$$

(5.31)

where the empirical numerical coefficient in the right part of the equation is obtained by equaling d_f values, calculated from equations (5.4) and (5.31), at $v_{nw} = 10^{27}$ m^{-3} [32].

Figure 5.4 compares d_f values, calculated by equations (5.4) and (5.31) as the function of v_{nw}, for studied EPs under the conditions $l_0 = 1.25$ Å [5] and f = 4. Clearly good correspondence of two selections of the fractal dimensionalities was obtained. This means that the increase of the carcass fractal dimensionality before the gel formation point [29] induces a decrease of the fractal dimensionality d_f of the supermolecular structure of polymer network in the glassy state. It should be noted also that knowledge of d_f allows the forecast of EP properties [35, 42]. Thus varying K_{st} or terminating the reaction at a definite stage, one can obtain polymers with required properties. However, from positions of practical significance, application of stoichiometric values K_{st}

= 1.0 is more preferable. It gives the most stable systems [18], including polymer networks [35]. In this case, properties of the final product can be varied by changing functionality of the cross-linking agent, *f*, or molecular characteristics of epoxy oligomer.

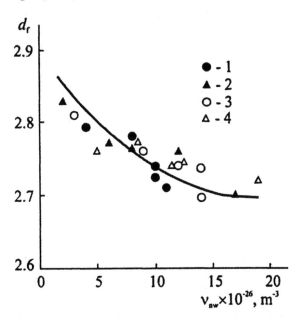

Figure 5.4. Dependence of fractal dimensionality, d_f, of EP structure on cross-linking frequency, v_{nw}, calculated as follows: by equation (5.4) for EP-1 (1), EP-2 (2), EP-1-200 (3) and EP-2-200 (4), and by equation (5.31) for all EPs [32]

Thus fractal analysis application allowed clearing up an interrelation between polymer network characteristics before the gel formation point (or at it) and in the glassy state. Note that this can be made for just synthesized polymer only. Further on, d_f will vary due to physical aging processes, unavoidable due to thermodynamically non-equilibrium origin of the polymer glassy state. Nevertheless, the methodology suggested allows forecast of properties of glassy polymer networks already at the cross-linking stage. Moreover, the model considered forecasts qualitatively previously expected effects, namely, the existence of unstable clusters [36, 37], self-similarity of the cluster structure [5, 34] and d_f decrease with the local ordering increase.

For a brief resume, it should be noted that in the case of glassy polymers, notions of the local order and fractality are tightly connected basing on the fundamental ideas of physics. This interrelation is expressed by simple analytical equations (5.5) - (5.8). It is also shown that combined application of these complement concepts allows significant broadening and improvement of the analytical description of the polymer structure and properties. The mentioned combination will be multiply used in the future discussion.

5.2. CLUSTERS AS DISSIPATIVE STRUCTURES

Obtaining of relations between structure and properties of polymers was always the foreground task of the polymer physics. However, solution of this problem is significantly complicated by the presence of a multilevel structure in polymers – molecular, topological, supermolecular (super-segmental), etc. That is why a polymer structural model is required, which would allow association of structural parameters of the mentioned levels by quantitative relations. Obviously such relations should better be deduced on the basis of fundamental physical principles. Recent rapid development of synergism of non-equilibrium systems [43, 44] and fractal analysis allows application of these fundamental concepts to solution of the tasks mentioned, because a solid-phase polymer represents thermodynamically highly non-equilibrium system.

The basic feature differing polymers from other classes of solids is their structure, composed of long chain molecules. That is why the assumption seems natural that the structural hierarchy of polymers must be based on their molecular structure. Recently, the characteristic relation C_∞ [45 - 47] is more and more frequently used as the molecular parameter for description of polymer properties, hence substituting, for example, such characteristics as relation a/δ (where a in the chain "thickness" and δ is its rigidity [48]). To the point of view of Boyer and Miller [48], this is stipulated by higher uncertainty of δ compared with C_∞, because the former depends upon valence angles of bonds in the macromolecule and the latter on the length of these bonds. Usually, bond lengths are determined with significantly higher accuracy than the valence angles. Not contradicting to the rightfulness of such conclusion, nevertheless, the authors assume that the reasons for broad application of C_∞ parameter are much deeper and fundamental; one of the objectives of the current Section is demonstration of declared fundamentality of C_∞ in the framework of thermodynamics of non-equilibrium processes [43, 44].

The behavior of a deformable solid subject to mechanical impact is determined by formation (decomposition) and evolution of dissipative structures (DS) providing for the optimal mode of supplied energy dissipation [18, 25, 49]. This general statement is also true for polymers. Broad application of polymers as construction materials requires the detail study of their mechanical properties which is impossible without understanding of fundamental processes, on which their behavior under load is based. At present, it is known well that polymers possess a local order that exists in them in the form of dissipative structures (DS) [50 – 52]. However, direct application of fundamental principles of thermodynamics and statistical physics of non-equilibrium processes to the study of behavior of any class of materials is often quite difficult or even impossible without knowing particular quantitative model of the structure of this material. For such model, the cluster model of the polymer amorphous state will be used below.

As shown [18, 49], the feature of DS, also formed in polymers [50 – 52], is existence of a universal hierarchy of spatial scales (structural levels), which is reflected in the hierarchy of linear sizes of the structure. For example, even in elastically isotropic bodies, at least, three independent length scales do exist: l_p, l_ε, and l_e, bound with one another by the following relations [49]:

$$l_p = a_M; \quad l_\varepsilon = a_M \Lambda_0; \quad l_e = l_\varepsilon \Lambda_1, \tag{5.32}$$

where a_M is the linear size; Λ_0 and Λ_1 are self-similarity coefficients.

At personification of these length scales related to polymers the atomic level must be neglected, because polymer represents a solid consisting of long chain macromolecules which, in turn, are composed of various groups of atoms. That is why the minimum scale, a_M, in equation (5.32) for polymers must be the length of the macromolecule backbone, l_0, varied within the range of several angstroms [45, 53]. Then the next higher structural (segmental) level will be determined by the statistical segment length, l_{st}, determined from the equation (2.8). As a consequence, for this structural level, the self-similarity coefficient $\Lambda_0 = C_\infty$ under the condition $l_p = l_0$ [5, 41, 54, 55].

Topological level is the consequent structural level in polymers. Its parameters are determined by characteristics of macromolecular entanglement ("hooking") network in linear polymers and by parameters of chemical cross-link network in polymer networks. For this level, the distance between entanglements or chemical cross-link points, R_{nw}, is the typical linear scale, l_e, determined from the equation (2.14).

As a consequence, it follows from the above discussion that if for the consequent structural levels the self-similarity coefficient value ($A_0 = C_\infty$) is preserved, then for polymers the equation (5.32) can be reduced to the following form [5, 55]:

$$l_p = l_0; \quad l_\varepsilon = l_{st} = l_0 C_\infty; \quad l_e = R_{nw} = l_0 C_\infty^2. \tag{5.33}$$

Figure 5.5 shows dependence $R_{nw}\left(C_\infty^2\right)$, which is linear and passes through the origin of coordinates, though it is fractioned into two straight lines. We are not to analyze the latter circumstance, though it should be mentioned that there are some obvious reasons for it, among which the following are: firstly, broad dispersion of parameters from the literature (l_0, C_∞, v_{nw}) [45, 53], required for calculation; secondly, there are different approaches to interpretation of the value $l_\varepsilon = l_{st}$ [45, 53]. Nevertheless, proportionality of R_{nw} and C_∞^2, shown in Figure 5.5, proves that C_∞ is the self-similarity coefficient at DS-size scale transition from l_{st} to R_{nw}.

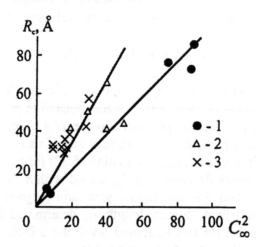

Figure 5.5. Dependence of the distance between entanglements (chemical cross-link points), R_{nw}, on the square characteristic relation, C_∞, for amorphous glassy (1) and amorphous-crystalline (2) polymers, and polymer networks (3) [5]

Figure 5.6 compares R_{nw} values, calculated from equations (2.14) and (5.33) (R_{nw1} and R_{nw2}, respectively). Data from the Figure indicate tight quantitative conformity between R_{nw1} and R_{nw2} for the majority of polymers. Possible reasons for occurrence of two lines have been already discussed. Thus C_∞ is regulated by the polymer topology, which can be changed by variation of v_{nw}. For a great number of linear polymers, Aharoni [53] has observed R_{nw} increase with the chain flexibility (C_∞ increase), expressed by the following equation:

$$R_{nw} = 10C_\infty. \tag{5.34}$$

For cross-linked polymers, similar equation as follows was obtained [56]:

$$R_{nw} = 5C_\infty. \tag{5.35}$$

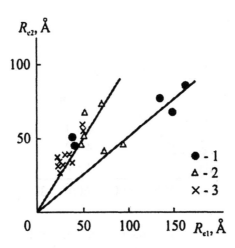

Figure 5.6. The relation between inter-entanglement (between chemical cross-link points) distances, R_{nw1} calculated from the equation (2.14) and R_{nw2} calculated from the equation (5.33), for amorphous glassy (1) and amorphous-crystalline (2) polymers, and polymer networks (3) [5]

One more specific length scale, which is the chain part length between entanglements or chemical cross-link points, L_c, does exist in polymers. It can be determined from the equation as follows [5]:

$$L_c = \frac{F}{2v_c S}.$$

(5.36)

Figure 5.7 shows the dependence of "transitional" chain length between entanglements (cross-link points), L_c, calculated by equation (5.36), on C_∞^3. It was also found linear and passing through the origin of coordinates. Expectedly, the straight line slope equals ~1 Å, i.e. possesses the order of magnitude of l_0 [53]. This indicates that for this structural level, typical of polymers, C_∞ represents the self-similarity coefficient [54, 55].

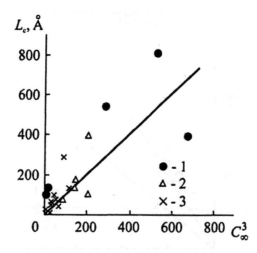

Figure 5.7. Dependence of macromolecule segment length between entanglements (chemical cross-link points), L_c, on cubic characteristic relation, C_∞^3, for amorphous glassy (1) and amorphous-crystalline (2) polymers, and polymer networks (3) [5]

DS evolution on the example of polymer networks during their deformation will be considered below using the cluster model for the structure personification. It has been shown [21] that elastic deformation of amorphous polymers is specifically realized in clusters, and the yield stress is the function of entanglement cluster network frequency, v_{cl}, only, i.e. the number of closely packed segments per specific polymer volume [57]. In the light of these facts the yielding can be considered as stability loss by clusters in the field of mechanical stresses. Cold flow of the polymer is considered as deformation of devitrified packless matrix, in which chemical cross-links are represented by clusters [57]. Such interpretation allows personification of structural changes proceeding at polymer deformation in the framework of the model [18, 21, 25].

As mentioned above, the feature of DS formed (or existing) in deformed solid is existence of the universal hierarchy of structural levels [25]. As Balankin has suggested [18, 49], the hierarchic structure of DS is based on the fundamental property of solids, the shearing stability that determines the difference of typical spatial scales of localization zones and dissipation of energy "injected" to deformed body under external impact. Typical scale of the zones (l_s), in which excessive energy is dissipated, is proportional to the shear modulus, G. Since in accordance with the above discussion, in the framework of the cluster model, these zones can be associated with the clusters, it is suitable to use their characteristic size, l_{st}, as l_s. Figure 5.8 shows relation between G and l_{st}, which is approximated by a straight line passing through the origin of coordinates. This proves correctness of l_{st} use for l_s.

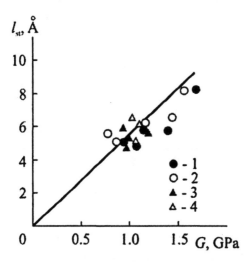

Figure 5.8. Relation between statistical segment length, l_{st}, and shear modulus, G, for EP-1 (1), EP-2 (2), EP-1-200 (3) and EP-2-200 (4) [5]

Concerning the next stage in the hierarchy of the structural levels, taking into account the above-mentioned interpretation of the cold flow process and equation (5.33), the distance between clusters, R_{cl}, should be taken for l_s. The typical size of injected energy localization zones, l_e, is proportional to $(B + 4/3G)$, where B is the uniform compression modulus [49]. The possibility of associating R_{cl} with l_e can be checked as follows. It is known [49] that the relation of DS spatial scales of neighbor structural levels, L_i^{DS} and L_{i+1}^{DS}, can be expressed by the following equation:

$$\frac{L_{i+1}^{DS}}{L_i^{DS}} = \frac{l_e}{l_s} = \frac{B + 4/3G}{G} = \Lambda_i, \qquad (5.37)$$

where Λ_i is the self-similarity coefficient.

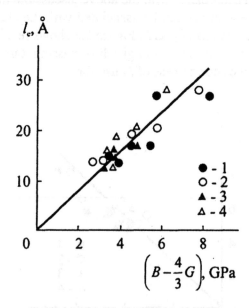

Figure 5.9. Correlation between characteristic dimensions of injected energy localization zones, l_e, and parameter $(B + 4/3G)$. For the legend, refer to Figure 5.8 [5]

Obviously l_e can be obtained by multiplying $(B + 4/3G)/G$ by l_s, which equals l_{st}. The relation between l_e and $(B + 4/3G)$, shown in Figure 5.9, was also found linear that proves principal possibility of using parameter l_e as the characteristic dimension of injected energy localization zones. However, at the structural level this assumption can be finally proved by comparison of l_e and R_{cl}. It displayed the presence of correspondence between these parameters, but also indicated systematic exceeding of R_{cl} over l_e by, approximately, 8 Å. The relation observed between l_e and R_{cl} is explained on the basis of schematically depicted cluster structure of amorphous polymers, shown in Figure 5.10. Obviously for obtaining l_e, the mentioned decrease of R_{cl} by 8 Å, which is approximately equal l_{st}, means that injected energy is localized not in the whole volume of the polymer, but in the packless matrix, because cluster with the characteristic dimension l_{st} is excluded from the consideration.

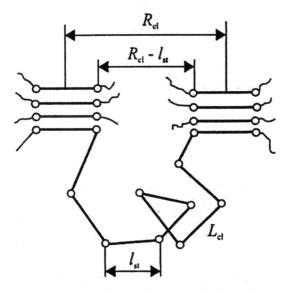

Figure 5.10. Schematic presentation of the cluster structure of amorphous polymer [5]

Comparison of l_e and $R_{cl} - l_{st}$ values, shown in Figure 5.11, proves this conclusion. As a consequence, structural characteristics l_{st} and $R_{cl} - l_{st}$ estimated in the framework of the cluster model, considered as spatial scales of DS in neighboring levels, allow assessment of the self-similarity coefficient of polymer mechanical deformation, Λ_i, from the equation (5.37). The interrelation of DS scales and injected energy localization processes allow an assumption of a definite correlation between Λ_i and dimensionality, D_f, of excessive energy localization zones in elastically deformable medium [18]. Parameter D_f value can be determined from the following equation [18]:

$$D_f = \frac{2(1-v)}{1-2v}.$$ (5.38)

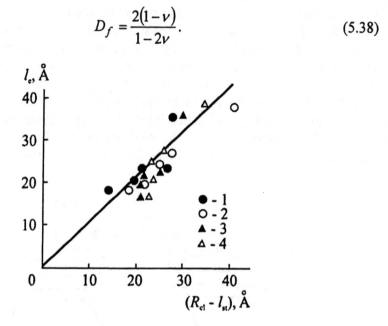

Figure 5.11. Correlation between characteristic size of injected energy
localization zone, l_e, and $R_{cl} - l_{st}$ value. For the legend, refer to
Figure 5.8. Straight line marks 1 : 1 relation [5]

Figure 5.12 shows comparison of D_f and the self-similarity coefficient,
Λ_i, for polymer networks. Clearly both absolute values and tendencies of
changes of these parameters conform well. Analogous correlation between R_{cl}
and $D_f - l_{st}$ was obtained [58] for amorphous, amorphous-crystalline and
oriented polymers. As a consequence, in the framework of thermodynamics of
non-equilibrium processes, yielding and cold flow processes are described well;
the cluster model describes structure of these materials properly. Obviously
clusters are identified as DS [5, 54, 55].

Basing on the above-stated, one may suggest that Λ_i in the equation
(5.37) also equals C_∞. Then this equation can be reduced to the following form
[5, 41]:

$$\frac{B}{G} + \frac{4}{3} = C_\infty.$$ (5.39)

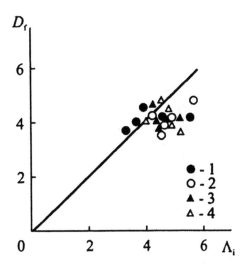

Figure 5.12. Correlation between dimensionality of injected energy localization zones, D_f, and self-similarity coefficient, Λ_i. For the legend, refer to Figure 5.8. Straight line marks 1 : 1 relation [5]

In their turn, B and G can be expressed via Young's modulus, E, and fractal dimensionality, d_f, of the polymer structure as follows [18]:

$$B = \frac{E}{d(d - d_f)},$$
(5.40)

$$G = \frac{(d-1)E}{2d_f}.$$
(5.41)

Substitution of equations (5.40) and (5.41) into (5.39) gives quantitative correlation between the molecular parameter C_∞, which characterizes statistical flexibility of the macromolecule [60] and compactness of macromolecular coil [53], and supermolecular structure characteristic d_f as follows [41]:

$$C_\infty = \frac{2f}{d(d-1)(d-d_f)} + \frac{4}{3}.$$
(5.42)

For the Euclidean three-dimensional space ($d = 3$), this equation is reduced to the form (5.30).

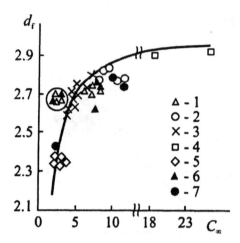

Figure 5.13. Dependence of the polymer structure fractal dimensionality d_f on characteristic relation C_∞. Experimental data: 1 – linear polymers (blow tests); 2 – linear polymers (d calculation by data from [59]); 3 – EP networks; 4 – ultrahigh-molecular polyethylene, produced by solid extrusion; 5 – polyarylate produced by solid extrusion at $T > T_g$; 6 – linear polymers (tensile tests); 7 – linear polymers (d_f calculation from small-angle X-ray diffraction results [15]). Continuous line presents calculation by the equation (5.42). The circle encloses experimental points with the highest deviation from the theoretical curve. For samples 7, experimental errors are indicated [41]

Figure 5.13 presents comparison of $d_f(C_\infty)$ dependencies, calculated by the equation (5.42) and obtained by different methods [41] for polymers, enumerated in the Figure legend. It is indicated that almost all samples shown display good conformity between the theory and the experiment, except for samples 5, experimental points for which in Figure 5.13 are enclosed in the circle. All these polymers (PC, PSF, and PAr) represent heterochain polymers with similar structures and properties, possessing relatively low (close to the minimum, as shown below) C_∞ values (2.2 – 2.4) [45, 53]. Samples of these polymers were produced by press molding (for blow tests) or cut from films, prepared by solution pouring on a glass support (for tensile tests). At the same time, data for PAr, prepared by solid extrusion at temperatures above T_g, conform much better with the calculation by equation (5.42). In this method of PAr sample preparation [61] devitrified polymer relaxes rapidly at the extrusion nozzle output that gives high v_{cl} value and induces formation of relaxed packless matrix. A state formed is much closer to the equilibrium, than for press molded or film samples [61]. As a consequence, equations (5.30) and (5.42) give the more accurate d_f values, the closer the structure of the polymeric sample approaches thermodynamically equilibrium one. To put it differently, for the current polymers, equations (5.30) and (5.42) give minimal d_f values [41].

Further on, variation limits for C_∞, determined from the equation (5.42), will be appraised. As is known [1, 18], d_f values for bulky samples at $d = 3$ vary within the limits as follows: $2 \leq d_f < 3$. At $d_f = 2$, parameter C_∞ possesses the minimal value equal 2. As observed from the literature [45, 53, 59], many polymers do not display $C_\infty < 2$. Concerning the minimal value of C_∞, at $d_f = 3$ it approaches the infinity. Actually, it has been indicated [53] that for some polymers C_∞ can reach ~700, which is equivalent of the mentioned tendency in practice, if the asymptotic run of $d_f(C_\infty)$ curve at large C_∞ is considered (Figure 5.13).

The expression binding dimensionalities D_f and d_f is the following [18]:

$$D_f = 1 + \frac{1}{d - d_f}.$$ (5.43)

For $d = 3$, it is reduced to the following form:

$$D_f = \frac{4 - d_f}{3 - d_f}.$$ (5.44)

Substitution of equation (5.44) into (5.30) gives the accurate result:

$$C_\infty = D_f$$ (5.45)

that proves the empirical correlation shown in Figure 5.12.

Two-level DS (micro- and macro-DS) are formed with the polymer network structure [62]. Micro-DS represent the local order zones (clusters), and their formation is due to high viscosity of the reaction mixture during the gel formation. This induces turbulence of viscous media and formation of oriented zones [18].

One more fundamental property of turbulent flows is fractality of the structures formed by them [18]. It is accepted that in three-dimensional turbulent flows energy dissipation mainly proceeds on a multitude with the fractional (fractal) dimensionality. However, experimental data on the rate fluctuation moments testify about impossibility of describing small-scale properties of the turbulent flow with the help of a self-similar fractal. That is why "inhomogeneous fractals", the rules of formation of which at every stage of the scale hierarchy are random in accordance with some probability distribution, are applied to description of turbulent DS. Hence, the energy transfer is described with the help of a random fragmentation model under the assumption that different stages of the process do not correlate. Therefore, the fractal volume does not possess the global invariance property in relation to similarity transformation. Nevertheless, one can calculate fractal dimensionality d_f determined by the relation as follows [18]:

$$\langle N_n \rangle \sim L_n^{-d_f},$$ (5.46)

where N_n is the number of active whirls at the n-th fragmentation step; β_n is the part of the volume occupied by whirls of the L_n scale. Therefore:

$$d_f = 3 + \log_2\{\beta\}.$$ (5.47)

The symbol $\{...\}$ indicates averaging by $\rho(\beta)$ distribution. It shall be expected that due to the above-mentioned connection between the turbulence and the order, a definite part of active whirls is finally transformed into clusters. That is why the relative part of clusters, φ_{cl}, must be proportional to β. Figure 5.14 shows the relation $\varphi_{cl}(\beta)$ that proves the effect of turbulence of the local order zone formation in epoxy polymers. The dependence presented does not pass through the origin of coordinates. This means that not all active whirls are transformed into clusters, by just a definite part of them. The part of transformed whirls depends on β: the higher β is, the greater the part of such whirls is. Note also that the reverse interrelation does also exist: the presence of the local order in the structure of polymers induces deformation turbulence at the cold flow plateau [63, 64].

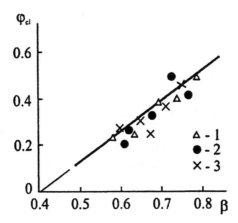

Figure 5.14. Dependence of the relative part of clusters, φ_{cl}, on the part of volume, β, occupied by active whirls, for sulfur-containing epoxy polymers cured by isomethyltetrahydrophthalic anhydride (1), 4,4'-diaminodiphenylmethane (2), and 4,4'-diaminodiphenylsulfone (3) [62]

The amorphous solid is considered as a fractal non-equilibrium nonergodic system [65] with hierarchically subordinate levels of the structural organization of self metastable defective states. Remind that in [34] the parts of chains participating in the clusters are taken for linear defects of the structure, analogous to dislocations in metals. In this case, dependencies of the cluster ensemble realization probability on the distance l between clusters in the ultrametric space can be presented as follows [65]:

$$\omega_\omega(l) \sim \exp(-l/l_n), \tag{5.48}$$

$$\omega_s(l) \sim l^{-D}, \tag{5.49}$$

where $\omega_\omega(l)$ describes low hierarchic systems; $\omega_s(l)$ describes highly hierarchic systems; l_n and D are positive parameters.

In this connection, two main questions occur: how good concordance between the concepts of hierarchic organization of the polymer amorphous state structure presented in [34] and [65] (equations (5.33) and (5.48), (5.49)) is, and what type of hierarchy structure is observed in polymers. The answer to these questions is given [66] on the example of 16 amorphous glassy and amorphous-crystalline polymers and polymer networks.

Similar to the works [67, 68], it is suggested that relative part of the local order zones (clusters), φ_{cl}, organizing the hierarchy of the polymer amorphous state structure must be a function of ensemble realization probability for such clusters ($\omega_\infty(l)$ or $\omega_s(l)$). To calculate probabilities $\omega_\infty(l)$ and $\omega_s(l)$ by equations (5.48) and (5.49), respectively, the following assessments were used. The value l is accepted equal R_{cl} and is calculated as shown in Chapter 2; $l_n =$ const $= 100$ Å, and this selection is stipulated by the fact that 100 Å size represents the upper linear scale of the fractal behavior of the polymer condensed state [4, 5]. Finally, index D in the equation (5.49) is equaled to the fractional part of the fractal dimensionality d_f of the polymer structure as follows:

$$D = d_f - 2, \tag{5.50}$$

because the fractional part, d_f, characterizes the structure change ($2 \le d_f < 3$ [5]).

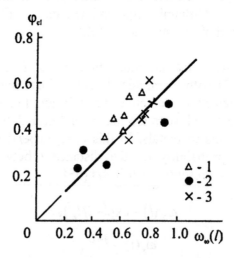

Figure 5.15. Dependence of the relative part of clusters, φ_{cl}, on low hierarchy system realization probability, $\omega_\infty(l)$, for amorphous glassy (1) and amorphous-crystalline (2) polymers, and polymer networks (3) [66]

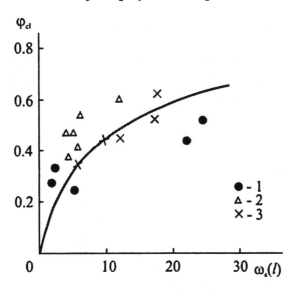

Figure 5.16. Dependence of the relative part of clusters, φ_{cl}, on the high hierarchy system realization probability, $\omega_s(l)$, for amorphous glassy (1) and amorphous-crystalline (2) polymers, and polymer networks (3) [66]

Figures 5.15 and 5.16 show dependencies of φ_{cl} on cluster ensemble realization probabilities, $\omega_\omega(l)$ (low hierarchy systems) and $\omega_s(l)$ (highly hierarchy systems), respectively. Clearly for all three classes of polymers under consideration, the correlation $\varphi_{cl}(\omega_\omega(l))$ is approximated by a linear dependence passing though the origin of coordinates, whereas the correlation $\varphi_{cl}(\omega_s(l))$ clearly represent the curve. Linearization of the latter correlation in the form as follows: $\varphi_{cl} = f[(\omega_s(l))^{1/2}]$, is shown in Figure 5.17, which indicates satisfactory linear approximation of this function by a straight line passing though the origin of coordinates. As a consequence, the dependence $\varphi_{cl}(\omega_\omega(l))$ is much stronger (magnitude 1) than the dependence $\varphi_{cl}(\omega_s(l))$ (magnitude 1/2). To put it differently, in the second case, the hierarchy structure is formed much less intensively than allowed by the probability $\omega_s(l)$. These results assume that hierarchic organization of the polymer supermolecular (supersegmental) structure occupies the intermediate position between high and low hierarchy systems, shifted towards the latter. This situation is explained by the macromolecular origin of polymers. Segments forming the hierarchy, which participate in the clusters, are linked by strong covalent bonds to the polymeric chain that strictly limits their ability to form similar structures. Moreover, the above-mentioned data have proved the hierarchy of supermolecular (supersegmental) structure of the amorphous state and conformity of models [34] and [65] for three main classes of polymers. Note also that the basic reason for distribution of data in Figures 5.15 – 5.17 is application of C_∞ values from the literature, which can vary in significantly broad range for the same polymer [59].

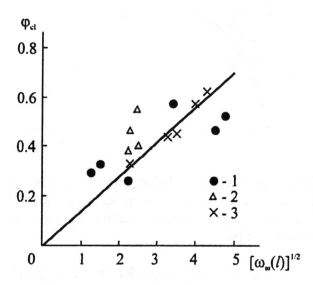

Figure 5.17. Dependence of the relative part of clusters, φ_{cl}, on parameter $[\omega_s(l)]^{1/2}$ for amorphous glassy (1) and amorphous-crystalline (2) polymers and polymer networks (3) [66]

To complete this Section, the authors discuss the way of obtaining two fundamental conclusions in the framework of synergism and fractal analysis: dependence of the "hooking" network frequency on statistical flexibility of the chain and increase of the physical entanglement network frequency (e.g. cluster network formation) in the polymer glassy state. The idea of macromolecular entanglements' network is one of the basic positions in the rheology of polymers [45], though attempts to apply other approaches shall also be mentioned [69]. Multiple attempts were also made to apply a model of such network, cross-link points of which are interpreted as macromolecular "hookings" [45], to description of glassy polymer properties (for example, [70, 71]). This has induced the increased interest to investigation of possible correlations between molecular characteristics of polymers and topology of macromolecular entanglement network [48, 53, 72]. Essentially, the authors of these works have come to the single point of view that macromolecular "hooking" network frequency, v_{nw}, increases with polymer chain rigidity. However, it should be noted that parameter v_{nw} is experimentally determined for the rubber-like state of polymers (at $T > T_g$); it is assumed that at transition to the glassy state v_{nw} does not change. Disadvantages of this interpretation became clear very soon. Actually, if cross-link points of such molecular hookings are considered as the points of topological chain fixation, then the assumption of equality of their frequencies in the rubbery and glassy states is very ambiguous. For example, polymer deformability decrease at transition from the rubbery to the glassy state is known well, but according to the indicated assumption this change must not happen, and for both cases, the stretch rating λ is determined from the known formula [73]:

$$\lambda = n_{st}^{1/2}, \tag{5.51}$$

where n_{st} is the number of statistical segments in the chain intersect between entanglements.

To explain the observed change in properties at coming over T_g, a series of models has been suggested. For example, it is assumed [74] that chain sliding through entanglements, which explains λ increase with temperature, is intensified with temperature. In the models [34, 75] increase of the number of chain fixation points is assumed at transition from the rubbery to the glassy state. These fixation points are called the "sliding units" [75] or clusters [34]. In any case, they determine limitations of molecular mobility of the chain in the glassy state [42].

In the framework of deformable solid synergism, it has been shown [76] that the part of dissipating energy, η, can be estimated from the following equation:

$$\eta = 1 - \Lambda^{\beta_p}, \tag{5.52}$$

where Λ is the structure self-similarity coefficient; β_p is the fractional part of the fractal dimensionality of the fracture surface, d_p.

In accordance with the above-shown, for the structure of polymers, the following equality is true: $\Lambda = C_\infty$, and in the case of quasi-friable destruction, d_p value can be assessed as follows [18]:

$$d_p = \frac{10(1+v)}{7-3v}. \tag{5.53}$$

Fractal dimensionality, D, of the chain part between entanglements is the structural parameter characterizing molecular mobility level and, consequently, the energy dissipation level [77]. Techniques for D assessment are given in the work [5]. Good correlation between $(D-1)$ and η parameters was obtained for 15 amorphous and amorphous-crystalline polymers [78] (remind that $1 \le D \le 2$ and $0 \le \eta \le 1$). Physical basis for this correlation is obvious. As mentioned above, D parameter characterizes molecular mobility, and intensification of the latter always induces increase of mechanical energy dissipation applied to the polymer [79].

The length of the chain intersect between "hookings", L_c, is calculated for the rubbery state that allow modification of the equation (5.52) as follows. For rubbers, v is close to 0.5 (but not equals it [18]). That is why dimensionality, d_p, must be calculated for viscous degradation under the condition $v \ge 0.375$ [18]. In this case, the equation for d_p calculation is as follows [18]:

$$d_p = \frac{2(1+4v)}{1+2v}. \tag{5.54}$$

As $v \approx 0.45$ is chosen for rubbers [80], calculation by the equation (5.54) gives $d_p \approx 2.95$, i.e. in the first approximation one can accept $\beta \approx 1$ in equation (5.52). Then under the above-mentioned conditions ($\eta = D - 1$ and $\Lambda = C_\infty$) modified equation (5.52) is obtained:

$$D = 2 - \frac{1}{C_\infty}. \tag{5.55}$$

In its turn, dimensionalities D and d_f are interrelated by the following expression [5]:

$$D = \frac{\ln n_{st}}{\ln(4-d_f) - \ln(3-d_f)}. \tag{5.56}$$

Table 5.2

Values of characteristic relations C_∞, backbone real bond lengths l_0, experimental L_c^e and theoretical L_c^t lengths of the chain intersects between macromolecular "hookings" for some polymers [78]

Polymer	C_∞ [45,53]	l_0 [45,53], Å	L_c^e, nm [81]	L_c^t, nm
Poly-*n*-dodecylmethacrylate	13.4	1.54	226	208
Poly-*n*-octylmethacrylate	10.0	1.54	177	149
Polystyrene	10.0	1.54	104	149
Polyvinylacetate	9.4	1.54	88	136
Poly-*n*-butylmethacrylate	9.1	1.54	130	131
Poly(methyl methacrylate)	8.6	1.54	97	92
Polyethylene	6.8	1.54	56	88
Polyvinylchloride	6.7	1.54	31	87
Polytetrafluoroethylene	6.3	1.54	41	80
Polyamide-6	5.3	1.49	46	61
Polypropylenoxide	5.1	1.49	60	58
Polyethylenoxide	4.2	1.49	45	43
Poly(ethyleneterephthalate)	4.2	1.25	21	36
Poly(ester terephthalate)	3.3	1.25	28	30
Polycarbonate	2.4	1.25	29	17

Combined equations (5.54) – (5.56) allow calculation of n_{st} for rubbers practically by known C_∞ values only, because v (or d_f and d_p) values are fixed. With respect to equation (2.8), the equation for L_c assessment is deduced as follows [78]:

$$L_c = n_{st} l_{st} = n_{st} l_0 C_\infty. \qquad (5.57)$$

Table 5.2 shows parameters l_0 and C_∞ required for calculation and accepted in accordance with [45, 53], $L_c \left(L_c^e \right)$ values from the literature (in accordance with [81]), and $L_c \left(L_c^t \right)$ values calculated by the method above. The Table indicates good conformity between theoretical and experimental results.

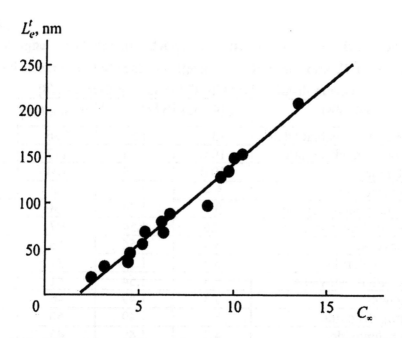

Figure 5.18. Dependence of theoretical lengths of the chain intersect between entanglements L_c^t on characteristic relation C_∞ for polymers listed in Table 5.2 [78]

Figure 5.18 shows correlation $L_c^t(C_\infty)$ for polymers, mentioned in Table 5.2. Good linear correlation obtained proves already known principle of entanglement network frequency increase with statistical chain rigidity [45]. At $L_c^t = 0$, the dependence $L_c^t(C_\infty)$ is extrapolated to $C_\infty \approx 2$ that corresponds to tetrahedral valence angle θ ($\approx 70.5°$) in the chain model with fixed valence angles [60]. However, this value of C_∞ cannot be considered the border one. As follows from the equation (5.55), $C_\infty = 1$ at the border value $D = 1$ [5]. In the mentioned model of the chain with fixed valence angles C_∞ is determined as follows [60]:

$$C_\infty = \frac{1+\cos\theta}{1-\cos\theta}. \tag{5.58}$$

Obviously at $C_\infty = 1$, $\cos\theta = 0$ and $\theta = 90°$. Thus the fractal analysis assumes C_∞ falling within the range of $1-2$ and arbitrary sharp valence angles.

It is known [80, 82] that transition from the rubbery to the glassy state is accompanied by the Poisson coefficient decrease, which value below T_g depends on the test temperature. This means decrease of d_f in accordance with the equation (5.4) and n_{st} (i.e. decrease of L_c) in accordance with the equation (5.56) in the glassy state rather than in the rubbery state even at invariable D. This effect is intensified by the appropriate decrease of D that follows even from the well known fact as follows: the decrease of molecular mobility in the polymer glassy state [59], and is proved experimentally [5].

Thus the fractal analysis suggested enables forecast of the topological chain fixation point number increase at $T < T_g$, suggested before [75, 83].

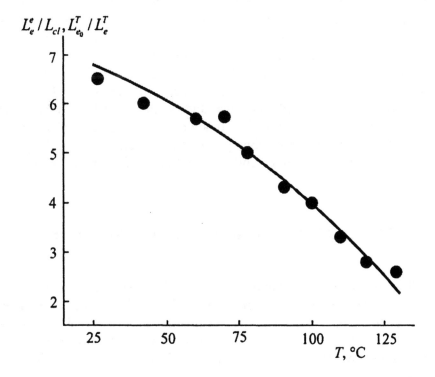

Figure 5.19. Dependencies of L_c^e/L_{cl} (1) and L_{c0}^t/L_c^t (2) (see text for details) on the test temperature for PC [78]

Quantitatively, this forecast can be checked by temperature dependence of two relations. The first of them is the relation of chain intersect lengths between "hookings" L_c^e (the rubbery state) [81] and between clusters L_{cl} (the glassy state) [83]. The second one is the relation between lengths, L_c^t, obtained by the above-mentioned method at $v = 0.45$ (L_{c0}^t, the rubbery state) and at v values, calculated by the equation (3.12) as the function of T (the glassy state). The first relation $\left(L_c^e/L_{cl}\right)$ gives results, obtained within the framework of the cluster model [83]; the second one $\left(L_{c0}^t/L_c^t\right)$ - in the framework of the fractal analysis. Figure 5.19 shows temperature dependencies of both relations for PC, for which good conformity was observed. Values of L_c^t were calculated at randomly chosen $D = 1.75$, but any other choice in the above-mentioned interval will not change symbate type of dependencies, shown in Figure 5.19, but just slightly deviate their quantitative conformity.

Thus in the framework of the fractal analysis the increase of macromolecular "hookings" network frequency with statistical chain rigidity was proved. In contrast with previous studies, the fractal analysis forecasts increase of the number of topological fixation points of macromolecules in the glassy state rather than in the rubbery one. The cluster model of the polymer amorphous state structure gives correct quantitative description of this effect [78].

5.3. CLUSTER AND FLUCTUATION FREE VOLUME MODELS

The fluctuation free volume model is widely spread in descriptions of the physical meaning of liquids and glasses [84 – 86]. Hence, the physical meaning of the model parameters is yet mainly unclear [87, 88]. Moreover, difficulties appear in the free volume theory concerning explanation of the temperature dependence of kinetic properties, in particular, viscosity at fixed volume of the system.

The relative part of the fluctuation free volume, f_f, is determined, for instance, by data in the glass transition zone [85, 86]. If f_f is known, one can calculate the formation energy of the minimal free volume microcavity, ε_h, by the formula as follows [86]:

$$\varepsilon_h = kT_g \ln(1/f_f). \qquad (5.59)$$

Let us now turn to comparison of the cluster model with the fluctuation free volume for solid polymers [83, 89]. Obviously in accordance with the cluster model the fluctuation free volume is accumulated in the packless matrix. In terms of the free volume theory, detachment of a kinetic unit (an atom or a group of atoms) from a cluster means formation of a random microcavity, and addition of a kinetic unit to a cluster means its "implosion" (annihilation). As a consequence, random change of the free volume with temperature can be realized in principle at a constant volume of the system due to microcavity exchange between clusters and the packless matrix. This approach to the mechanism of generation and migration of random microcavities overcomes the above-mentioned difficulty in explanation of the temperature dependence of viscosity at fixed volume of the system [89].

Figures 2.1 and 2.2 show symbate decrease of the cluster network frequency, v_{cl}, and cluster functionality, F, with temperature increase for amorphous glassy polymers [90]. Analysis of these data induces a conclusion that the number of clusters per specific volume of the polymer remains practically unchanged with temperature increase, but the number of segments in each cluster decreases. However, at $T = T_g$ in the glass transition zone clusters are completely decomposed ($v_{cl} = 0$), and excessive fluctuation free volume reaches the critical value.

Thus from these positions, devitrification of a glassy polymer is caused by cluster decomposition (i.e. annihilation of the local order). Therefore, energy of the cluster formation or decomposition, U_{cl}, must possess the magnitude of the average energy of kinetic unit (statistical segment) heat motion at the glass transition temperature: $U_{cl} \cong (i/2)kT_g$, where i is the number of degrees of freedom of the kinetic unit. If one takes into consideration that a segment consists of many atoms, then in the first approximation i must equal the number of degrees of freedom in a multiatomic molecule, $i = 6$ [89].

Expectedly, the energy of segment detachment from the cluster, determined in the equation (3.19), was found proportional to the glass transition temperature of polymers (Figure 5.20). It is noteworthy that the slope of $U_{cl}(kT_g)$ line is practically coincident with $i/2$ value and the coefficient of proportionality between energy of the minimal random microcavity formation (see equation (5.59)): $U_{cl} = 3.3\ kT_g$ and $\varepsilon_h = 3.5\ kT_g$. It is taken into account here that common value of f_f for glassy polymers equals 0.025 and $\ln(1/f_f) = 3.5$. As a

Kozlov G.V. and Zaikov G.E.

consequence, the energy of random microcavity formation is practically coincident with the energy of segments' interaction in the cluster ($\varepsilon_h = U_{cl}$).

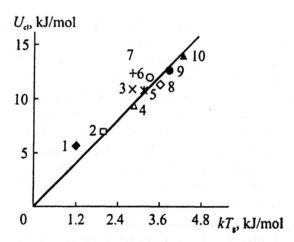

Figure 5.20. Correlation between the segment interaction energy in the cluster, U_{cl}, and kT_g value for the polymer solid phase at $T = 293$ K: 1 – HDPE, 2 – PP, 3 – PA-6, 4 – PETP, 5 – PVC, 6 – PHE, 7 – PTPE, 8 – PC, 9 – PSP, 10 – PAr [89]

From this point of view, generation and implosion of a random microcavity in amorphous polymers are stipulated by dissociation and segment association into a cluster, respectively. Microcavity formation energy equals the work for collinear segments detachment from one another in the cluster by breaking intermolecular bonds between them, and the magnitude of the microcavity minimum size ($\approx 4 - 5$ Å [89]) equals the border elastic deformation of the intermolecular bond between segments.

The equation (3.19) suggests U_{cl} decrease with temperature due to the presence of temperature dependencies of G and v [91]. It can be shown that for glassy polymers, f_f increases (though slowly) with temperature [92] that induces the corresponded decrease of ε_h. However, it should be noted that temperature dependencies of energies ε_h and U_{cl} are very weak. Apparently, this is explained by the fact that parameters ε_h and U_{cl} represent characteristics of the short-range (local) order of glasses, weakly responding to temperature variation (anyway, at $T < T_g'$ [90]).

Besides the above-considered correspondence between the cluster model and the kinetic theory of the fluctuation free volume, application of the former one allowed obtaining of a series of new ideas on the interconnection between structure and free volume [93]. For example, Figure 5.21 shows dependencies of the relative random volume, f_f, on the volumetric part of packless matrix, $\varphi_{p.m.}$, for two linear amorphous polymers, PC and PAr. The dependencies are linear, and their extrapolation to $\varphi_{p.m.} = 0$ gives $f_f = 0$. The latter means that in this case, an amorphous polymer represents a huge cluster without a fluctuation free volume. Clearly this circumstance does not exclude the presence of a configurative free volume [94]. At $\varphi_{p.m.} = 1.0$ f_f is extrapolated to a finite value, approximately equal 0.14 and corresponded to f_f at the glass transition temperature [95].

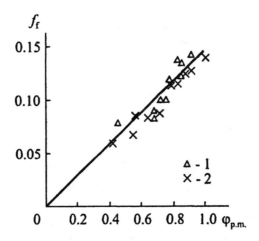

Figure 5.21. Dependence of the relative fluctuation free volume, f_f, on volumetric part of the packless matrix, $\varphi_{p.m.}$, for linear amorphous polymers: PC (1) and PAr (2) [93]

For glassy polymer networks, the type of dependencies $f_f(\varphi_{p.m.})$ differs from the above-considered one (Figure 5.22). Similar to the case of linear amorphous polymers, they are approximated well by a straight line. However, the line does not pass through the origin of coordinates, but at $\varphi_{p.m.} = 0$ cuts an intersect $f_f \approx 0.024$ on the axis of ordinates. One can assume that this value of f_f is associated with the presence of a network of chemical cross-link points. Thus in contrast with linear polymers, in polymer networks the relative fluctuation free volume consists of two components. One of them is constant and connected with the zones of chemical cross-link points; the second one is the linear function of $\varphi_{p.m.}$ and is determined by supersegmental (cluster) structure.

Various components of f_f determine different properties of epoxy polymers: variable part of f_f regulates elasticity and local plasticity, and total value of f_f regulates properties in the zone of macroscopic flow behavior [96].

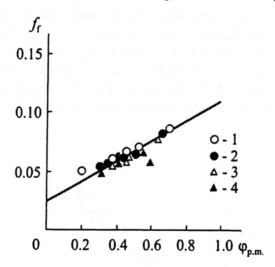

Figure 5.22. Dependence of the relative fluctuation free volume, f_f, on volumetric part of the packless matrix, $\varphi_{p.m.}$, for epoxy polymers: EP-1 (1), EP-2 (2), EP-1-200 (3), and EP-2-200 (4) [93]

Thus correlations obtained using the cluster model of the polymer amorphous state structure allow an observation that the fluctuation free volume, and the type of its variation with respect to structural factors are significantly determined by the type of polymer structure [93].

5.4. THE CLUSTER MODEL AND THE ANHARMONICITY CONCEPT

In recent 20 years, increased interest to the Grüneisen parameter, γ, is observed in physics of liquids and glasses. This parameter participates in the equation of state and represents the anharmonicity index of quasi-lattice oscillations and nonlinearity of interatomic interaction forces [59, 97]. The Grüneisen parameter was successfully applied to analysis of the cluster formation event and orientation of molecules in liquids [98 – 100]. Since a liquid and the glassy state are structurally indistinguishable [101], of interest is the Grüneisen parameter application to studies the local order in glassy polymers. In the framework of the cluster model temperature dependence of γ for amorphous polymer was interpreted [83].

Knopoff and Shapiro [98] have indicated that for water and mercury, γ decrease with the specific volume increase can be assigned to the change of the average number of molecules in the cluster, N. They have suggested a semi-quantitative one-dimensional model, which enables description of γ dependence on volume. It is assumed that γ represents the sum of two components [98]:

$$\gamma = \gamma_D + \xi \frac{\partial \ln N}{\partial \ln V}\bigg|_T , \qquad (5.60)$$

where γ_D is the component of γ corresponded to the Debye heat capacity theory; ξ is a factor dependent on N.

The equation (5.60) describes γ variation with pressure, P, but due to equivalence of P and T variables in the equation of state it can be used for assessment of the temperature dependence of γ under constant pressure. In this case, index T in the equation (5.60) is replaced by P.

The Debye component γ_D in the Grüneisen parameter can be assessed as follows [98]:

$$\gamma = -\xi \frac{\partial \ln \omega_D}{\partial \ln V}\bigg|_P , \qquad (5.61)$$

where $\omega_D = \pi \omega_0$ is the Debye frequency of polymeric chain oscillations; ω_0 is determined as follows:

$$\omega_0 = \frac{U}{M} . \qquad (5.62)$$

Here M is the mass of oscillating kinetic unit of the chain; U is the potential energy of interaction between two kinetic units, accepted equal U_{cl} in the framework of the cluster model (equation (3.19)).

Temperature dependence of the specific volume, V, is calculated by the following equation [100]:

$$\left(\tilde{V}^{1/3} - 1\right) = \frac{\alpha T}{3(1 + \alpha T)}, \tag{5.63}$$

where $\tilde{V} = V/V^*$ is the reduced specific volume; V^* is the characteristic specific volume; α is the volumetric coefficient of heat expansion, temperature dependence of which is determined as follows [83]:

$$\left(\frac{\partial \alpha}{\partial T}\right)_P = (7 + 4\alpha T)\left(\frac{\alpha^2}{3}\right). \tag{5.64}$$

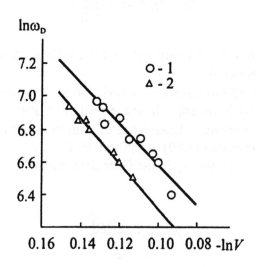

Figure 5.23. Variation of the Debye frequency, ω_D, with the specific volume, V, present in double logarithmic coordinates for PAr (1) and PSF (2) [83]

Accepting α value at $T = 293$ K [83] and characteristic volume as the reverse density, ρ, one can obtain correlation $\ln\omega_D(\ln V)$, shown in Figure 5.23 for PAr and PSF. Clearly the correlation between ω_D and V in double logarithmic coordinates is well approximated by a linear correlation. In accordance with the equation (5.61), this means that γ_D constancy in the considered temperature interval (293 – 453 K for PAr and 293 – 433 K for PSF). Absolute values of γ_D can be estimated from the slope of lines in Figure 5.23, which are independent of V^* selection: these values ($\gamma_D = 12$ for PAr and

γ_{D} = 13.75 for PSF) are close to the ones previously obtained for polymers of this type [102].

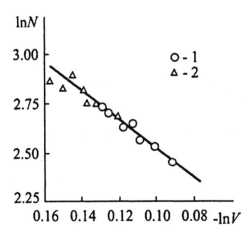

Figure 5.24. Correlation between the average number of segments in the cluster, n_{seg}, and the specific volume, V, present in double logarithmic coordinates for PAr (1) and PSF (2) [83]

Thus for the case of temperature independence of the segments' number in the cluster, the above-shown results indicate obligatory fulfillment of the following condition (equation (5.60)) [83]:

$$\gamma = \gamma_{D} = \text{const.} \tag{5.65}$$

Indicated in the above discussion was γ decrease with temperature [59]. It is assumed that $dN \neq 0$, whence the necessity of assessment of the second term in the equation (5.60) follows. As indicated in Chapter 2, $N = n_{seg} = F/2$. Figure 5.24 shows dependence $\ln n_{seg}(\ln V)$ for PAr and PSF; both dependencies are described by single linear relation that gives an opportunity to assess the relation $\partial \ln n_{seg}/\partial \ln V \approx 6.5$. For large n_{seg}, parameter ξ in equation (5.60) at temperatures above the Debye point is approximated by the following series expansion [98]:

$$\xi = 0.6942 + \frac{0.03780}{n_{seg}} + ... \tag{5.66}$$

Figures 5.25 and 5.26 show temperature dependencies of γ, calculated by equation (5.60) for PAr and PSF, respectively.

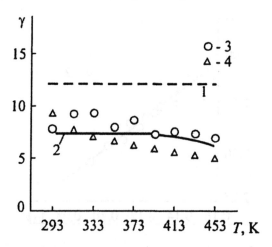

Figure 5.25. Temperature dependence of the Grüneisen parameter γ for PAr: the Debye component γ_D (1), calculations by equations (5.60) (2), (5.67) (3), and (5.68) (4) [83]

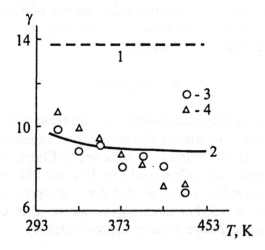

Figure 5.26. Temperature dependence of the Grüneisen parameter γ for PSF: the Debye component γ_D (1), calculations by equations (5.60) (2), (5.67) (3), and (5.68) (4) [83]

The value of γ can also be calculated by two independent methods as follows. The first method uses simple relation for temperature dependence of the elasticity modulus, E [103]:

$$E = E_0(1 - \gamma\alpha T), \qquad (5.67)$$

where E_0 is the value of E extrapolated to 0 K. The second method is based on application of the Sharma equation [100]:

$$\gamma = 1 + \left(\frac{1}{2\alpha T}\right) - \frac{1}{3}\left(5\tilde{V}^{1/3} - 2\right). \qquad (5.68)$$

The values of γ calculated for PAr and PSF by equations (5.67) and (5.68) are also shown in Figures 5.25 and 5.26, respectively. Clearly calculation results by three equations mentioned display high coincidence.

Previously, γ decrease with temperature was attributed to accompanying the elasticity modulus decrease; in its turn, the latter was attributed to decrease of the cluster network frequency, e.g. n_{seg} decrease [57]. Thus the above-shown results conform to previously suggested explanation of the effect observed. Equations (5.60), (5.67) and (5.68) give adequate description of $\gamma(T)$ dependence by both the general tendency and absolute values. In its turn, this proves correctness of the cluster model and suitability of application of the anharmonicity concept to description of the local orientation of the polymer amorphous state structure [83].

5.5. CLUSTER STRUCTURE AS A PERCOLATION SYSTEM

Percolation models are widely used for description of many physical problems [22], including polymers [104 – 108]. These models are differed by simplicity and demonstrativeness [22], and their application to description of the structure and properties of polymers allows the use of the well-developed mathematical apparatus of the percolation theory. The cluster structure formation at T_g sharply changes properties of amorphous polymers attributing their rigidity typical of solids [90]. That is why one may suggest T_g representing the percolation threshold on the temperature scale in this case, at which an infinite cluster (in the sample borders) is formed. Relating to description of polymers, the "carcass coherency" concept [109, 110], suggested 35 years ago, can be accepted to be the predecessor of percolation models. This concept assumes that the existence of carcass coherency of macromolecules in polymers is the necessary condition of their rigidity and strength. It also gives a relation for assessing probability of the carcass bond formation, η_x. The most probable candidate for the carcass role in linear amorphous and amorphous-crystalline polymers is the network of macromolecular entanglements. As mentioned above, two basic concepts of the structure of such network do exist now: the "hooking" network and the cluster network (refer to Figure 1.2). The contribution of these networks to the carcass coherency of amorphous and amorphous-crystalline polymer and, consequently, their mechanical properties were determined [111].

If the cluster network of macromolecular entanglements is accepted to be such carcass, it is natural to assume equality of the probability of this "carcass" formation, η_x, to the relative part of clusters, φ_{cl}. The value of η_x can be assessed as follows [109]:

$$\eta_x = \frac{S}{a_c E},$$ (5.69)

where S is the cross-section of the macromolecule; a_c is the analogue to Young's modulus for elastic stretch of macromolecules; E is the elasticity modulus, determined experimentally. The value of a_c can be assessed in accordance with recommendations in the ref. [109], which gives the following results [11]: 2.22×10^{-1} N for LDPE; 4.5×10^{-10} N for HDPE; 7.5×10^{-10} N for PC and PAr.

Figure 5.27 shows comparison of φ_{cl} and η_x for PC and PAr. Clearly a linear correlation between them is observed, but η_x values regularly exceed appropriate φ_{cl} values. At $\varphi_{cl} = 0$ η_x is extrapolated to $\eta_{x0} \approx 0.18$ that assumes existence of two carcasses in amorphous PC and PAr. Moreover, frequency of one of these carcasses (the cluster one) is the function of temperature, whereas the second carcass (macromolecular hookings) possesses constant frequency. Macromolecular hookings' carcass frequency, η_{x0}, is assessed from the following equation [109]:

$$\eta_{x0} = n_{x0} S.$$ (5.70)

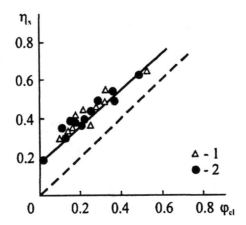

Figure 5.27. Dependence of the coherency carcass formation probability, η_x, on the relative part of clusters, φ_{cl}, for PC (1) and PAr (2) [111]

Then from the equation (2.3) the molecular mass, M_{x0}, of the chain segment between neighbor hookings can be calculated. For PC and PAr, $M_{x0} \approx 1,610$ g/mol. Values of this parameter shown in the literature are as follows: $1,780 - 2,250$ g/mol [45] and $1,400 - 2,426$ g/mol [112], which correlate well with the assessment by equation (2.3). Thus for amorphous glassy PC and PAr polymers, the cluster network of macromolecular entanglements represents the "carcass", and contribution of the macromolecular "hooking" network is rather low (especially at low temperatures) and constant [111].

Figure 5.28 shows the same comparison of φ_{cl} and η_x for LDPE and HDPE. Similar to two amorphous polymers (Figure 5.27), linear correlation between φ_{cl} and η_x is observed. However, for LDPE and HDPE linear approximation passes through the origin of coordinates, i.e. $\varphi_{cl} \approx \eta_x$. This fact is explained by the devitrified state of the polyethylenes' amorphous phase in the temperature interval studied; thus the elasticity modulus of the phase (and, respectively, η_x) is very low (~2.1 MPa [45]). That is why η_{x0} value is much lower than inaccuracy of the graphical correlation $\eta_x(\varphi_{cl})$.

Thus the above results allow the statement that the macromolecular entanglement cluster network represents the "carcass" for amorphous and amorphous-crystalline polymers. As a consequence, this very network causes the main impact on mechanical properties of solid polymers (especially amorphous ones [57]). Contribution of the macromolecular hooking network is independent of temperature and is sufficient in the glassy state of polymer only. Estimation performed by the equation (5.69) has indicated that the presence of the macromolecular hooking network only in PC and PAr would give, approximately, three-fold lower elasticity modulus compared with the experimentally observed one, and polyethylenes would completely lose rigidity [111, 113].

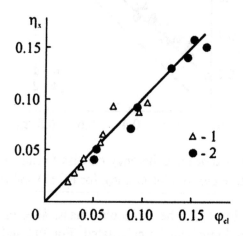

Figure 5.28. Dependence of the probability of coherence carcass formation, η_x, on the relative part of clusters, φ_{cl}, for LDPE (1) and HDPE (2) [111]

It is known [22] that at approach of percolation threshold, x_c, critical behavior of the infinite cluster cardinality, P_∞ (the probability of cross-link point belonging to the current cluster), is described by the scaling relation as follows:

$$P_\infty \sim (x - x_c)^{\beta_n} .\qquad(5.71)$$

In the framework of the cluster model, φ_{cl} is the obvious selection for P_∞, and the glass transition temperature, as mentioned above, is accepted for x_c [67, 68]. In this case, the test temperature, T, is accepted as the current probability, X, and the relation (5.71) can be reduced to as follows [67]:

$$\varphi_{cl} \sim (T_g - T)^{\beta_n} = (\Delta T)^{\beta_n} ,\qquad(5.72)$$

where changing over T_g and T is caused by the inequality $T_g > T$. Figure 5.29 shows dependence $\varphi_{cl}(\Delta T)$ in double logarithmic coordinates, corresponded to correlation (5.72). The graph of this dependence is linear that enables calculation of $\beta_n \approx 0.46$.

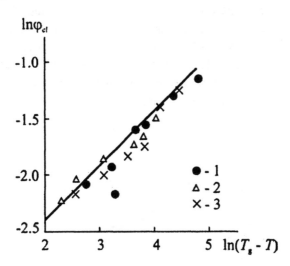

Figure 5.29. Dependence of the relative part of clusters, φ_{cl}, on temperature difference $(T_g - T)$, presented in double logarithmic coordinates for PMMA (1), PC (2), and PAr (3) [67]

This value is close to theoretical "classical" one $\beta_n = 0.40$ for a three-dimensional percolation cluster [22]. The graph in Figure 5.29 is plotted for the temperature range $T_g - (T_g - 80\ \text{K})$, because correlation (5.71) is true under the condition $x \to x_c$ [22] (or $T \to T_g$). Thus the amorphous polymer structure represents a percolation cluster, which can be described by general laws of the percolation theory [67]. The same result was obtained for glassy polymer networks [68] and amorphous-crystalline polymers [114]. For example, in the framework of these general laws, the number of cross-link points, S, in a finite cluster, which can be the local order zone consisting of $F/2$ closely packed segments (where F is the cluster functionality [90]), depends upon dimensionless deviation of concentration, τ, from its critical value ($\tau = (x - x_c)/x_c$ [22]) at $\tau \to 0$ as follows:

$$S \sim |\tau|^{-\gamma_n}. \tag{5.73}$$

As mentioned above, the number of cross-link points, S, in the local order zone (classified as the cluster [34, 90]) equals $F/2$, and for τ parameter $(T_g - T)/T_g$ should be accepted. Figure 5.30 shows dependencies of $F/2$ on $(T_g - T)/T_g$ in double logarithmic coordinates for epoxy-polymers. The dependencies were found linear that allowed estimation of the index in correlation (5.73) as 1.28 for EP-1 and 2.28 for EP-2. For amorphous glassy polymers the following values of γ_n were obtained: ~1.45 for PMMA, ~1.36 for PC, and ~1.14 for PAr. All calculated values of γ_n satisfactorily correlate with its theoretical value (= 1.84) for a three-dimensional cluster [22].

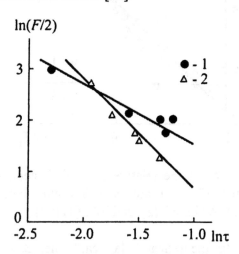

Figure 5.30. Dependencies of the number of cross-link points (segments) in a finite cluster, $F/2$, on dimensionless temperature deviation, $\tau = (T_g - T)/T_g$, presented in double logarithmic coordinates for EP-1 (1) and EP-2 (2) [68]

As a consequence, the results obtained [67, 68, 114] indicate that at temperature T_g (T_{melt}) a structure of statistical segments is formed in amorphous (amorphous-crystalline) polymers representing a percolation cluster and being described properly in the framework of the cluster model. This circumstance assumes the polymer glass transition represents the phase transition, and φ_{cl} value is the orientation parameter [22]. Moreover, this is one more proof for fractality of the polymer structure in the range of ~3 – 100 Å (refer to Section 5.1).

Of special interest is the fact that critical indices of the orientation parameter, β_n, obtained in refs. [67, 68, 114], though are very close to the classical percolation indices ($\beta_n = 0.37 - 0.40$ [115]), but differ from them by the absolute value. The reason for such inconsistency is the percolation cluster representation as a pure geometrical structure, which is too simplified simulation of natural amorphous polymers, which possess thermodynamically non-equilibrium structure. That is why the structure of the mentioned polymers is affected by thermal interactions (remind that the cluster structure is postulated as a thermofluctuational system [34]). This means that the percolation cluster (carcass) of the polymer structure consists of elements (statistical segments) possessing molecular mobility, e.g. they oscillate around some quasi-equilibrium position. Such elements form a virtual volume, which exceeds the volume of a cluster from static elements that makes the percolation process simpler and allows formation of the percolation carcass from smaller number of elements or with higher β_n value. Moreover, cluster formation is studied by the scale of relative temperatures, but not by the scale of concentrations as for the percolation cluster [22, 115]. As a consequence, thermal cluster, i.e. the cluster which equilibrium configuration is determined by both geometrical and thermal interactions [116], is more correct simulation of the polymer amorphous state. In this case, the orientation parameter φ_{cl} of the polymer structure is determined from the following equation [116]:

$$\varphi_{cl} = \left(\frac{T_g - T}{T_g} \right)^{\beta_T}, \qquad (5.74)$$

where index of the orientation parameter, β_T, is not necessarily equal to the appropriate critical index, β_n, in pure configuration percolation models. An attempt was made [117] to estimate variation of β_T absolute values, detect possible connection between this index and structural characteristics of polymers, and determine factors affecting β_T value. This was performed on the example of three series of epoxy polymers: amine (EP-1) and anhydride (EP-2) cross-linking type, and EP-1 after natural aging under atmospheric conditions during 3 years (EP-3) [18].

It has been shown [119] that universality of critical indices of the percolation system is directly bound to the fractal dimensionality of this system. Self-similarity of the percolation system assumes the presence of a series of subsets possessing the order n ($n = 1, 2, 4, ...$), which in the case of the amorphous polymer structure are identified as follows [117]. The first subset (n = 1) is the carcass of the percolation cluster or, as shown above, cluster network of the polymer. The second subset ($n = 2$) represents the packless matrix, to which cluster network is buried. In this interpretation, critical indices β_n, ν_n, and t_n are given as follows (in the three-dimensional Euclidean space) [119]:

$$\beta_n = \frac{1}{d_f}, \tag{5.75}$$

$$\nu_n = \frac{2}{d_f}, \tag{5.76}$$

$$t_n = \frac{4}{d_f}, \tag{5.77}$$

where d_f is the fractal dimensionality of the polymer structure, determined by the equation (5.4).

Thus β_n, ν_n, and t_n parameters represent border values for β_T pointing out the structural component of the polymer that determines its behavior. As $\beta_T = \beta_t$, this component are clusters or, more accurately, the percolation cluster carcass identified with the cluster network. As $\beta_n < \beta_T < \nu_n$, the polymer behavior is determined by joint influence of clusters and packless matrix. As $\beta_T = \nu_n$, the determining structural component will be the packless matrix, and at $\beta_T = t_n$ the chemical cross-link network; at $\nu_n < \beta_T < t_n$ a joint influence of two above-mentioned structural components is observed.

For three series of epoxy polymers under consideration, assessments of β_T parameter have indicated their possible approximation by the following values: ≈ 0.54 for EP-1, ≈ 0.55 for EP-2, and ≈ 0.38 for EP-3. Figure 5.31 compares φ_{cl} values as the function of K_{st}, experimental [5] and calculated by the equation (5.74), using the above β_T values. In the framework of the thermal cluster model, experimental and calculated values conform well. The average D_f value for three studied series of epoxy polymers equals 2.613. This means that according to equations (5.75) and (5.76) $\beta_n \approx 0.38$ and $\nu_n \approx 0.77$. Comparison of these assessments and β_T values, calculated from the equation (5.74), indicates that the cluster network is the basic structural factor for EP-3, and behavior of EP-1 and EP-2 is regulated by the joint effect of cluster network and packless matrix. Thus for aged epoxy polymers (EP-3) only indices of the orientation parameter of percolation and thermal clusters coincide. For the case under consideration, this indicates insufficiency of thermal interaction for the polymer structure and that its equilibrium configuration is determined by geometrical interactions only. It has been shown [118] that during heat aging the structure of epoxy polymer EP-3 reaches its quasi-equilibrium state, characterized by the balance between the tendency to local order increase, the index of which is φ_{cl}, and entropic chain straining. Hence chain intersects between clusters are stretched, and their fractal dimensionality D approaches 1. This indicates [5, 77] full suppression of molecular mobility of these parts of the chains.

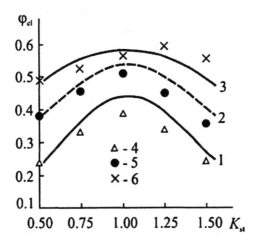

Figure 5.31. Dependencies of the relative part of clusters, φ_{cl}, obtained experimentally (1 – 3) and calculated by equation (5.74) (4 – 6), on K_{st} for epoxy polymers: EP-1 (1), EP-2 (2), and EP-3 (3) [117]

For EP-1 and EP-2, $D > 1$ [5] and, as follows from Figure 5.31, $\beta_T > \beta_n$. Consequently, it may be assumed that the condition $\beta_T > \beta_n$ is induced by non-zero molecular mobility, e.g. $1 < D \leq 2$ [5]. To check this assumption, one shall calculate D value that can be made differently, (for example, using equation (5.55) or (5.56)), but the authors in the work [117] have used a simple formula as follows [77]:

$$\frac{2}{\varphi_{cl}} = C_\infty^D . \tag{5.78}$$

Figure 5.32 shows the dependence $\beta_T(D)$ for three series of studied epoxy polymers. Expectedly, increase of D leads to β_T growth. As $D = 1$, $\beta_T \approx 0.38$, i.e. the condition $\beta_T = \beta_n$ is fulfilled. The condition $\beta_T = v_n$ is reached at $D \approx 1.42$, and extrapolation of the dependence $\beta_T(D)$ shown in Figure 5.32 to $D = 2$ gives $\beta_T \approx 1.35$. As known [120], $D = 2$ indicates a rubber-like state of polymer ($d_f \approx 3$), characterized by complete degradation of the cluster structure [121]. In accordance with the equation (5.77), at $d_f \approx 3$ $t_n \approx 1.33$, i.e. for rubber-like polymers the condition $\beta_T \approx t_n$ and, consequently, its behavior is completely determined by the chemical cross-link network, which is the well-known fact [122]. Taking into account that for glassy epoxy polymers $\beta_n \leq \beta_T \leq t_n$, one should assume incorrectness of the latter conclusion for them and general regulation of their behavior by the totality of three structural components: cluster network (local order zones), packless matrix and partly (at high D) chemical cross-link network (topological structure level).

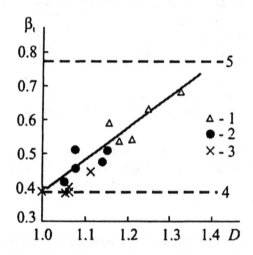

Figure 5.32. Dependence of index β_t on the fractal dimensionality D of the chain segment between clusters for EP-1 (1), EP-2 (2), and EP-3 (3). Horizontal dashed lines indicate β_n (4) and v_n (5) values for the percolation cluster [117]

As a consequence, the above-discussed results indicate that the structure of polymer networks can be simulated by a thermal cluster. Hence, correspondence between thermal and percolation clusters, expressed by the equality of critical indices of the orientation parameter $\beta_T = \beta_n$, is reached for quasi-equilibrium state of the polymer structure only. Actually, β_T value is the function of molecular mobility, characterized by the fractal dimensionality (D) of the chain segment between clusters [117].

Some other important points should be noted, associated with the possibility of describing the polymer structure in the framework of equation (5.72). For the "rigidity segment", the statistical segment with the length l_{st}, determined from the equation (2.8), is accepted in the cluster model. The value of l_{st} is several times lower than the Kuhn segment length, A, which characterizes rigidity of freely joint chain [60]. Selection of A as the "rigidity segment" mostly gives φ_{cl} above the unit, which is physically meaningless. The following correlation binds parameters l_{st} and A [60]:

$$l_{st} = A\cos(\theta/2), \qquad (5.79)$$

where θ is the valence angle.

Thus shorter length of the chain "rigidity segment", l_{st}, in the glassy state is caused by conformational restructurings typical of this state [9].

For amorphous-crystalline polymers, reduced relative part of clusters, φ_{cl}^r, is used. It characterizes this parameter in relation to the entire volume of the polymer, but not its amorphous phase exclusively [123, 124]:

$$\varphi_{cl}^r = \frac{\varphi_{cl}}{1-K}, \qquad (5.80)$$

where K is the crystallinity degree.

Thus for amorphous-crystalline polymers, the equation (5.72) can be reduced to the form as follows [9]:

$$\varphi_{cl} \sim (1-K)(T_g - T)^{\beta_T}. \qquad (5.81)$$

Figure 5.33 shows correlation between D_f and $\varphi_{p.m.}$ for three polymers. Clearly at $\varphi_{p.m.} = 0$ the value of D_f does not approach its border value equal 2.0 [5], namely, d_f^0 value, which is different for different polymers. Obviously, this is the minimum of D_f that can be reached for current polymers. The equation (5.42) binding parameters D_f and C_∞ and gives D_f values close to the minimum. Table 5.3 shows comparison of d_f^T values, calculated by the equation (5.42), and d_f^0 (Figure 5.33), which indicate their good conformity. Since the equation (5.42) gives the better correspondence to the experiment, the closer the polymer structure approaches thermodynamic equilibrium [41], dimensionality of d_f^0 characterizes some quasi-equilibrium state of the structure. As follows from this equation, $D_f = 2$ can be approached for polymers possessing $C_\infty = 2$, i.e. for chains with tetrahedral valence angle [60]. Thus substitution of d_f^0 or d_f^T into equation (5.5) or (5.7) can help in the local order assessment, φ_{cl}^0, maximum approachable for the current polymer, as the function of C_∞ and S, the most important characteristics of the polymer. As temperature decreases below the point, at which φ_{cl}^0 is approached, the local order remains unchanged, so this temperature T_0 can be assumed to be the analogue of T_∞, at which the polymer reaches constant free volume level [125]. In its turn, this temperature can be determined from the equation (5.42) under the condition $\varphi_{cl} = \varphi_{cl}^0$. The values of T_0 and T_∞, obtained in this manner [125], are compared in Table 5.4.

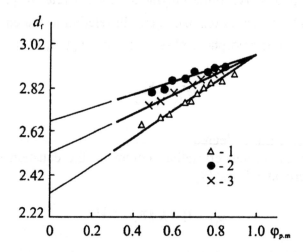

Figure 5.33. Dependencies of the structure fractal dimensionality, d_f, on the relative part of packless matrix, $\varphi_{p.m.}$, for PC (1), PMMA (2) and HDPE (3) [9]

Good conformity was obtained for all polymers, except for PC, for which calculation by the equation (5.42) gave physically meaningless $\Delta T \approx 670$ K. The reason for this seeming non-correspondence is as follows. Previously [5], the notion of quasi-equilibrium state of the polymer structure was introduced, to which $D_f \approx 2.5$ is corresponded. For real polymers, the tendency of transition to thermodynamically more equilibrium state (i.e. in the state with higher local order degree, φ_{cl}) at a definite stage must be balanced by entropic chain straining forces. To put it differently, for the majority of polymers thermodynamic balance by means of the local orientation cannot be reached. That is why $D_f = 2.5$ instead of theoretical value d_f^T should be used. Moreover, data from Figure 5.33 indicate that the value $\varphi_{p.m.} = 1.0$ (i.e. $\varphi_{cl} = 0$) is approached at $D_f = 2.92$, but not at $D_f = 3$ as suggested by the equation (5.5). With respect to these adjustments, calculation by the equation (5.42) gives $T_0 = 318$ K for PC that properly correlates with $T_\infty = 323$ K (Table 5.4).

Table 5.3

Critical fractal dimensionalities of polymer structures: $\left(d_f^T\right)$, calculated by equation (5.42), and d_f^0, estimated from graphs in Figure 5.33 [9]

Polymer	$\left(d_f^T\right)$	d_f^0
Poly(methyl methacrylate)	2.83	2.66
Polyethylene	2.82	2.50
Epoxy polymer	2.58	2.51
Polycarbonate	2.29	2.31

Table 5.4

Transition temperatures to "frozen" free volume (T_∞) and structure (T_0) [9]

Polymer	φ_{cl}^0	T_∞, K	T_0, K
Poly(ethylene terephthalate)	0.24	280	300
Polyamide-6	0.14	277	304
Polyethylene	0.10	227	243
Polypropylene	0.24	227	199
Polystyrene	0.23	324	320
Poly(methyl methacrylate)	0.36	290	269
Polyvinylchloride	0.17	270	333
Polycarbonate	$\dfrac{1.0}{0.36}$	323	– 318

It is known [126] that the fluctuation glass transition theory assumes the following temperature dependence of the characteristic length of cooperatively restructured zones (CRZ), ξ_A:

$$\xi_A \sim (T - T_0)^{2/3}, \tag{5.82}$$

where T_0 is the Foegel temperature in the VLF equation.

Actually, the relation (5.82) is the percolation one or, more exactly with respect to the above discussion, the thermal cluster model relation. As indicated in Section 2.3, ξ_A value is equivalent to the local order diameter, D_{cl} (Figure 2.26). That is why Figure 5.34 shows the dependence $D_{cl}(T_g - T)$ plotted in double logarithmic coordinates, corresponded to the relation (5.82) (under the condition $\xi_A = D_{cl}$ and $T_0 = T_g$) for the initial PC and the one, annealed at 393 K during 6 hours [127]. Clearly in both cases dependencies of $\ln D_{cl}$ on $\ln(T_g - T)$ were found linear, i.e. are described by the relation (5.82), by the slope of this dependence is lower than that for annealed PC [128]. Thus the index in the relation (5.82) is smaller for annealed PC compared with the initial one. It is known [127, 129] that annealing of amorphous polymers shifts their structure towards thermodynamically higher balanced state. As a consequence, data in Figure 5.34 allow an assumption about connection between the index in the relation (5.82) with the polymer structure thermodynamic state: the more unbalanced the structure is, the higher the index is [128].

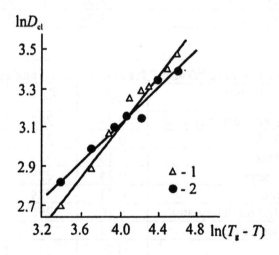

Figure 5.34. Dependencies of the cluster diameter, D_{cl}, on temperature difference $(T_g - T)$ in double logarithmic coordinates,

corresponded to the relation (5.82), for initial (1) and annealed (2) PC [128]

For a hypothetical polymer existing in the quasi-equilibrium state, D_{cl} can be estimated with the help of the above-mentioned thermal cluster model, namely, by the equation (5.74) at $\beta_T = 0.40$. Figure 5.35 shows temperature dependencies of D_{cl} for three types of PC polymer: initial, annealed and existing in a hypothetical quasi-equilibrium state. Clearly linear correlations of $D_{cl}(T)$ conforming with the data in Figure 2.26 and ref. [126] are observed for all three cases. However, the slope of linear graphs is decreased with the approach of the polymer structure to thermodynamic equilibrium: from initial (0.63) to hypothetical (0.47) PC.

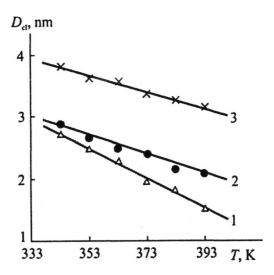

Figure 5.35. Temperature dependencies of the cluster diameter, D_{cl}, for PC: initial (1), annealed (2) and existing in a quasi-equilibrium state (3) [128]

Discussed below are data from the work [126], which indicate that though linear decrease of ξ_A with temperature responds to Adams and Gibbs theory [130], the slope of these dependencies is much higher than that forecasted theoretically: by three times, approximately, for PMMA and by 6 times for PS. For one of the possible reasons for this deviation (Bittrich, [126]) the non-equilibrium origin of heat fluctuations is assumed. The above-obtained results on PC allow more exact explanation of this deviation.

Figure 5.36 compares of temperature dependencies of ξ_A in accordance with the data from [126] and D_{cl}, calculated for the hypothetical quasi-equilibrium state for PMMA and PS structures. Clearly the slope of $D_{cl}(T)$ dependencies is much smaller than for $\xi_A(T)$: by three times, approximately, for PMMA and by 3.5 times for PS. This approximately conforms to the deviation in slopes of theoretical and experimental $\xi_A(T)$ dependencies for the mentioned polymers, obtained in the work [126].

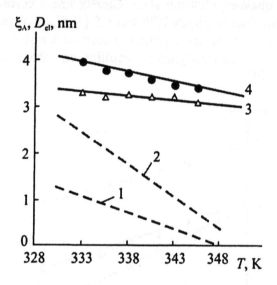

Figure 5.36. Temperature dependencies of CRZ characteristic length, ξ_A, (1, 2) and cluster diameter, D_{cl}, for quasi-equilibrium state (3, 4) for PMMA (1, 3) and PS (2, 4) [128]

As a consequence, the above results indicate that the thermal cluster model correctly describe both structure of the entire amorphous polymer and its basic element (cluster) separately that might be expected due to thermofluctuation origin. Moreover, the cluster model is more general than the fluctuation glass transition theory, because it allows correct theoretical calculation of structural parameters both in thermodynamically non-equilibrium and quasi-equilibrium states, whereas Adams and Gibbs theory allows it for the latter case only [128].

5.6. CLUSTER MODEL AS THE MODEL OF IRREVERSIBLE AGGREGATION

In 1981, Witten and Sander developed the model of the particle-cluster (P-Cl) diffusion-limited aggregation (DLA) [131]. Imagine a particle present in space. Another particle making the Brownian motion appears remote from the first one. Meeting the first particle, it adheres to it, the adhesion being irreversible. Then another particle appears in the distance, diffuses and touching the primary two particles adheres to them. This process is multiply repeated. Some particles do not touch the propagating cluster during random walk, but diffuse infinitely far from it. These particles are neglected from the future consideration.

Figure 5.37. Two-dimensional Witten-Sander cluster composed of 11,260 particles. The adhesion coefficient (steric factor) $p = 1$ [132]

Such cluster is shown in Figure 5.37. It was obtained using computerized Monte-Carlo method. In this method, all the totality of events is simulated in detail (the Brownian motion is given by a random number generator). Particularities of computer simulation are discussed in the work

[132]. The reason for formation of branched structures (refer to Figure 5.37) in the P-Cl DLA model is as follows. A particle walking randomly will most probably contact cluster branches first thus increasing its size. Therefore, the probability of its penetration inside the cluster that would cause formation of a compact structure with $D_f = D$ is infinitely low. Thus cluster branches shield the internal structure of it, somewhat freezing it. As mentioned above [133], propagation of such a cluster should be considered as the phase transition between finite and infinite clusters. This formalization allows comparison of the P-Cl DLA model with others which simulate formation of an infinite system (a polymer gel). Mikin [132] has performed computerized simulation of P-Cl DLA for the case of $D = 2 - 6$. As detected, the fractal (Hausdorff) dimensionality $D_f \approx 2.5$ for the Witten-Sander cluster with $D = 3$.

As indicated [8], the Witten-Sander (WS) P-Cl DLA model allows description of a great number of real physical processes and structures formed by them. Computer simulation gives an opportunity, at least, to determine the tendencies in structure variation of final aggregates under the effect of one or another factor, which is of importance for WS model application to description of the real objects. Nevertheless, under consideration of the latter computerized simulation schemes require clear physical particularization in relation to structure of the studied object. Advantages of the description of one or another process (aggregate) in the framework of P-Cl DLA model (or other aggregation models [8]) are obvious. If the cluster model of the polymer amorphous state structure can be described as WS cluster, this means their belonging to the same class of universality of physical events. Remind that the essence of the universality hypothesis is the following: if the same limiting conditions (interactions between system components) are typical of different systems, these systems belong to the same class of universality of physical events [133]. The more so this means that the polymer amorphous state structure belongs to much broader class of physical events, but not represents something isolated from other braches of physics. In this case, it is subordinate to the general regularities of systems from the current universality class that proves theoretically the description of modern physical concepts within its framework, and the universality class is characterized by a definite selection of scaling indices [133]. It has been shown on this basis [134] that model presentations of WS clusters allow description of polymer network structures using the same scaling index, for which the fractal (Hausdorff's) dimensionality is taken.

There are many reasons (one of which is the cluster appearance, Figure 5.3) for association of the polymer amorphous structure in the framework of the cluster model and WS cluster. Here are three of them. The first is correspondence of the scaling index (fractal dimensionality, D_f) in both models [67, 131]. The second reason is the well-known fact [135] that transition from one universality class to another is observed in the gel formation point, e.g. from cluster-cluster (Cl-Cl) to P-Cl DLA. This means that after the gel formation point the polymer represents the WS cluster. Finally, the third reason is the mechanism of local order zone formation in the cluster model. It is indicated [89] that detachment of a single segment from such zone means formation of one microcavity of fluctuation free volume, whereas addition of a segment induces "implosion" of the microcavity. In the framework of DLA model such mechanism can be identified with P-Cl aggregation that, besides D_f parameter, is the basic characteristic of this class of universality [133].

Note also that the meaning of the "cluster" term in models [34] and [131] are different. In the WS model by the cluster both the local order zone (or close packing zone [136]) and chains (branches) coming out of it are meant. In the cluster model by "cluster" statistical segments only participating in the local order zone are meant. Obviously, the polymer sample structure should not be considered as WS cluster, because in such scales it represents the Euclidean object. That is why by analogy with the modified WS model [137], the polymer network structure was associated [134] with a selection of WS clusters, where every cluster is sized about $2R_{cl}$. In accordance with the model [137] WS cluster propagation is realized on a great number of "seedings", and R_{cl} parameter is bound to concentration of freely diffusing particles, c, by the following equation [137]:

$$c = R_{cl}^{d_f - d} .$$

$$(5.83)$$

Let us discuss the physical meaning of parameter c. Figure 5.3 indicates that the only statistical segments of the macromolecule have the opportunity to be added to the local order zone, which exist in the packless matrix. Thus c decreases with the local order, and under the condition $R_{cl} = $ const this means decrease of $(D_f - D)$ or D_f. Parameter c value was accepted as multiplication of the number of particles (statistical segments) per specific volume and relative part of the packless matrix. Figure 5.38 shows comparison of the fractal dependencies of EP structure, d_f^e [63], and WS cluster, d_f^t, calculated by the equation (5.83). The data in this Figure indicate good conformity of d_f^e and d_f^t values. As a consequence, modified P-Cl DLA aggregation model [137] gives good description of the polymer network structure using the expected scaling indices D_f. This allows belonging of the polymer network structure to the mentioned universality class of physical systems and describes its properties in the framework of this class of relations [134].

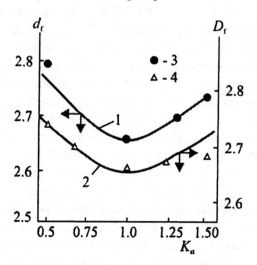

Figure 5.38. Dependencies of structure fractal dimensionalities, experimentally determined d_f^e (1, 2) and calculated by the equation (5.83) d_f^t, on the curing agent:oligomer relation, K_{st}, for EP-1 (1, 3) and EP-2 (3, 4) [134]

Basically, cluster growth is induced by diffusive aggregation of particles, whereas under the stationary mode (c_r = const) the following correlation is valid [138]:

$$\frac{dR}{dt} \sim KR, \quad K = \frac{c_r}{R^{d_f}}, \tag{5.84}$$

where t is time; c_r is the rate of particle delivery to the propagation zone; R is the typical linear size of the cluster.

If l_{st} is accepted for R, the correlation (5.84) can be transformed to the following form [39]:

$$\frac{d\left(l_{st}^{d_f}\right)}{d\left(T^{-1}\right)} \sim \frac{c_r}{l_{st}^{d_f-1}},\qquad(5.85)$$

where T is the cluster formation temperature.

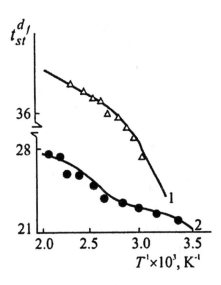

Figure 5.39. Dependencies of parameter $l_{st}^{d_f}$ on reverse temperature for PC (1) and PAr (2) [39]

Figure 5.39 shows $l_{st}^{d_f}\left(T^{-1}\right)$ dependencies, calculated in the above-described manner, for two amorphous glassy polymers (PC and PAr). Clearly the plots show that the assumed correlation does exist, and its nonlinearity is caused by the presence of $l_{st}^{d_f-1}$ term in the right part of equation (5.85). This means that for glassy amorphous polymers, aggregation of segments into clusters is caused by its diffusion mechanism. It should be noted that determination of segments participating in the cluster as linear defects is proved in detail in Chapter 3. Curvature of $l_{st}^{d_f}\left(T^{-1}\right)$ graphs means decrease of the defect generation rate with temperature, and linearity of correlations $l_{st}^{d_f}(T)$,

Kozlov G.V. and Zaikov G.E.

shown in Figure 5.40, proves this conclusion followed from consideration of the correlation (5.85).

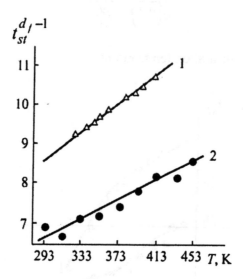

Figure 5.40. Dependencies of parameter $l_{st}^{d_f-1}$ on test temperature T for PC (1) and PAr (2) [39]

Power-series size distribution of objects is widely spread in the nature. Some examples of the power-series correlation between the number, N, and size, r, of defects are shown in the work [138]. General form of the correlation is as follows:

$$N(r) \sim r^{-\alpha}, \qquad (5.86)$$

where α is a constant.

Figure 5.41 shows the dependence $v_{cl}(l_{st})$ in double logarithmic coordinates for 16 polymers of different classes that corresponds to the correlation (5.86). This dependence is linear, which is typical fractal behavior proving the above-mentioned results [39].

For the cluster WS, the following correlation characterizing variation of its density, ρ_{WS}, with the cluster radius, R_{WS}, and fractal dimensionality, D_f, is presented [136]:

$$\rho_{WS} = \rho_{c.p.} \left(\frac{R_{WS}}{a} \right)^{d_f - d},$$ (5.87)

where $\rho_{c.p.}$ is the substance density in closely packed WS cluster; a is the lower border of WS cluster fractal behavior.

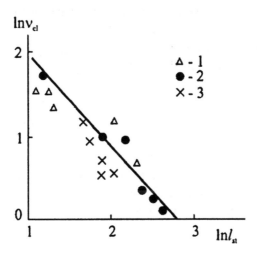

Figure 5.41. Dependence of entanglement cluster network frequency, v_{cl}, on the length of statistical segment, l_{st}, in double logarithmic coordinates for amorphous (1) and amorphous-crystalline (2) polymers and polymer networks (3) [39]

Let us consider equation (5.87) in more detail. For WS cluster, correlations $D_f < D$ and $\rho_{WS} < \rho_{c.p.}$ (change of signs in these two inequalities by the equality sign means cluster transition to another universality class, for example, transition to the Iden cluster [133]) are determined by the condition $a < R_{WS}$. At the same time, since $\rho_{WS} \neq 0$ and $\rho_{c.p.} \neq 0$, a cannot equal zero. Thus a compact zone ($D_f \approx D$) sized a does always exist in the center of WS cluster. For the polymer amorphous state structure, this compact zone is identified as the local order zone (Figure 5.3). This circumstance is one of the conditions of WS cluster existence [139]. Two remarks should be presented on this point. Firstly, the reason for formation of the compact zone on primary stages of WS cluster formation is easy for understanding. The fractal origin of WS cluster is determined by the screening effect of its "branches" preventing access of particles to internal cluster zones [139, 140]. However, on the primary propagation stages these "branches" are low developed, and their screening effect is also low. Secondly, the structure of polymers, as any real (physical) fractal, possesses fractal properties in a definite range of linear scales only [14]. Thus, identification of the polymer amorphous state structure as WS cluster (or, more exactly, the totality of huge amount of clusters) and the necessary existence of local order zones in the amorphous state [141]. This conclusion has been confirmed [141] by results of a value comparison in accordance with the equation (5.87) and sizes of the local order zones, $r_{c.p.}$, estimated from the equation (3.18), shown in Figure 5.42. For ρ_{WS}, density of the polymer packless matrix was accepted that coincides with the polymer density above T_g [93]; for $\rho_{c.p.}$, density of crystalline zones in PC and HDPE are accepted [142, 143]. As indicated by data from Figure 5.42, both absolute values and tendencies of temperature variations of $r_{c.p.}$ and a parameters display good conformity. A definite spread in a values in relation to $r_{c.p.}$ is induced by inaccuracy of D_f assessment due to power-series dependence of a on D_f (equation (5.87)) that causes a significant effect of the comparison accuracy. Data from Figure 5.42 prove the above assumption about local order zones' necessary existence in the structure of the polymer amorphous state [141].

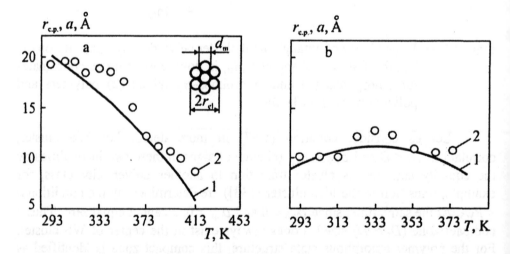

Figure 5.42. Temperature dependencies of sizes of local order zones $r_{c.p.}$ (1) in the model [34] and compact zones a (2) in the model [131, 139] for PC (a) and HDPE (b) [141]

As mentioned above, one of the advantages of WS model application to simulation of the polymer amorphous state structure is its thoroughly developed theoretical basis, proved by great number of computerized simulations. WS cluster growth on many "seedings" and their interaction are discussed in detail in [137]. In the framework of this model, on the example of PC and PAr, temperature variations of the structure fractal dimensionalities, D_f, and the chain segment between entanglements, D, were forecasted. Two alternatives of WS cluster growth on many "seedings" (models A and B) were considered [137]. These models are thoroughly described in the works [8, 137]. Here it should be mentioned that the model A possesses two alternatives, too: with low and high seeding concentrations, c_0, the border between which is the concentration R_c^{-d}, where R_c is the critical (border) radius of the WS cluster. Values of R_c parameter for PC and PAr polymers are taken from Table 2.2. The border value c_0, calculated in this manner, was found equal 3.36×10^{25} and 6.40×10^{25} m^{-3} for PC and PAr polymers, respectively [144].

Let us consider physical interpretation of "seedings" for the case of the polymer amorphous state. In accordance with the model [145], an avalanche-like formation and "freezing" of local order zones proceed with temperature, $T > T_g$, at transition over T_g. These very local order zones are the basic structural feature of glassy polymers [34, 90]. At $T > T_g$ macromolecular "hookings" and short-living local order zones exclusively exist in the polymer structure [125]. As shown above, the "frozen" local order zones represent the central oriented part of the WS cluster in relation to the polymer amorphous state structure. That is why, essentially, in the case under consideration, the only candidates for the role of "seedings" are macromolecular "hookings" [144]. For the polymers under consideration, the "hooking" network frequency equals, approximately, the magnitude of 10^{27} m^{-3}, more precisely, 7×10^{26} m^{-3} [112]. Thus in this case, the frequency of "seedings", c_0, is greater than R_c^{-d}, and WS cluster propagation is terminated, when the amount of initial diffusing particles becomes equal $R_c^{-d_f}$ [137]. Analytically, this condition is expressed as follows [133]:

$$c = R_c^{d_f - d} . \tag{5.88}$$

As before, for diffusing particles a statistical segment is accepted. In this case, c value equals [144]:

$$c = \frac{L}{l_{st}} = \frac{1}{Sl_0 C_\infty},$$
(5.89)

where L is total length of macromolecules per specific polymer volume, determined by the equation (2.7).

For PC and PAr polymers, calculation of parameter D_f by the equation (5.88) gives good conformity to experimental results [144]. As shown in Chapter 2, injection of relatively low concentrations ($\sim 0.01 - 1.0$ wt.%) of highly disperse Fe/FeO (Z) mixture into HDPE causes significant changes in the structure and properties of HDPE + Z composites [146]. It is also indicated [146] that these effects are stipulated by formation of local order zones (clusters) by local magnetic fields of superparamagnetic domains Z. Thus Z domains represent somewhat "seedings" for amorphous phase structure formation in HDPE + Z. In its turn, this affects the structure of crystalline and interphase zones in HDPE + Z during their formation from melt. That is why the method different from the above-described one was suggested for determination of the "seedings" density, c_0 [147]. Accepting in the first approximation for HDPE that $R_c = 50$ Å [147], the border "seeding" concentration equal $\sim 0.8 \times 10^{25}$ m^{-3} is obtained. In a natural polymer, the "seeding" density must equal the amount of clusters per specific polymer volume, N_{cl}, determined by the following relation: $2 v_{cl}/F$ [144].

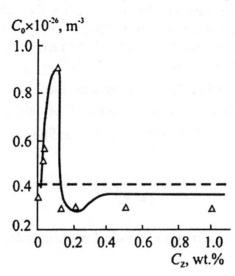

Figure 5.43. Dependence of the "seeding" density, c_0, on Z concentration, C_Z, for HDPE + Z composites. The dashed line shows frequency of macromolecular "hooking" network for HDPE [147]

Figure 5.43 shows the dependence $c_0(N_{cl})$ at Z concentration, C_Z, for HDPE + Z composites. Dashed line in the Figure shows frequency of macromolecular "hooking" network, ν_h, calculated in the work [148]. Clearly the dependence $c_0(C_Z)$ is of clear extreme type, corresponded to the type pf HDPE + Z composite properties variation [146]. At $C_Z = 0.05$ wt.%, the number of "seedings" by, approximately, 2.5-fold exceeds their concentration for unmodified HDPE. For HDPE and HDPE + Z composites with $C_Z \geq 0.10$ wt.%, the value of ν_h parameter is approximately equal c_0. Thus data from Figure 5.43 induce an assumption that previously detected extreme variation of HDPE + Z composite properties in C_Z range of $0.01 - 0.10$ wt.% is induced by the same change of the "seeding" density. In the rest cases, e.g. when Z particles do not form superparamagnetic domains, $c_0 = \nu_h$.

Assessment of the structure change induced by c_0 variation can be performed in the framework of the above-described WS model [137]. Since the minimal value of c_0 (Figure 5.43) is, approximately, five-fold higher than the border concentration of "seedings", the alternative of high "seedings" concentration (alternative **A** [137]), the basic correlation of which is equation (5.88), was used for assessment of structural changes [147]. Figure 5.44 shows comparison of fractal dimensionalities, D_f, of HDPE + Z composite structure, calculated by equations (5.4) and (5.88). Clearly good correspondence by run of D_f concentration dependence and absolute values of this parameter (maximal deviation does not exceed 1.5%) was obtained.

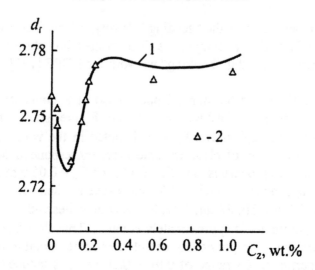

Figure 5.44. Comparison of fractal dimensionalities, d_f, of the structure, calculated by equations (5.4) (1) and (5.88) (2) as the function of C_Z for HDPE + Z composites [147]

The number of particles (statistical segments), n_{seg}, in the central closely packed zone of WS cluster can be determined by the following equation [119]:

$$\frac{R_{cl}}{2} = \left(\frac{n_{seg} S}{\pi \varphi_{cl}}\right)^{1/2}$$
(5.90)

in accordance with equations (2.5) and (5.90) (under the condition $F = 2n_{seg}$), i.e. in accordance with the cluster model of the polymer amorphous state structure and WS model, respectively. Both good qualitative and quantitative conformity is obtained again.

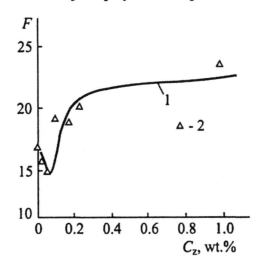

Figure 5.45. Comparison of local order zone functionality, F, calculated by equations (2.5) (1) and (5.90) (2) as the function of Z concentration, C_Z, for HDPE + Z composites [147]

As a consequence, the results have displayed [147] that besides molecular parameters S and C_∞, "seeding" density, c_0, is one more the important factor determining structure of the polymer amorphous state. It should be noted that the crystallinity degree is the function of D_f and, consequently, c_0, too [123, 124]. Application of modified WS model gives an opportunity of forecasting the structure of polymers as the function of density of artificially injected "seedings" (Z in the present case) or frequency of macromolecular "hookings" [144] (chemical cross-links [134]).

The energy of intermolecular interaction characterized either by molar cohesive energy, E_c, or cohesive energy density, W_c, is (besides the glass transition temperature, T_g) one of the most important characteristics of the polymer condensed state [149]. However, at present, correlations of these parameters with polymer properties are usually considered [150, 151], and there is no description of E_c (or W_c) effect on the polymer structure, though such influence is assumed quite frequently [149, 152]. Using the above-discussed models, the effect of cohesive parameters on amorphous polymer structure formation was cleared out on the example of two heterochain polyesters [153, 154]: aromatic copolyester sulfone formals (APESF) and 2,2-di-(4-oxyphenyl)propane oligoformal, phenolphthalein and isophthalic dichloranhydride (SP-OFD-10/F-1) diblock-copolyesters.

Molar cohesion, E_c, can be estimated by the following equation [149]:

$$T_g = T_{g0} \frac{2RT_{g0}^2 n}{E_c},$$ (5.91)

where T_g is the glass transition temperature of copolymer; T_{g0} is the glass transition temperature of more rigid-chain component (polysulfone (PSF) for APESF and polyarylate (PAr) for SP-OFD-10/F-1); n is the part of more flexible-chain component (polyformal) in block-copolymer.

However, in the framework of models applied to description of the polymer structure characteristics calculated per specific volume of the polymer are used. That is why to obtain correct correlations between cohesive characteristics and parameters of the structure, the use of density of cohesive energy, W_c, possessing "energy/volume" dimensionality (usually, MJ/m^3) instead of E_c is required [155].

Figure 5.46 shows W_c dependencies on the composition of copolyesters, namely, on concentration of rigid blocks (PSF or PAr: C_{PSF} and C_{PAr}, respectively) in them. The data shown indicate that increase of rigid-chain component concentration leads to W_c increase. As a rule, such correlations are explained by formation of domains from closely packed segments in the polymer structure [149, 152]. Figure 5.47 shows dependencies of T_g on the formal block concentration, C_{form}, comparison of which with the data from Figure 5.46 gives interesting results. It is known [156] that disposition of T_g dependencies with respect to composition of copolymers allows identification of their type. For example, for APESF, dependence $T_g(C_{form})$ is located above the additive glass transition temperature, T_g^{ad}. That is why for this polymer a tendency to regular alternation of units shall be expected. For SP-OFD-10/F-1, $T_g < T_g^{ad}$, i.e. the copolymer is inclined to formation of long sequences of the same blocks [156]. Nevertheless, for both these copolymers dependencies of W_c on composition occur beneath the appropriate additive dependence. This circumstance induces two conclusions. Firstly, in contrast with T_g dependencies on composition, the shape of analogous W_c dependence is independent of the type of copolymer. Secondly, T_g and W_c values characterize different properties of copolymers, though interdependent ones. For example, T_g is determined by rigidity of the polymer chain [157], and W_c by intensity of intermolecular interactions. In this connection, of interest is comparison of two polymers: polyamide-6 and polyvinylchloride, for which T_g are approximately equal (~323 and 343 K, respectively [155]). This is caused by approximately equal flexibility of their chains, determined by characteristic relation C_∞ (6.2 and 6.7, respectively [45, 53]). However, W_c values for these polymers differ by two times (783 and 389 MJ/m^3, respectively [155]).

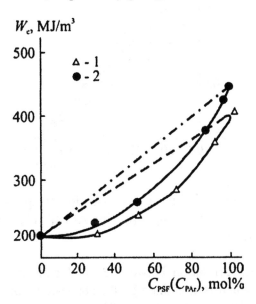

Figure 5.46. Dependencies of cohesive energy density, W_c, on concentration of rigid-chain blocks, C_{PSF} for APESF (1) and C_{PAr} for SP-OFD-10/F-1 (2) [153]

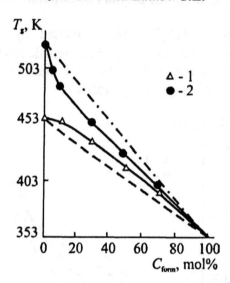

Figure 5.47. Dependencies of the glass transition temperature, T_g, on formal block concentration, C_{form}, for APESF (1) and SP-OFD-10/F-1 (2) [153]

Nevertheless, correlation between T_g and cohesive characteristics are known well [158], which is explained by the above-mentioned interdependence of these parameters (refer to equation (5.91)). Simultaneously, the increase of the polymer chain rigidity with T_g promotes formation of closely packed domains that increases W_c (or E_c) [152]. Figure 5.48 shows correlation $W_c(T_g)$ for APESF and SP-OFD-10/F-1, which is linear and reflects the above-considered interrelation. Since all data were obtained at $T = 293$ K, extrapolation of linear dependence $W_c(T_g)$ to $T_g = 293$ K means polymer transition to devitrified state, and at this temperature $W_c = 0$ (anyway, in the scale of Figure 5.48). Broad dispersion usually observed for such correlations is caused by the fact that T_g and W_c (or E_c) characterize different polymer properties that follows from comparison of data from Figures 5.46 and 5.47.

The effect of attraction and repulsion forces between particles forming WS cluster on its structure was studied [159]. It is shown that attraction force increase makes the cluster, more precisely, its external zones more packless. This aspect of amorphous polymer structure formation can also be considered in the framework of the cluster model [8].

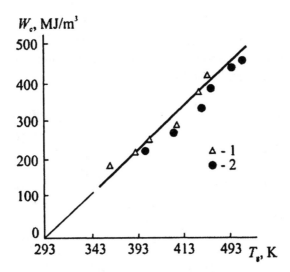

Figure 5.48. Correlation between cohesive energy density, W_c, and glass transition temperature, T_g, for APESF (1) and SP-OFD-10/F-1 (2) [153]

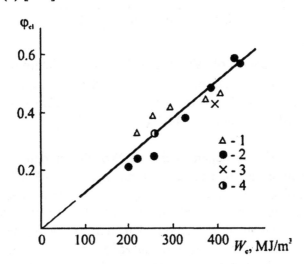

Figure 5.49. Correlation between relative part of clusters, φ_{cl}, and cohesive energy density, W_c, for APESF (1), SP-OFD-10/F-1 (2), PC (3), and HDPE (4) [153]

Figure 5.49 shows dependence $\varphi_{cl}(W_c)$ for APESF and SP-OFD-10/F-1. Expectedly, W_c or intermolecular attraction force intensification causes increase of the relative part of closely packed local order zones in the structure of copolymers. Data in the Figure on PC and HDPE [5] conform to $\varphi_c(W_c)$ correlation, obtained for copolymers under consideration. This indicates quite high generality of the correlation [8].

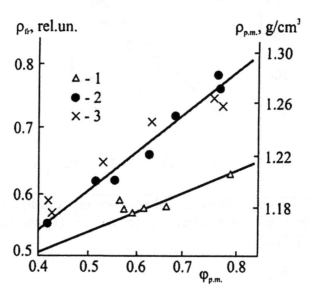

Figure 5.50. Dependencies of WS cluster density, ρ_{fr}, determined by equation (5.2) (1, 2) and packless matrix density, $\rho_{p.m.}$, determined by equation (5.92), on the relative part of packless matrix, $\varphi_{p.m.}$, for APESF (1, 3) and SP-OFD-10/F-1 (2) [154]

Figure 5.50 shows dependencies of WS cluster density, ρ_{fr} (equation (5.2)) on the relative part of packless matrix, $\rho_{p.m.}$, for APESF and SP-OFD-10/F-1, which indicate ρ_{fr} increase with $\rho_{p.m.}$, i.e. φ_{cl} or W_c decreased (Figure 5.49). Thus similar to computerized simulation [159], attraction forces increase between particles (statistical segments), characterized by W_c increase, induces higher loosening (ρ_{fr} decrease) of external zones in WS clusters, identified as the packless matrix [154]. Note also that at T_g relative free volume, f_f, increases with the polymer chain rigidity (or W_c, refer to Figure 5.48) [95]. Clearly W_c increase induces φ_{cl} growth (Figure 5.49) and $\rho_{p.m.}$ decrease (Figure 5.50) and consequently, f_f increase. Thus application of the WS model does also explain this feature of amorphous polymer structure [154].

Free particles are used in computerized simulation of WS clusters, whereas the principal difference of the polymer structure is linking of statistical segments forming it to the chain by strong covalent bonds. This restricts their motion. That is why φ_{cl} and $\varphi_{p.m.}$ variations can be accompanied by redistribution of frequencies of these structural components only. This effect can be estimated using simple rule of mixtures:

$$\rho = \rho_{cl}\varphi_{cl} + \rho_{p.m.}\varphi_{p.m.}, \tag{5.92}$$

where ρ is the polymer density; ρ_{cl} and $\rho_{p.m.}$ are densities of clusters and the packless matrix, respectively.

Taking for ρ experimentally determined densities of copolyesters and assuming $\rho_{cl} = 1.08\rho$ [142], one can estimate $\rho_{p.m.}$ variation with $\varphi_{p.m.}$, (under the assumption that $\rho_{fr} \sim \rho_{p.m.}$). Figure 5.50 shows the dependence $\rho_{p.m.}(\varphi_{p.m.})$, which can be coincident with the correlation $\rho_{fr}(\varphi_{p.m.})$ by simple adjustment of the scale, because ρ_{fr} values are given in relative units.

Thus application of the WS model gives quite full picture of intermolecular interaction forces effect on formation of the amorphous polymer structure. The increase of W_c characterizing intensity of intermolecular attraction forces induces growth of the local order degree, φ_{cl}. This effect is accompanied by loosening of the packless matrix by means of segments redistribution in structural components of the amorphous polymer [154].

It has been shown [131] that autocorrelation function:

$$c(r) = (p(r') \cdot p(r' + r)), \tag{5.93}$$

where $p(r')$ and $p(r' + r)$ are probabilities of the presence of particles, from which the cluster is composed (the probability equals 1 in the presence and 0 in the absence of the particle), in points r' and $r' + r$, can be presented by the following scaling relation:

$$c(r) \sim r^{-\alpha}, \tag{5.94}$$

for distances r greater than several steps of the lattice. In the relation (5.94), α is determined by the difference as follows [139]:

$$\alpha = D - D_f. \tag{5.95}$$

As a consequence, by determining the slope, α, of linear autocorrelation function, one can estimate the fractal dimensionality of the object, D_f, using relations (5.94) and (5.95) [160].

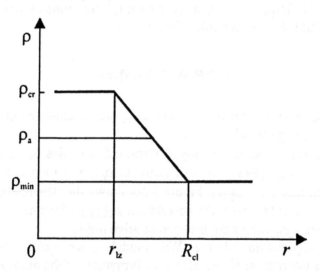

Figure 5.51. Schematic presentation of the function $\rho(r)$ for amorphous polymer [160]

Figure 5.51 shows schematic presentation of the autocorrelation function for the structure of amorphous polymer. Here variation of density is presented by the function of distance, r, and the origin of coordinates ($r = 0$) is disposed in the center of the local order zone (refer to Figures 5.2 and 5.3). In some interval from 0 to r_{lz} (where r_{lz} is the radius of the local order zone) parameter ρ is constant and equals density of the local order zone, which in the first approximation is accepted equal the appropriate ρ_{cr} value for crystalline polymer (particularly, for PC [160]). Then in the interval from r_{lz} to R_{cl} a linear decrease of ρ to some ρ_{min} value is assumed, assessed as follows. The average density of amorphous PC, ρ_a, was calculated [142], corresponded to the middle of the sloped linear part of $\rho(r)$ dependence in Figure 5.51.Then basing on simple geometrical ideas, one may obtain the following [160]:

$$\rho_{min} = \rho_a - (\rho_{cr} - \rho_a).$$
(5.96)

Further on, parameter α can be assessed from the following correlation [160]:

$$(\rho_{cr} - \rho_{min}) \sim (R_{cl} - r_{lz})^{-\alpha}, \tag{5.97}$$

and then D_f value by equation (5.95) can be calculated. Figure 5.52 shows comparison of temperature dependencies of the fractal dimensionality values of PC structure: experimental values $\left(d_f^e\right)$ calculated by the equation (5.4) and theoretical ones within the framework of the model suggested $\left(d_f^t\right)$ calculated by equations (5.95) and (5.97).

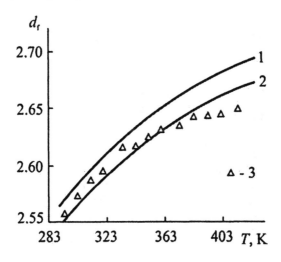

Figure 5.52. Dependence of fractal dimensionality, d_f, of PC structure on test temperature, T. Calculation was performed by equation (5.4) using σ_y (1) and σ_y^r (2) in equation (3.12); by equations (5.95) and (5.97) (3) [160]

Clearly good conformity between theoretical and experimental results is observed (deviation below 10%). However, this conformity can be improved using the following methodological method. To assess d_f^e in accordance with the equation (5.4), ν parameter was determined from the equation (3.12). If in this equation the real yield stress, σ_y^r, estimated as follows, is used instead of technical one, σ_y, [59]:

$$\sigma_y^r = \sigma_y(1 + \varepsilon_y), \tag{5.98}$$

where ε_y is the yield deformation, somewhat lower v value will be obtained. Data in Figure 5.52 indicate that the adjustment gives much better conformity between d_f^e and d_f^t for PC. For PC also at $T = 293$ K D_f value was determined [67] by five different methods that gave, on average, $D_f = 2.51$. This result conforms well to D_f value at $T = 293$ K, obtained in the framework of the WS model ($D_f = 2.56$, Figure 5.53) [160].

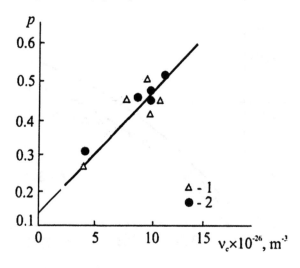

Figure 5.53. Dependence of steric factor, p, on cross-linking frequency, v_c, for EP-2 (1) and EP-3 (2) [162]

As observed in the framework of computerized simulation of DLA clusters, the steric factor p ($p \le 1$) indicating that not all collisions of particles happen with orientation necessary for particle "adhesion" to the cluster plays the important role basically defining the aggregation mechanism and structure of the final aggregate, characterized by its fractal dimensionality, D_f [159, 161]. For example, at high p approaching 1 the diffusion-limited mechanism of the aggregate propagation with relatively low values D_f is realized. On this basis an attempt was made [162] to clear out the effect of cross-linking frequency, v_c, on p and conditions of local order zone formation for polymer networks, EP-2 and EP-3.

For WS clusters, one can write down as follows [133]:

$$p \sim \frac{1}{r_{lz}}, \qquad (5.99)$$

where r_{lz} is determined from the equation (3.18).

The proportionality factor in the correlation (5.99) can be approximately assessed as follows. The distance between chemical cross-link points, R_c, was calculated by the equation (2.14), and then the proportionality factor was determined from the correlation (5.99) under the condition $r_{lz} = R_c$ and the appropriate minimal value $p = 0.1$.

Figure 5.53 shows dependence $p(v_c)$, found linear. Clearly v_c increase leads to p growth, which is accompanied by D_f decrease [68]. This conforms to general statements of the model [161]. There are data [163, 164] displaying that v_c increase induces cluster network frequency increase, v_{cl}, and decrease of the number of segments, n_{seg}, in a single cluster. This means formation of a large number of small clusters, N_{cl} ($= v_{cl}/n_{seg}$). Figure 5.54 shows dependence $N_{cl}(p)$ that indicates linear increase of N_{cl} with p (or v_c, refer to Figure 5.53). However, dependence $N_{cl}(p)$ does not pass through the origin of coordinates, and at $N_{cl} = 0$ value of p is finite and equals ~0.215. This means that at $p \le 0.215$ the cluster network is not formed or, to put it differently, polymer network does not transit to the glassy state. Figure 5.53 indicates that $p = 0.215$ corresponds to $v_c = 2.2 \times 10^{26}$ m^{-3}, and at v_c below the mentioned value the cross-linkable polymer does not transit to the condensed state. The critical value $v_c \left(v_c^{cr} \right)$ can also be estimated theoretically. The condition $N_{cl} = 0$ and, consequently, $v_{cl} = 0$ is the criterion of polymer transition to the rubbery state [121]. For such state, fractal dimensionality D of a part of chain between cross-link points that characterizes molecular mobility equals 2 [120]. Moreover, for the series of epoxy polymers under consideration, the maximum of C_∞ (at the minimum of v_c) equals 4.8 [165]. This allows determination of maximal D_f for epoxy polymers by equation (5.42) and then the number of segments, n_{seg}, between chemical cross-link points by equation (5.56). In accordance with the relation (5.42) and at $D = 2$ and $D_f = 2.737$, $n_{seg} = 23.3$ was obtained from equation (5.56) and the chain length between cross-link points, L_c, was calculated by equation (5.57). After assessment of the total length of macromolecules, L, per specific volume of the polymer by the equation (2.7) one can determine v_c^{cr} via L/L_c, whence it follows that $v_c^{cr} = 2.1 \times 10^{26}$ m^{-3}. This theoretical value of v_c^{cr} conforms properly to the appropriate experimental one, graphically estimated from the data in Figure 5.54 ($v_c^{cr} = 2.2\times10^{26}$ m^{-3}).

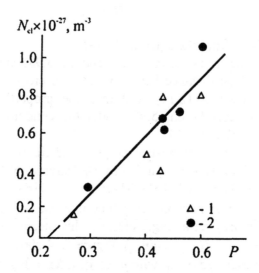

Figure 5.54. Dependence of the number of local order zones, N_{cl}, per specific volume of the polymer on steric factor, p, for EP-2 (1) and EP-3 (2) [162]

Extrapolation of $p(v_c)$ curve (Figure 5.53) to the maximum $p = 1$ gives the maximum of v_c $\left(v_c^{max}\right)$ equal ~26×10^{26} m^{-3}. At this v_c value the following fractal dimensionalities shall be obtained: $D = 1$ and $D_f = 2$ [5]. Under these conditions, $n_{st} = 2$ from equation (5.56) and $L_c = 12$ Å from equation (5.57) are obtained. Then theoretical value of v_c^{max} may be estimated as ~24.5×10^{26} m^{-3}, which again conforms well to the above experimental assessment. Moreover, the minimum of $n_{st} = 2$ conforms completely to the analogous assessment for completely cross-linked polymers [125].

Thus the above-considered results indicate that chemical cross-link point frequency, v_c, regulating conformation of the chain part between them determines the steric factor, p: the higher v_c is, the greater p is. This means that v_c increase forms the chain conformation favorable for statistical segments' packing in the local order zone. Therefore, this circumstance determines formation of a large number of small clusters at high v_c values [162].

In Section 5.5 the notion of quasi-equilibrium state of the polymer structure corresponded to the balance of tendencies to local order and entropic chain strain increase was introduced [9]. Generally, this state is characterized by $D_f = 2.5$. It should be noted that at $N \to \infty$ (where N is the number of particles in the cluster) this value of D_f coincides with the fractal dimensionality of the WS cluster for three-dimensional space [132]. In contrast with the primary Witten-Sander model [131], in which WS cluster growth is unlimited ($N \to \infty$), an interaction between WS clusters occurs in the modified simulation at multiple growth points [137], and the condition of the process termination is obtained. That is why the conformity of quasi-equilibrium state formation conditions in cluster and WS models was considered [166].

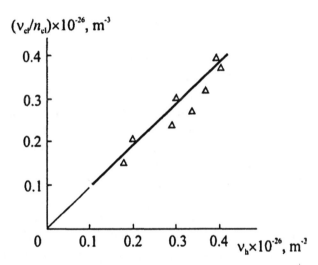

Figure 5.55. Correlation between cluster density, N_{cl}, and macromolecular "hooking" network frequency, v_h, for 8 amorphous and amorphous-crystalline polymers. The continuous line marks 1 : 1 ratio [166]

Data from the literature for a series of amorphous and amorphous-crystalline polymers were considered [166]. For these polymers, the "seeding" density $c_0 = N_{cl} = 2 v_{cl}/F$ and macromolecular "hooking" network frequency, v_h, were determined by the equation (2.??) using M_h values from the literature [45, 53]. Figure 5.55 shows comparison of N_{cl} and v_h values for 8 polymers under the strain and stress relaxation tests. Clearly quite good conformity of these parameters by both absolute values and the tendency of their variations was obtained. Quite broad dispersion is explained by corresponded deviation in M_h values from different literature sources, obtained for one and the same polymer (for example, for polyethylene M_h = 1,900 g/mol [53] and 1,390 g/mol [45], respectively). Nevertheless, Figure 5.55 indicates that the condition $N_{cl} = \nu_h$ allows identification of ν parameter as the "seeding" density [166].

The simulation [137] assumes the presence of such parameter as the diffusion length, ξ, depended on the local environing of every "seeding". Fractal propagation of WS cluster lasts until the condition $\xi = R_c$ is reached [137]. Parameters ξ and R_c are bound to the concentration of free particles, c, as follows [137]:

$$\xi^2 = c^{-1} R_c^{d_f - d + 2}. \qquad (5.100)$$

Substituting the above condition of WS cluster termination into equation (5.100), the criterion (5.88) is obtained that gives concentration of free particles, c, at the moment of WS cluster termination, when it reaches radius R_c (hereinafter, c_{WS}), or as in the framework of the WS model the quasi-equilibrium state is reached, because in this case, too, the structure invariance is determined by the balance of tendencies: diffusive propagation is limited by interaction of neighboring WS clusters [166].

In accordance with the relation (5.88) the use of dimensional R_c values induces c_{SW} dependence on dimensionality of the length, selected for R_c. That is why for normalizing factor the lower scale of the fractal behavior (b_l = 5 Å) was chosen [166], and dimensionless parameter (R_c/b_l) was used in the equation (5.88). Selection of another b_l value will change quantitative estimations of c_{SW} insignificantly and is negligible for tendencies in their variations.

In the framework of the cluster model (Figure 5.3), the ability to be added to the local order zone is inherent to the only segments, present in the packless matrix. As a consequence, in terms of the WS model these very segments represent free particles. That is why in the framework of the cluster model under reaching a quasi-equilibrium state, the concentration of the free particles, c_{cr}, will equal the relative part of the packless matrix, $\varphi_{p.m.}^{cr}$, in this state. This parameter was calculated as follows. First, D_f corresponded to the quasi-equilibrium state [41] was determined from the equation (5.42); sequentially, φ_{cl} and $\varphi_{p.m.}^{cr}$ as (1 - φ_{cl}) (or as (1 - φ_{cl}^{max}) for amorphous-crystalline polymers, equation (5.80)), which equals c_{cr} (see above) for the quasi-equilibrium state were determined from equation (5.7) [166].

Figure 5.56 gives comparison of c_{WS} and c_{cr} for 12 amorphous and amorphous-crystalline polymers. Clearly correspondence of these parameters indicating, in its turn, the conformity between interpretations of quasi-equilibrium states of the polymer structure in the cluster and WS models is observed. Somewhat overestimated c_{WS} values over c_{cr} are induced by arbitrary choice of b_1 value.

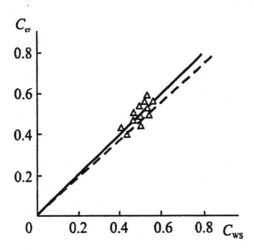

Figure 5.56. Correlation between concentrations of free particles, c_{WS}, and statistical segments in the packless matrix, c_{cr}, for quasi-equilibrium states for 12 amorphous and amorphous-crystalline polymers. Dashed line marks 1 : 1 ratio [166]

Briefly resuming results of the present Chapter, note that the cluster model of the polymer amorphous state conforms both qualitatively and quantitatively to the modern physical concepts, which can be applied to description of the polymer structure. Applications of one or another alternative of the combined models allows significant expansion of the analytical abilities of the structure description and obtaining of principally new ideas on it.

REFERENCES

1. Feder J., *Fractals*, New York, Plenum Press, 1988, 254 p.
2. Ivanova V.S., Balankin A.S., Bunin I.Zh., and Oksogoev A.A., *Synergism And Fractals In The Material Science*, 1994, Moscow, Nauka, 383 p. (Rus)
3. Novokov V.U and Kozlov G.V., *Uspekhi Khimii*, 2000, vol. **69**(4), pp. 378 - 393. (Rus)
4. Novokov V.U and Kozlov G.V., *Uspekhi Khimii*, 2000, vol. **69**(6), pp. 572 - 599. (Rus)
5. Kozlov G.V. and Novikov V.U., *Synergism And Fractal Analysis Of Polymer Networks*, 1998, Moscow, Klassika, 112 p. (Rus)
6. Novikov V.U. and Kozlov G.V., *Analysis of Polymer Destruction In The Framework Of The Concept Of Fractals*, 2001, Moscow, Izd. MGOU, 136 p. (Rus)
7. *Fractals And Local Order In Polymer Materials*, Ed. G.V. Kozlov and G.E. Zaikov, 2001, New York, Nova Science Publishers. Inc., 190 p.
8. Shogenov V.N. and Kozlov G.V., *Fractal Clusters In Physicochemistry Of Polymers*, 2002, Nalchik, Poligrafservis i T., 268 p. (Rus)
9. Kozlov G.V. and Zaikov G.E., In Coll.: *Fractals And Local Order In Polymer Materials*, Ed. G.V. Kozlov and G.E. Zaikov, 2001, New York, Nova Science Publishers, Inc., pp. 55 - 64.
10. Gladyishev G.P. and Gladyishev D.P., *On Physicochemical Theory Of Bilogical Evolution* (Preprint), 1993, Moscpw, Olimp, 24 p. (Rus)
11. Gladyishev G.P. and Gladyishev D.P., *Zh. Fiz. Khim.*, 1994, vol. **68**(5), pp. 790 - 792. (Rus)
12. Hornbogen E., *Int. Mater. Rev.*, 1989, vol. **34**(6), pp. 277 - 296.
13. Bessendorf M.H., *Int. J. Eng. Sci.*, 1987, vol. **25**(6), pp. 667 - 672.
14. Zemlyanov M.G., Malinovsky V.K., Novikov V.N., Parshin P.P., and Sokolov A.P., *Zh. Eksper. Teor. Fiz.*, 1992, vol. **101**(1), pp. 284 - 293. (Rus)
15. Kozlov G.V., Ozden S., Krysov V.A., and Zaikov G.E., In Coll.: *Fractals And Local Order In Polymeric Materials*, Ed. G.V. Kozlov and G.E. Zaikov, 2001, New York, Nova Science Publishers, Inc., pp. 83 - 88.
16. Kozlov G.V., Ozden S., and Dolbin I.V., *Russian Polymer News*, 2002, vol. **7**(2), pp. 35 - 38.
17. Bagryansky V.A., Malinovsky V.K., Novikov V.N., Pushchaeva L.M., and Sokolov A.P., *Fizika Tverdogo Tela*, 1988, vol. **30**(8), pp. 2360 – 2366. (Rus)

18. Balankin A.S., *Synergism of Deformable Body*, 1991, Moscow, MO SSSR, 404 p. (Rus)

19. Kozlov G.V., Belousov V.N., and Mikitaev A.K., *High Pressure Physics and Technology*, 1998, vol. 8(1), pp. 64 - 70. (Rus)

20. Flory P.J., *Brit. Polymer J.*, 1976, vol. 8(1), pp. 1 - 10.

21. Balankin A.S., Bugrimov A.L., Kozlov G.V., Mikitaev A.K., and Sanditov D.S., *Doklady AN*, 1992, vol. 326(3), pp. 463 - 466. (Rus)

22. Sokolov I.M., *Uspekhi Fizicheskikh Nauk*, 1986, vol. 150(2), pp. 221 - 256. (Rus)

23. Song H.-H. and Roe R.-J., *Macromolecules*, 1987, vol. 20(11), pp. 2723 - 2732.

24. Rammal R. and Toulouse G., *J. Physiq. Lettr.*, 1983, vol. 44(1), pp. L13 - L22.

25. Balankin A.S., *Pis'ma V ZhTF*, 1991, vol. 17(6), pp. 84 - 89. (Rus)

26. Novikov V.U., Kozlov G.V., and Boronin D.V., *Materialovedenie*, 1999, No. 3, pp. 25 – 28. (Rus)

27. Meakin P., *Phys. Rev. Lett.*, 1983, vol. 51(13), pp. 1119 - 1122.

28. Kolb M., Botet R., and Jullien R., *Phys. Rev. Lett.*, 1983, vol. 51(13), pp. 1123 - 1126.

29. Family F., *Phys. Rev. Lett.*, 1983, vol. 51(23), pp. 2112 - 2115.

30. Irzhak V.I., Rozenberg B.A., and Enikolopyan N.S., *Polymer Networks: Synthesis, Structure, Properties*, 1978, Moscow, Nauka, 248 p. (Rus)

31. Kozlov G.V., Beloshenko V.A., Stroganov I.V., and Lipatov Yu.S., *Doklady NAN Ukrainy*, 1995, No. 10, pp. 117 - 118. (Rus)

32. Kozlov G.V., Burya A.I., and Zaikov G.E., *Inzhenerno-Fizicheskii Zhurnal*, 2003, vol. 77, No. 1 (in press). (Rus)

33. Katz A.J. and Tompson A.H., *Phys. Rev. Lett.*, 1985, vol. 54(12), pp. 1325 - 1328.

34. Kozlov G.V. and Novikov V.U., *Uspekhi Fizicheskikh Nauk*, 2001, vol. 171(7), pp. 717 - 764. (Rus)

35. Kozlov G.V., Beloshenko V.A., Varyukhin V.N., and Lipatov Yu.S., *Polymer*, 1999, vol. 40(4), pp. 1045 - 1051.

36. Beloshenko V.A. and Kozlov G.V., *Mekhanika Kompozitnyikh Materialov*, 1994, vol. 30(4),pp. 451 - 454. (Rus)

37. Kozlov G.V., Belousov V.N., Serdyuk V.D., and Kuzhetsov E.N., *High Pressure Physics and Technology*, 1995, vol. 5(3), pp. 59 - 64. (Rus)

38. Smirnov B.M., *Uspekhi Fizicheskikh Nauk*, 1986, vol. 149(2), pp. 177 - 219. (Rus)

39. Kozlov G.V., Beigelzimer Ya.E., Gazaev M.A., Novikov V.U., and Mikitaev A.K., *Fractal Physics Of The Polymer Amorphous State*, Manuscript deposited to VINITI RAS, Moscow, December 05, 1995, No. 3221-V95. (Rus)

40. Flory P.J., *J. Chem. Phys.*, 1977, vol. **66**(12), pp. 5720 - 5729.

41. Kozlov G.V., Aloev V.Z., and Lipatov Yu.S., *Doklady NAN Ukrainy*, 2001, No. 3, pp. 150 - 154. (Rus)

42. Kozlov G.V., Beloshenko V.A., Gazaev M.A., and Varyukhin V.N., *High Pressure Physics and Technology*, 1995, vol. **5**(1), pp. 74 - 80. (Rus)

43. Careri G., *Ordine E Disordine Nella Materia*, Laterza, 1982, 228 p.

44. Ivanova V.S., *Synergism. Strength And Destruction Of Metal Materials*, 1992, Moscow, Mir, 160 p. (Rus)

45. Wu S., *J. Polymer Sci., Part B, Polymer Phys.*, 1989, vol. **27**(4), pp. 723 - 741.

46. Wu S., *J. Appl. Polymer Sci.*, 1992, vol. **46**(4), pp. 619 - 624.

47. Boyer R.F., *Macromolecules*, 1992, vol. **25**(20), pp. 5326 - 5330.

48. Boyer R.F. and Miller R.L., *Macromolecules*, 1977, vol. **10**(5), pp. 1167 - 1169.

49. Balankin A.S., *Pis'ma V ZhTF*, 1990, vol. **16**(7), pp. 14 - 20. (Rus)

50. Aliguliev R.M., Khiteeva D.M., and Oganyan V.A., *Vysokomol. Soedin.*, 1988, vol. **B30**(4), pp. 268 - 272. (Rus)

51. Aliguliev R.M., Khiteeva D.M., Oganyan V.A., and Nuriev R.A., *Doklady AN AzSSR*, 1989, vol. **45**(5), pp. 28 - 31. (Rus)

52. Perepechko I.I. and Maksimov A.V., *Vysokomol. Soedin.*, 1989, vol. **B31**(1), pp. 54 - 57. (Rus)

53. Aharoni S.M., *Macromolecules*, 1983, vol. **16**(9), pp. 1722 - 1728.

54. Novikov V., Gazaev M., and Kozlov G., *Proc. Second Symposium "Advances in Structured and Heterogeneous Continua"*, August 14 - 16, 1995, Moscow, Russia, p. 45.

55. Kozlov G.V., Gazaev M.A., Novikov V.U., and Mikitaev A.K., *Behavior Of Dissipative Structures In Deformable Polymer Matrices*, Manuscript deposited to VINITI RAS, Moscow, November 03. 1995, No. 2944-V95. (Rus)

56. Kozlov G.V., Beloshenko V.A., Novikov V.U., and Lipatov Yu.S., *Ukrainsky Khimichesky Zhurnal*, 2001, vol. **67**(3), pp. 57 - 60. (Rus)

57. Shogenov V.N., Belousov V.N., Potapov V.V., Kozlov G.V., and Prut E.V., *Vysokomol. Soedin.*, 1991, vol. **A33**(1), pp. 155 - 160. (Rus)

58. Kozlov G.V., Shustov G.B., and Temiraev K.B., *Vestnik KBGU, Ser. Khimicheskie Nauki*, 1997, No. 2, pp. 50 - 52. (Rus)

59. Kozlov G.V. and Sanditov D.S., *Anharmonic Effects And Physicomechanical Properties Of Polymers*, 1994, Novosibirsk, Nauka, 261 p. (Rus)
60. Budtov V.P., *Physical Chemistry Of Polymer Solutions*, 1992, Saint-Petersburg, Khimia, 384 p. (Rus)
61. Kozlov G.V., Temiraev K.B., and Shustov G.B., *Plast. Massy*, 1998, No. 9, pp. 27 - 29. (Rus)
62. Kozlov G.V., Burmistr M.V., Korenyako V.A., and Zaikov G.E., *Voprosy Khimii I Khimicheskoi Tekhnologii*, 2002, No. 6, pp. 77 – 81. (Rus)
63. Kozlov G.V., Beloshenko V.A., Gazaev M.A., and Novikov V.U., *Mekhanika Kompozitnyikh Materialov*, 1996, vol. **32**(2), pp. 270 - 278. (Rus)
64. Gazaev M.A., Kozlov G.V., Milman L.D., and Mikitaev A.K., *High Pressure Physics and Technology*, 1996, vol. **6**(1), pp. 76 - 81. (Rus)
65. Lyubchenko L.S. and Lyubchenko M.L., *Khimicheskaya Fizika*, 1997, vol. **16**(8), pp. 24 - 51. (Rus)
66. Kozlov G.V., Burya A.I., Sviridenok A.I., and Zaikov G.E., *Doklady NAI Belarusi*, 2003, vol. 47 (in press).
67. Kozlov G.V., Gazaev M.A., Novikov V.U., and Mikitaev A.K., *Pis'ma V ZhTF*, 1996, vol. **22**(16), pp. 31 - 38. (Rus)
68. Kozlov G.V., Novikov V.U., Gazaev M.A., and Mikitaev A.K., *Inzhenerno-Fizicheskyi Zhurnal*, 1998, vol. **71**(2), pp. 241 - 247. (Rus)
69. Porter D., *Polymer*, 1987, vol. **28**(8), pp. 1652 - 1656.
70. Donald A.M., Kramer E.J., and Bubeck R.A., *J. Polymer Sci., Polymer Phys. Ed.*, 1982, vol. **20**(7), pp. 1129 - 1141.
71. Wu S., *"Benibana" Int. Symp. October 8-11, 1990, Yamagata, Japan*, Abstract, Yamagata, Japan, 1990, pp. 130 - 137.
72. Lipatov Yu.S. and Privalko V.P., *J. Macromol. Sci.-Phys.*, 1973, vol. **B7**(3), pp. 431 - 444.
73. Haward R.N. and Thackray G., *Proc. Royal Soc. Lond.*, 1968, vol. **A302**(1471), pp. 453 - 472.
74. Plummer C.J.G. and Donald A.M., *Macromolecules*, 1990, vol. **23**(7), pp. 3929 - 3937.
75. Edwards S.F. and Vilgis T., *Polymer*, 1987, vol. **28**(3), pp. 375 - 378.
76. Balankin A.S., Ivanova V.S., and Breusov V.P., *Doklady AN*, 1992, vol. **322**(6), pp. 1080 - 1085. (Rus)
77. Kozlov G.V., Temiraev K.B., Shetov R.A., and Mikitaev A.K., *Materialovedenie*, 1999, No. 2, pp. 34 - 39. (Rus)

78. Kozlov G.V., Ozden S., and Zaikov G.E., In book: *Fractal Analysis of Polymers: From Synthesis to Composites*, Ed. Kozlov G., Zaikov G., and Novikov V., New York, Nova Science Publishers, Inc., 2003, p. 113-121.
79. Caush G., *Destruction Of Polymers*, 1981, Moscow, Mir, 440 p. (Rus)
80. Kozlov G.V., Belousov V.N., Sanditov D.S., and Mikitaev A.K., *Izvestiya VUZov, Severo-Kavkazsky Region, Estestvennye Nauki*, 1994, No. 1 – 2(86), pp. 54 - 58. (Rus)
81. Aharoni S.M., *Macromolecules*, 1985, vol. **18**(12), pp. 2624 - 2630.
82. Hong S.-D., Chung S.Y., Fedords R.F., and Moacanin J., *J. Polymer Sci.: Polymer Phys. Ed.*, 1983, vol. **21**(9), pp. 1647 - 1660.
83. Sanditov D.S., Kozlov G.V., Belousov V.N., and Lipatov Yu.S., *Ukrainian Polymer J.*, 1992, vol. **1**(3-4), pp. 241 - 258.
84. Frenkel Ya.I., *Kinetic Theory Of Liquids*, 1945, Moscow-Leningrad, Izd. AN SSSR, 424 p. (Rus)
85. Ferry J.D., *Viscoelastic Properties of Polymers*, 3rd Ed., New York, Wiley and Sons, 1980, 522 p.
86. Sanditov D.S. and Bartenev G.M., *Physical Properties of Non-Oriented Structures*, 1982, Novosibirsk, Nauka, 256 p. (Rus)
87. Sanditov D.S. and Kozlov G.V., *Vysokomol. Soedin.*, 1996, vol. **A36**(8), pp. 1389 - 1393. (Rus)
88. Sanditov D.S., Sangadiev S.Sh., and Kozlov G.V., *Fizika I Khimia Stekla*, 1998, vol. **24**(6), pp. 758 - 766. (Rus)
89. Sanditov D.S., Kozlov G.V., Belousov V.N., and Lipatov Yu.S., *Fizika I Khimia Stekla*, 1994, vol. **20**(1), pp. 3 - 13. (Rus)
90. Belousov V.N., Kozlov G.V., Mikitaev A.K., and Lipatov Yu.S., *Doklady AN SSSR*, 1990, vol. **313**(3), pp. 630 - 633. (Rus)
91. Sanditov D.S. and Kozlov G.V., *Fizika I Khimia Stekla*, 1993, vol. **19**(4), pp. 593 - 601. (Rus)
92. Kozlov G.V., Beloshenko V.A., and Lipskaya V.A., *Ukrainsky Fizichesky Zhurnal*, 1996, vol. **41**(2), pp. 222 - 225. (Rus)
93. Belousov N.V., Beloshenko V.A., Kozlov G.V., and Lipatov Yu.S., *Ukrainsky Khimichesky Zhurnal*, 1996, vol. **62**(1), pp. 1 - 4. (Rus)
94. Lipatov Yu.S., *Uspekhi Khimii*, 1978, vol. **47**(2), pp. 332 - 366. (Rus)
95. Sanchez I.C., *J. Appl. Phys.*, 1974, vol. **45**(10), pp. 4204 - 4215.
96. Beloshenko V.A., Kozlov G.V., and Varyukhin V.N., *High Pressure Physics And Technology*, 1994, vol. **4**(2), pp. 70 -74. (Rus)
97. Sanditov D.S. and Kozlov G.V., *Fizika I Khimia Stekla*, 1995, vol. **21**(6), pp. 547 - 576. (Rus)

98. Knopoff L. and Shapiro J.N., *Phys. Rev. B*, 1970, vol. 1(10), pp. 3893 - 3895.

99. Sharma B.K., *Acoust. Lett.*, 1980, vol. 4(2), pp. 19 - 23.

100. Sharma B.K., *Acoust Lett.*, 1980, vol. 4(1), pp. 11 - 14.

101. Olemsky A.I., Katsnelson A.A., and Umrikhin V.V., *Fizika Tverdogo Tela*, 1985, vol. 27(2), pp. 318 - 321. (Rus)

102. Kozlov G.V. and Mikitaev A.K., *Vysokomol. Soedin.*, 1987, vol. B29(7), pp. 490 - 492. (Rus)

103. Kozlov G.V., Shetov R.A., and Mikitaev A.K., *Doklady AN SSSR*, 1986, vol. 290(4), pp. 885 - 888. (Rus)

104. Cohen M.H. and Grest G.S., *Phys. Rev. B*, 1979, vol. 20(3), pp. 1077 - 1098.

105. Chen H.S., *J. Non-Cryst. Solids*, 1981, vol. 46(3), pp. 289 - 305.

106. Manevich L.I., Oshmyan V.G., Gai M.I., Akopyan E.L., and Enikolopyan N.S., *Doklady AN SSSR*, 1986, vol. 289(1), pp. 128 - 131. (Rus)

107. Margolina A. and Wu S., *Polymer*, 1988, vol. 29(12), pp. 2170 - 2173.

108. Bridge B., *J. Mater. Sci. Lett.*, 1989, vol. 8(2), pp. 102 - 103.

109. Patrikeev G.A., *Doklady AN SSSR*, 1968, vol. 183(9), pp. 636 - 639. (Rus)

110. Patrikeev G.A., *Mekhanika Polimerov*, 1971, No. 2, pp. 221 - 231. (Rus)

111. Kozlov G.V., Sanditov D.S., Milman L.D., and Serdyuk V.D., *Izvestiya VUZov. Severo-Kavkazskyi Region, Estestvennye Nauki*, 1993, No. 3 – 4(83-84), pp. 88 - 92. (Rus)

112. Prevorsek D.C. and De Bona B.T., *J. Macromol. Sci.-Phys.*, 1981, vol. B19(4), pp. 605 - 622.

113. Shustov G.B., Belousov V.N., Kozlov G.V., and Mikitaev A.K., *Carcass Coherency And Mechanical Properties Of Extruded Amorphous Polymers*, Manuscript disposed to VINITI RAS, Moscow, February 26, 1997, No. 626-V97. (Rus)

114. Kozlov G.V., Aloev V.Z., Bazheva R.Ch., and Zaikov G.E., In Coll.: *IV All-Russian Sci.-Tech. Conf. "New Chemical Technologies: Production and Application"*, Penza, PGU, 2002, pp. 59 – 61. (Rus)

115. Shklovsky B.I. and Efros A.L., *Uspekhi Fizicheskikh Nauk*, 1975, vol. 117(3), pp. 401 - 436. (Rus)

116. Family F., *J. Stat. Phys.*, 1984, vol. 36(5/6), pp. 881 - 896.

117. Kozlov G.V. and Dolbin I.V., *Mater. Intern. Conf. "Physics of Electronic Materials"*, Kaluga, Russia, Oct. 1-4, 2002, p. 134 - 135.

118. Kozlov G.V., Beloshenko V.A., and Lipatov Yu.S., *Ukrainsky Khimichesky Zhurnal*, 1998, vol. **64**(3), pp. 56 - 59. (Rus)
119. Bobryishev A.N., Kozomazol V.N., Babin L.O., and Solomatov V.I., *Synergism Of Composite Materials*, 1994, Lipetsk, NPO ORIUS, 153 p. (Rus)
120. Kozlov G.V., Serdyuk V.D., and Dolbin I.V., Materialovedenie, 2000, No. 12, pp. 2 - 5. (Rus)
121. Beloshenko V.A., Kozlov G.V., and Lipatov Yu.S., *Fizika Tverdogo Tela*, 1994, vol. **36**(10), pp. 2903 - 2906. (Rus)
122. Bartenev G.M. and Frenkel S.Ya., *Physics Of Polymers*, 1990, Leningrad, Khimia, 432 p. (Rus)
123. Kozlov G.V., Beloshenko V.A., Varyukhin V.N., and Novikov V.U., *Zhurnal Fizicheskikh Issledovanii*, 1997, vol. **1**(2), pp. 204 - 207. (Rus)
124. Aloev V.Z., Kozlov G.V., and Beloshenko V.A., *Izvestiya KBNTs RAN*, 2000, No. 1(4), pp. 108 - 113. (Rus)
125. Berstein V.A. and Egorov V.M., *Differential Scanning Calorimetry In Physicochemistry Of Polymers*, 1990, Leningrad, Khimia, 256 p. (Rus)
126. Bittrich H.-J., *Acta Polymerica*, 1982, vol. **33**(12), p. 741.
127. Kozlov G.V., Burya A.I., Sanditov D.S., Serdyuk V.D., and Lipatov Yu.S., *Materialy, Tekhnologii, Instrumenty*, 1999, vol. **4**(2), pp. 51 - 54. (Rus)
128. Kozlov G.V. and Lipatov Yu.S., *Vysokomol. Soedin.*, 2003, vol. **B45**(4), pp. 660 - 664. (Rus)
129. Kozlov G.V., Sanditov D.S., and Lipatov Yu.S., In Coll.: *Fractals And Local Order In Polymeric Materials*, Ed. G.V. Kozlov and G.E. Zaikov, New York, Nova Science Publishers, Inc., 2001, pp. 65 - 82.
130. Adam G. and Gibbs J.H., *J. Chem. Phys.*, 1965, vol. **43**(1), pp. 139 - 146.
131. Witten T.A. and Sander L.M., *Phys. Rev. Lett.*, 1981, vol. **47**(19), pp. 1400 - 1403.
132. Meakin P., *Phys. Rev. A*, 1983, vol. **27**(3), pp. 1495 - 1507.
133. Kokorevich A.G., Gravitis Ya.A., and Ozon'-Kalnin V.G., *Khimia Drevesiny*, 1989, No. 1, pp. 3 - 24. (Rus)
134. Kozlov G.V., Beloshenko V.A., and Varyukhin V.N., *Ukrainsky Fizichesky Zhurnal*, 1998, vol. **43**(3), pp. 322 - 323. (Rus)
135. Botet R., Jullien R., and Kolb M., *Phys. Rev. A*, 1984, vol. **30**(4), pp. 2150 - 2152.
136. Brady R.M. and Ball R.C., *Nature*, 1984, vol. **309**(5965), pp. 225 - 229.

137. Witten T.A. and Meakin P., *Phys. Rev. B*, 1983, vol. **28**(10), pp. 5632 - 5642.
138. Mosolov A.B. and Dinariev O.Yu., *Problemy Prochnosti*, 1988, No. 1, pp. 3 - 7. (Rus)
139. Witten T.A. and Sander L.M., *Phys. Rev. A*, 1983, vol. **27**(9), pp. 5686 - 5697.
140. Meakin P. and Witten T.A., *Phys. Rev. A*, 1983, vol. **28**(5), pp. 2985 - 2988.
141. Kozlov G.V., Shogenov V.N., and Mikitaev A.K., *Inzhenerno-Fizicheski Zhurnal*, 1998, vol. **71**(6), pp. 1012 - 1015. (Rus)
142. Litt M.H., Koch P.J., and Tobolsky A.V., *J. Macromol. Sci.-Phys.*, 1967, vol. **B1**(3), pp. 587 - 594.
143. Wunderlich B. *Macromolecular Physics*, Volume 1, New York – London, Academic Press, 1973, 623 p.
144. Burya A.I., Shogenov V.N., Kozlov G.V., and Arlamova N.T., *Pridneprovsky Nauchny Vestnik*, 1998, No. 83, pp. 9 - 16. (Rus)
145. Kozlov G.V., Beloshenko V.A., and Gazaev M.A., *Ukrainski Fizicheski Zhurnal*, 1996, vol. **41**(11-12), pp. 1110 - 1113. (Rus)
146. Mashukov N.I., Gladyishev G.P., and Kozlov G.V., *Vysokomol. Soedin.*, 1991, vol. **A33**(12), pp. 2538 - 2546. (Rus)
147. Afaunov V.V., Shogenov V.N., Mashukov N.I., and Kozlov G.V., *Elektr. Zhurnal "Issledovano V Rossii"*, 1999, **33**. http://zhurnal mipt.rssi.ru/articles /1999/033.pdt. (Rus)
148. Graessley W.W. and Edwards S.F., *Polymer*, 1981, vol. **22**(10), pp. 1329 - 1334.
149. Bessonov M.I., Koton M.M., Kudryavtsev V.V., and Laius L.A., *Polyimides – The Class of Thermoresistant Polymers*, 1983, Leningrad, Nauka, 302 p. (Rus)
150. Willbourn A.H., *Polymer*, 1976, vol. **17**(10), pp. 965 - 976.
151. Mikitaev A.K., Kozlov G.V., and Shogenov V.N., *Plast. Massy*, 1983, No. 2, pp. 32 - 33. (Rus)
152. Koton M.M., Artemieva V.N., Kukarkina N.V., Kuznetsov Yu.P., and Dergacheva E.N., *Vysokomol. Soedin.*, 1987, vol. **B29**(8), pp. 571 - 575. (Rus)
153. Kozlov G.V., Temiraev K.B., and Afaunova Z.I., *The Effect of Intermolecular Interaction Energy On Structure Of Heterochain Polyesters*, Manuscript deposited to VINITI RAS, Moscow, June 15, 1999, No. 1918-V99. (Rus)

154. Kozlov G.V., Temiraev K.B., and Bedanokov A.Yu., *Vestnik Adyigeiskogo Universiteta*, 1999, No. 3, pp. 79 - 82. (Rus)
155. Kalinichev E.L. and Sakovtseva M.B., *Properties And Processing Of Thermoplasts*, 1983, Leningrad, Khimia, 288 p. (Rus)
156. Vasnev V.A. and Kuchanov S.I., *Uspekhi Khimii*, 1973, vol. **42**(12), pp. 2194 - 2220. (Rus)
157. Privalko V.P. and Lipatov Yu.S., *Vysokomol. Soedin.*, 1971, vol. **A13**(12), pp. 2733 - 2738. (Rus)
158. Lee W.A. and Sewell J.H., *J. Appl. Polymer Sci.*, 1968, vol. **12**(6), pp. 1397 - 1409.
159. Meakin P., *J. Chem. Phys.*, 1983, vol. **79**(5), pp. 2426 - 2429.
160. Kozlov G.V., Burya A.I., Sviridenok A.I., and Zaikov G.E., *Doklady NAI Belarusi*, 2003, vol. **47** (in press). (Rus)
161. Jullien R. and Kolb M., *J. Phys. A*, 1984, vol. **17**(12), pp. L639 - L643.
162. Kozlov G.V., *Doklady Adyigeiskoi (Cherkesskoi) Mezhdunarodnoi AN*, 2001, vol. **6**(1), pp. 83 – 87. (Rus)
163. Kozlov G.V., Beloshenko V.A., Gazaev M.A., and Lipatov Yu.S., *Vysokomol. Soedin.*, 1996, vol. **B38**(8), pp. 1423 - 1426. (Rus)
164. Kozlov G.V., Beloshenko V.A., and Varyukhin V.N., *Prikladnaya Mekhanika I Tekhnicheskaya Fizika*, 1996, vol. **37**(3), pp. 115 - 119. (Rus)
165. Kozlov G.V., Beloshenko V.A., and Varyukhin V.N., *Ukrainski Fizicheski Zhurnal*, 1996, vol. **41**(2), pp. 218 - 221. (Rus)
166. Kozlov G.V., Afaunova Z.I., and Zaikov G.E., *Elektron. Zh. "Made In Russia"*, 2003, Iss. 42, pp. 489 – 501; http: // Zhurnal.apl.relarn.ru/articles/2003/042.pdf. (Rus)

Chapter 6.

Description of impacts on the polymer structure in the framework of the cluster model

6.1. CRYSTALLIZATION

Actually, there are certain reasons why crystallization of polymers cannot be described in the framework of the cluster model. Nevertheless, there is a definite connection between short-range (local) order and long-range order formation [1]. Heat motions in a liquid induce continuous formation and termination of heterophase fluctuations, which represent potential focuses of future crystalline structures [2]. As temperature falls below the glass transition temperature, chemical potential of molecules in crystal is lower than in liquid. That is why in heterophase fluctuation the nucleation center can become stable and start propagating spontaneously [2].

It is assumed that configuration, morphology and mechanism of crystalline phase nucleation and propagation will be highly determined by the origin of heterophase fluctuations in the amorphous phase. For example, the assumption of heterophase fluctuation from folded chains (the Yech model [3]) leads to formation of crystallites with folded chains (CFC). An alternative model of heterophase fluctuation was suggested [2], where the presence of parallel chain segments of different macromolecules in it is assumed. The use of these models means the possibility of formation of CFC, crystallites with extended chains (CEC) or some intermediate crystalline morphology [4]. In its physical essence, the type of heterophase fluctuation suggested [2] is analogous to that of the local order zone in the cluster model [5]. Application of the mentioned model allows quantitative description of such heterophase fluctuations and, as a consequence, analysis of variations in the nucleation mechanism and crystallization morphology [6]. Such analysis has been performed on the example of oriented crystallization of two cross-linked polymers [1]: low density polyethylene (LDPE) and polychloroprene (PCP).

It has been shown [7] that the increase of stretching, λ, during oriented crystallization induces elongation of the statistical segment, l_{st}. Thus as the critical condition $l_{st} = l^*$ (where l^* is the crystallization nucleus length) is reached, the cluster becomes such nucleus and the crystallite morphology assessment is rather approximate, because curves 1 – 4 are plotted in accordance with the data on polyethylene [1]. Nevertheless, the data in Figure 6.1 demonstrate correctness of the cluster identification as heterophase fluctuations, which subsequently become crystalline phase growth nuclei. From these data it also follows that the increase of experimental λ_{cr}^{*} values means decrease of the part of fibrillar crystallites in mixed crystalline morphology due to increase of the CFC part. The data obtained prove conclusions [10] assuming formation of mixed-type crystallites at PCP oriented crystallization.

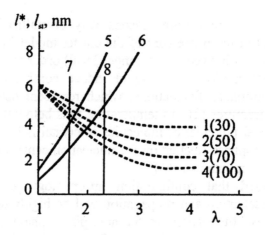

Figure 6.1. Dependencies of crystalline nucleus size, l^{*} (1 – 4), and length of statistical segment, l_{st} (5, 6), on the stretch degree, λ, for LDPE (1 – 5) and PCP (6). Straight lines (7) and (8) indicate critical values, λ_{cr}^{*}, for LDPE (7) [9] and PCP (8) [10]. Digits at $l^{*}(\lambda)$ curves give percentage of CEC content [1]

Besides longitudinal size of critical heterogeneous fluctuation, l^{*}, there is one more critical size of this fluctuation, the transversal one [4, 6]. The above-mentioned size is determined as follows:

$$(Sv^{*})^{1/2},$$

where S is the macromolecule cross-section; v^{*} is the functionality of critical nucleus. For polyethylene, dependence of $(Sv^{*})^{1/2}$ on λ is shown in the work [6]. For the cluster, analogous parameter can be determined as follows:

$$(SF^*)^{1/2},$$

where F is the cluster functionality. For LDPE, Figure 6.2 shows dependencies of $(Sv^*)^{1/2}$ in accordance with the data from [6] and $(SF^*)^{1/2}$ on the stretching, λ. Clearly in accordance with the condition of equality of transverse dimensions, the cluster becomes the critical nucleus of crystallization at $\lambda^*_{cr} \approx 1.75$, which is close to experimental value of this parameter (≈ 1.60) [9].

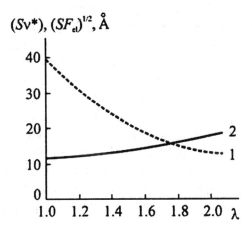

Figure 6.2. Dependencies of critical crystallization nucleus $(Sv^*)^{1/2}$ (1) and cluster $(SF^*)^{1/2}$ (2) transverse dimensions on the stretch rating, λ, for LDPE [1]

Thus clusters by both physical meaning (remind that they are of thermofluctuational origin [5]) and their critical dimensions correspond to heterophase fluctuations, which may become crystallite nuclei. It should be noted that in contrast with the glassy state, dynamic local order zones in the devitrified state may partly consist of folded chains. Results obtained for PCP indicate its mixed crystalline morphology including both folded and fibrillar crystallites. The increase of part of the latter leads to decrease of the critical stretch rating, λ^*_{cr}, at which transition from mono- to multimolecular nucleation is observed.

6.2. PLASTICIZATION

Plasticization is introduction of organic compounds into a polymer for the purpose of attaching elasticity and frost resistance to them, and decreasing processing temperature [11]. In recent 50 years, due to high practical importance of the plasticization effect, great attention was paid to the study of different aspects of it. The so-called anti-plasticization effect, which is the increase of polymer rigidity and strength with plasticizer concentration in them, has been discovered [11]. At present, the approach based on the assumption of occurrence (or modification) of ordering at the introduction of low amounts of plasticizer that induces polymer rigidity increase, is the most reasonable one [11 – 14]. Further increase of plasticizer concentrations does not cause ordering of the polymer structure: polymer rigidity and strength are decreased. However, so far these structural changes have been described qualitatively only. That is why structural changes proceeding in polymers at plasticizer injection and determining the well-known plasticization and anti-plasticization effects were described quantitatively [15]. The above-mentioned changes were described quantitatively in the framework of the cluster model on the example of polycarbonate (PC) plasticized by dibutylphthalate (DBP) [5].

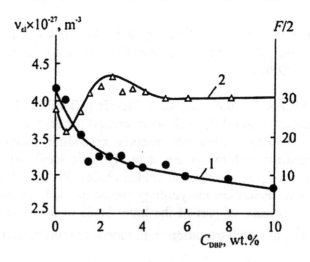

Figure 6.3. Dependencies of macromolecular entanglement cluster network frequency, v_{cl} (1), and the number of segments in the cluster, $F/2$ (2), on dibutylphthalate concentration in PC, C_{DBP} [15]

Figure 6.3 shows dependencies of v_{cl} and $F/2$ on DBP plasticizer concentration in PC, C_{DBP}. Of special interest is different behavior of v_{cl} and $F/2$ parameters with the concentration. The former decreases monotonously; to put it differently, local ordering in PC is monotonously decreased by the plasticizer. However, the extreme increase of the number of statistical segments in the cluster (equal $F/2$) indicates cluster restructuring. Simultaneous consideration of $v_{cl}(C_{DBP})$ and $F/2(C_{DBP})$ dependencies suggests that plasticizer injection leads to degradation of lower stable (with lower $F/2$) clusters and simultaneous formation of more stable (with higher $F/2$) local order zones [15].

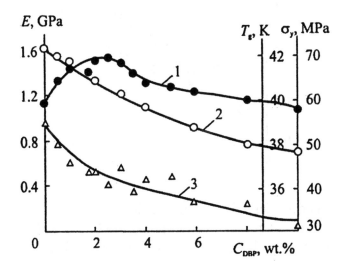

Figure 6.4. Dependencies of elasticity modulus E (1), glass transition temperature T_g (2) and yield stress σ_y (3) on dibutylphthalate concentration in PC, C_{DBP} [15]

Figure 6.4 shows dependencies of elasticity modulus E, glass transition temperature T_g and yield stress σ_y on the plasticizer concentration in PC, C_{DBP}. Two latter parameters decrease monotonously with C_{DBP} increase, and the former one (E) passes through a maximum at $C_{DBP} = 2.5$ wt.% that conforms to $F/2$ maximum in Figure 6.3.

Comparison of Figures 6.3 and 6.4 indicates complete determination of T_g and σ_y by v_{cl} parameter (e.g. the local order degree), and more complicated dependence between E and the structure. The first assumption is proved by plots in Figure 6.5 presenting dependencies of T_g and σ_y on the entanglement cluster network frequency v_{cl} (or the number of closely packed segments per specific volume). Clearly both plots are linear. Moreover, extrapolation of $T_g(\nu_{cl})$ dependence to $\nu_{cl} = 0$ demonstrates that the glass transition temperature of PC will equal the test temperature, i.e. under such conditions PC becomes the rubber-like composite. To put it differently, the polymer devitrification condition is disappearance of the local order. Naturally, under these conditions $\sigma_y = 0$, which is displayed by $\sigma_y(\nu_{cl})$ curve extrapolation to $\nu_{cl} = 0$ (Figure 6.5).

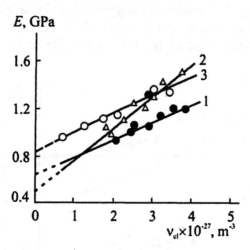

Figure 6.5. Dependencies of elasticity modulus E on entanglement cluster network frequency ν_{cl} for primary PC (1) and PC with DBP contents as follows: 2.5 (2) and 10 (3) wt.% [15]

Figure 6.6 shows dependencies of elasticity modulus (E) on entanglement cluster network frequency (ν_{cl}) for primary PC (line 1), PC containing 2.5 wt.% of DBP (anti-plasticized one, line 2), and the one containing 10 wt.% of DBP (plasticized one, line 3). In contrast with $\sigma_y(\nu_{cl})$ plot (Figure 6.5, these lines at $\nu_{cl} = 0$ are extrapolated to non-zero E_0 values. This indicates that E is determined by both ν_{cl} value and packless matrix rigidity [16]. Comparison of three plots shown in Figure 6.6 allows the following conclusions. Firstly, anti-plasticized PC possesses the minimum of E_0 at $\nu_{cl} = 0$. This assumes that anti-plasticization effect is induced by formation of more stable clusters and, the more so, by more loosening of the packless matrix that results in decrease of its rigidity. Secondly, anti-plasticized PC displays the greatest slope of $E(\nu_{cl})$ line. To put it differently, more stable clusters are the most rigid ones. Thirdly, at equal slopes of $E(\nu_{cl})$ dependencies E_0 at $\nu_{cl} = 0$ is greater for plasticized PC rather than for the primary one. It may be assumed that increased rigidity of plasticized PC packless matrix is stipulated by free volume microcavity filling in by plasticizer, which is indicated by monotonous decrease of gas permeability with C_{DBP} increase.

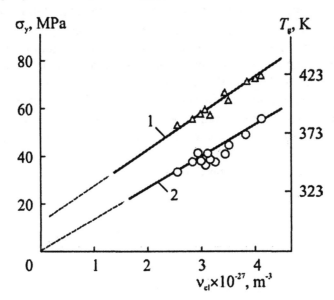

Figure 6.6. Dependencies of glass transition temperature T_g (1) and yield stress σ_y (2) on entanglement cluster network frequency ν_{cl} for PC, plasticized by DBP [15]

The mentioned curves (Figures 6.3 – 6.6) give demonstrative quantitative description of structural changes in PC under plasticizer injection into it. Note also that interpretation of plasticization and anti-plasticization effects of amorphous glassy-like PC in the framework of the cluster model correlates with the results obtained by different authors [12 – 14].

6.3. HYDROSTATIC PRESSURE

Usually, changes in the structure of amorphous polymers under hydrostatic pressure are interpreted in the framework of free volume models [17 – 19]. As shown before [20] for glassy-like epoxy polymer, dependence of the relative part of the fluctuation free volume, f_f, on external pressure, P, possesses three typical areas displaying different mechanisms of polymer structure consolidation. For example, in the pressure range $P = 0 - 200$ MPa fluctuation microcavities "implode", and in the range $P = 200 - 600$ MPa consolidation is stipulated by changes in macromolecule conformations (entropy decrease). Finally, at $P \geq 800$ MPa entire macromolecules are compressed that requires extremely high pressures. In this range, f_f reaches its maximum [20].

The same data on $f_f(P)$ dependence are considered [21] in the framework of the fractal analysis [22] and the cluster model of the polymer amorphous state [5]. For epoxy polymer derived from bisphenol A diglycidyl ester, experimental data obtained by the positron annihilation spectroscopy method were used [18].

Parameter f_f and fractal dimensionality of the structure d_f are determined from equations (2.33) and (5.4), respectively. Combination of these equations gives the following correlation between f_f and d_f [21]:

$$f_f \approx 8.5 \times 10^3 \left(\frac{d_f}{d - d_f} \right). \tag{6.3}$$

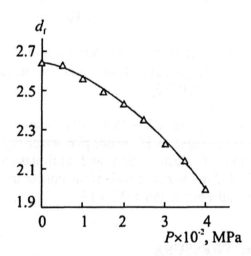

Figure 6.7. Dependence of epoxy polymer fractal dimensionality, d_f, on hydrostatic pressure, P [21]

Figure 6.7 shows dependence $d_f(P)$, calculated by the equation (6.3) using parameter f_f as the function of P with regard to previously obtained data [20]. Clearly increase of P induces monotonous decrease of d_f. From these values of d_f using equation (5.7) the relative part of clusters φ_{cl} was calculated as a function of P (Figure 6.8) under the conditions $S = $ const and $C_\infty = $ const. As indicated in the Figure, φ_{cl} is rapidly increased in the range of $P = 0 - 200$ MPa. At $P \approx 200$ MPa, the relative part of clusters reaches its maximum, $\varphi_{cl} = 1.0$, and is kept at this level during further pressure increase: $\varphi_{cl} = $ const $= 1.0$ [21].

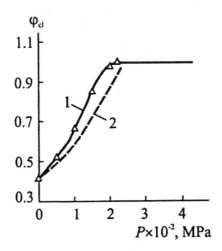

Figure 6.8. Dependence of the relative part of epoxy polymer volume, incapable of relaxation, φ_{cl}, on hydrostatic pressure, P: 1 – calculated by equation (5.7) at $C_\infty = $ const; 2 – calculated by equation (5.42 with regard to C_∞ assessment [21]

As accepted [5] from the cluster model, the entire fluctuation free volume, f_f, is concentrated in the packless matrix. Under the impact of hydrostatic pressure applied to the polymer, fluctuation free volume microcavities "implode" and φ_{cl} is increased. Obviously there are no reasons for the supposition that increase of epoxy polymer structure ordering is induced by pressure. In this case, more general definition of parameter φ_{cl} can be given: it is the part of the polymer volume incapable of relaxation. Primarily, this part of the volume includes local order zones. In its turn, incapability of relaxation is stipulated by an abrupt decrease of f_f or real absence of the fluctuation free volume. To put it differently, "porosity" degree of the packless matrix is determined by f_f value; as $f_f \approx 0$, the matrix obtains a new quality: it behaves itself similar to closely packed local order zones [21].

Revert to the condition $\varphi_{cl} = \text{const} = 1.0$ at $P > 200$ MPa. Obviously at $P > 200$ MPa S remains constant, too ($S = \text{const}$), because compression of macromolecules themselves requires much higher pressures. That is why C_∞ is the only variable parameter, remained in the equation (5.7). Figure 6.9 shows dependence $C_\infty(P)$ under the supposition of its constancy at $P \leq 200$ MPa and $\varphi_{cl} = \text{const} = 1.0$, and $S = \text{const}$ at $P > 200$ MPa. In the latter case ($P > 200$ MPa), an abrupt decrease of C_∞ is clearly observed, and at the limit C_∞ approaches 1. It is common knowledge [23] that in accordance with the equation (5.58) the characteristic relation C_∞ is the function of internal rotation angle of chemical valence bonds, θ. Obviously for $C_\infty \to 1$, $\cos\theta \to 0$ and $\theta \to 90°$. Thus as assumed before [20], pressure increase in the range $P > 200$ MPa will induce conformational changes in macromolecules, characterized by C_∞ decrease and θ increase.

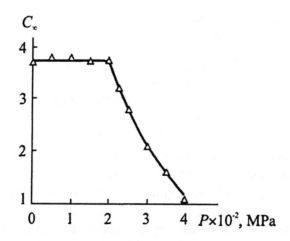

Figure 6.9. Dependence of characteristic relation, C_∞, of epoxy polymer on hydrostatic pressure, P, calculated by the equation (5.42) under the assumption $C_\infty = \text{const}$ at $P < 200$ MPa [21]

Assuming that a change of d_f induces variation of C_∞ equal that for the case without application of hydrostatic pressure, the interconnection between C_∞ and d_f is set by the equation (5.42). In Figure 6.8 dashed line presents dependence $\varphi_{cl}(P)$, calculated by the equation (5.7) under the condition of variable C_∞. Clearly qualitatively analogous result is obtained: the conditions $\varphi_{cl} = 1.0$ is reached at $P = 250$ MPa.

Thus the fractal analysis methods combined with the cluster model of the polymer amorphous state structure allow quantitative description of structural changes in epoxy polymer impacted by external hydrostatic pressure. The results obtained correlate well with the data followed from kinetic theory of the fluctuation free volume [20].

6.4. HEAT (PHYSICAL) AGING

Polymer heat aging processes are studied in many works. It is found that the effect of aging on structure and properties of these polymers is caused by thermodynamic non-equilibrium of the glassy state, and aging itself can be considered as slow approaching of the equilibrium state of a system displaying a broad distribution of relaxation times [24]. It has been assumed [25 – 27] that aging of amorphous polymers at temperatures below their glass transition points, T_g, induces consolidation of their liquid-like packing, which, in its turn, approaches their structure to the equilibrium glassy state. There are two points of view on particular ways of the mentioned approaching. The first is [28] that aging causes changes of the free volume in polymers. The second states [29] that the effects of aging are caused by thermally reversible molecular restructurings. These processes were specified by other authors [20]. It has been supposed that aging causes rather fine changes in conformations, smaller than, for example, transition from *cis-* to *trans*-conformation. The authors have also suggested that the free volume decrease is the demonstration of aging, but not the main reason for changes in physical properties of the polymer. The results of heat aging for PC were also described in the framework of the cluster model [31], which indicated that this interpretation generalizes both points of view.

Figure 6.10 shows dependencies of the entanglement cluster network frequency, v_{cl}, on test temperature, T, for PC, primary and annealed at 393 K during 6 hours. Clearly v_{cl} is higher for aged PC, i.e. it displays higher local order degree. This tendency conforms completely to the amorphous polymer structure approaching to the equilibrium state, because the latter is the ordered state. Moreover, increase of the number of segments in clusters correlates with the conclusions made in ref. [30] in the point that structural changes are stipulated by relatively fine conformational changes.

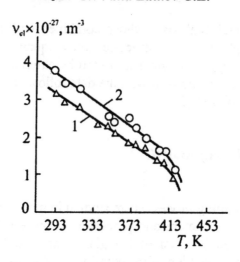

Figure 6.10. Dependencies of entanglement cluster network frequency, ν_{cl}, on test temperature, T, for PC: primary (1) and aged (2) [31]

Discussed below are changes of the relative fluctuation free volume, f_f, during heat aging of PC. Dependencies of f_f on test temperature T are shown in Figure 6.11. It is shown that in the whole temperature range f_f is higher for primary PC rather than for the aged one. This conforms to the conclusions made in ref. [29]. Much more interesting conclusions can be made on the basis of $f_f(\nu_{cl})$ dependencies, shown in the insertion to Figure 6.11. For primary and aged PCs, these dependencies do not coincide giving lower f_f for aged polymer. Since the entire fluctuation free volume is concentrated in the packless matrix, at heat aging the latter is consolidated. This conclusion conforms to positron annihilation spectroscopy data on polyarylate annealed at temperature below T_g [32].

Of special interest is also extrapolation of $f_f(\nu_{cl})$ dependencies to $f_f = 0$ and $\nu_{cl} = 0$. Obviously at $f_f = 0$ a closely packed polymer structure - the unique super-cluster, in principle, must be obtained. For this structure, ν_{cl} can be assessed as L/l_{st} (refer to equations (2.7) and (2.8)). In the case of PC (either primary or aged), this gives $\nu_{cl} = 7.04 \times 10^{27}$ m^{-3}. In the insertion to Figure 6.11, extrapolation of $f_f(\nu_{cl})$ dependence to $f_f = 0$ gives quite close ν_{cl} value, 6.8×10^{27} m^{-3}, approximately.

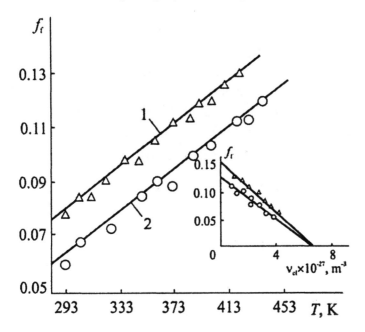

Figure 6.11. Dependencies of the relative fluctuation free volume, f_f, on test temperature, T, for primary (1) and aged (2) PC. In insertion: dependencies $f_f(v_{cl})$ for initial (1) and aged (2) PC [31]

Extrapolation to $v_{cl} = 0$ gives the fluctuation free volume for both primary (~0.150) and aged (~0.133) PC. Though these values exceed the universal Simhi-Boyer one for f_f (~0.122 [33]), it falls within the range of calculations by Sanchez (0.105 – 0.159 [34]).

It is common knowledge [35] that for the majority of amorphous glassy-like polymers, the Prigozhin-Defye criterion is not met. This means that for characterization of the structure of these polymers, the single order parameter is not enough. On the example of PC, the reasonableness of this argument is demonstrated by $f_f(v_{cl})$ dependence. Obviously different values of f_f at the same v_{cl} indicate that besides v_{cl}, the order parameter is required that characterizes structural state of the packless matrix. Thus in the framework of the cluster model, interpretation of physical aging of amorphous glassy-like PC allows quantitative description of structural changes happening.

As shown in Section 5.2, fractal dimensionality, D, of the chain segment between entanglement or cross-link points characterizes mobility of the chain. The interconnection of mobility and local order was studied on the example of two series of epoxy polymers (primary (EP-1) and aged during 3 years under natural conditions (EP-2)) [36]. Dependence $D\left(v_{cl}^{-1/3}\right)$ in Figure 6.12 indicates that increase of v_{cl} or local order degree induces decrease of D and, consequently, suppresses chain mobility. Any decrease of the chain mobility, independently of its reasons, induces polymer friability increase [37], which is always observed for physical aging of these materials. Comparison of dependencies $D\left(v_{cl}^{-1/3}\right)$ and $v_{cl}(K_{st})$, shown in Figure 6.13, allows a conclusion that in the present case, this process is stipulated by polymer aging [36].

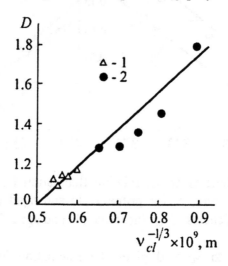

Figure 6.12. Dependence of the fractal dimensionality, D, of chain intersect between cross-link points on parameter $v_{cl}^{-1/3}$ for EP-1 (1) and EP-2 (2) [36]

As mentioned in Chapter 2, the initial temperature of the glass transition interval corresponds to the glass transition point of the packless matrix, and the final temperature corresponds to the glass transition point of the polymer [38]. Therefore, the glass transition interval width, ΔT_g, is determined by the number of segments in the cluster, n_{cl}. Based on these data and Figure 6.14, one can state that the aging of polymers promotes increase of the cluster structure homogeneity [36].

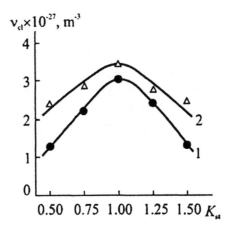

Figure 6.13. Dependence of the cluster network frequency, ν_{cl}, on the curing agent:oligomer ratio, K_{st}, for EP-1 (1) and EP-2 (2) [36]

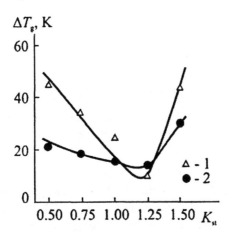

Figure 6.14. Dependence of the glass transition interval width, ΔT_g, on the curing agent:oligomer ratio, K_{st}, for EP-1 (1) and (EP-2) (2) [36]

Thus physical aging of amorphous glassy polymers represents changes in structure and properties with time and is the consequence of thermodynamically non-equilibrium origin of these polymers. At quite low (about room level) temperatures, this process proceeds slowly, and significant changes in properties are realized during several years [36, 39]. That is why such events must be forecasted with regard to both spatial and time disorder of the systems subject to aging [40]. Spatial structural inhomogeneity of amorphous glassy-like polymers makes the range of microscopic rates of relaxation transitions rather wide, i.e. spatial disorder induces the time one [40].

Relating to description of the physical aging in the framework of the fractal analysis, the following facts should be noted. The study of reaction dynamics in amorphous systems indicate [40] that both types of disorder must be considered. At long time periods for the majority of reactions the subordination expressed by the relation below [40] takes place:

$$\Phi(t) \sim f(t^{\alpha\gamma}),\tag{6.4}$$

where $\Phi(t)$ is an arbitrary function, independent of time t; α and γ are exponents taking into account the presence of spatial and time disorder, respectively.

Relating to the physical aging, structural changes with time can be described based on the general correlation (6.4) as follows [41]:

$$v_{cl} = cv_{cl}^0 t^{\alpha\gamma},\tag{6.5}$$

where v_{cl} and v_{cl}^0 are entanglement cluster network frequencies for aged and primary samples, respectively; c is the proportion coefficient considering replacement of the proportion sign in the correlation (6.4) by the sign of equality in equation (6.5).

In the equation (6.5) the physical meaning of exponents α and γ shall be determined. In the framework of the fractal analysis, exponent α characterizing spatial disorder can be determined as $(d_f - 2)$, where d_f is the fractal dimensionality of the primary epoxy polymer structure (EP-1), due to the following reasons. Variation of d_f in the range of 2 -3 characterizes the disorder degree of the polymer structure, and the basic information about this degree is carried by fractional part of d_f: the greater d_f is, the lower the local order degree is and, consequently, higher spatial disorder of the system is [42]. Essentially, the integer part of d_f carries the information about dimensionality of the Euclidean space, to which the fractal object is buried. Moreover, it should be noted that the exponent $\alpha = d_f - 2$ coincides with the fractional index in the theory of fractional integration-differentiation [43], in which this index characterizes the part of system states, not varied during evolution [44]. To put it differently, in the framework of the fractional integration-differentiation it is assumed that the ordered part of the polymer structure will be changed during physical aging, which is clearly displayed by the equation (6.5).

As shown in Chapter 5, there is a quasi-equilibrium state of polymers, approached by the polymer structure with increase of its thermodynamic equilibrium degree, characterized by the condition $d_f = 2.5$. This condition relates to completely relaxed polymer with a narrow distribution of free volume cavity sizes [45] and, consequently, to the shortest time disorder characterized by the relaxation time spectrum width. That is why, henceforth, the following approximation will be used [41]:

$$\gamma = d_f - 2.5. \tag{6.6}$$

Thus the equation (6.5) can be reduced to the form as follows [41]:

$$V_{cl} = c V_{cl}^0 t^{(d_f-2)(d_f-2.5)}, \tag{6.7}$$

where empirically determined constant c equals ~0.14, and the aging time t is given in seconds.

Figure 6.15 shows dependencies of φ_{cl}, calculated by equations (6.7) and (2.11), for primary (EP-1) and aged (EP-2) epoxy polymers on the curing agent:oligomer ratio, K_{st}, as well as experimental ones. The data shown indicate good correspondence between theory and experiment. Even at $K_{st} = 0.50$ and 1.50, at which deviation is maximal, it does not exceed ~25%. It has been concluded [41] that the mentioned deviation is associated with the calculation method for the Poisson parameter, v, and, consequently, D_f (equation (5.4)), namely, with application of equation (3.12)and strong dependence of v_{cl} on d_f in equation (6.7) for this purpose. It is known [46] that for epoxy polymers, inaccuracy of v assessment by the equation (3.12) increases with test temperature, T. At high temperatures this method can give v by ~7% higher than those, determined by other methods. Since at $K_{st} = 0.50$ and 1.50 the highest values of v are observed (0.364 and 0.370, respectively), the highest inaccuracy shall be expected for them. Decrease of both these values of v to 0.360 (i.e. by 1 – 3% only) and recalculation of φ_{cl} by equations (6.7) and (2.11) give φ_{cl} values much closer to experimental ones (the dashed line in Figure 6.15).

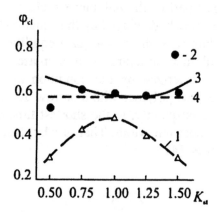

Figure 6.15. Dependencies of the relative part of clusters, φ_{cl}, on the curing
agent:oligomer ratio, K_{st}, for EP-1 (1) and EP-2 (2 – 4); 1, 2 –
experimental data; 3 – calculation by equations (6.7) and (2.11);
4 – calculation by the same equations with adjustment of v
values for EP-2 with $K_{st} = 0.50$ and 1.50 (refer to the text) [41]

The second important aspect of physical aging also follows from data of
the Figure 6.15. For example, for different K_{st} (and, consequently, for different
frequencies of chemical cross-link points v_c [47]) intensities of aging, expressed
by difference of φ_{cl} values for EP-2 and EP-1, are significantly different.
Expectedly [42], minimal intensity of aging is observed for the system with
stoichiometric value $K_{st} = 1.0$, and K_{st} deviation from this value leads to the
increase of physical aging intensity [36]. As follows from equation (6.7), the
effect observed is induced by increase of both α and γ for systems with highest
deviations from $K_{st} = 1.0$ or, to put it differently, with increase of fractal
dimensionality of the structure, D_f [41].

As shown in Chapter 5, correlations (5.33), (5.38) and (5.45) are true for
polymers, for which changes of supermolecular structure parameters (d_f and D_f)
and molecular index (statistical chain rigidity, characterized by C_∞) during
physical aging are assumed. Finally, physical aging levels the structure of epoxy
polymers independently of v_c, which varies by, approximately, an order of
magnitude for them [47]. Naturally, such structure leveling also determines a
significant approach of properties for EP-2. For example, the elasticity modulus
E in squeeze tests varies for EP-1 in the range of 2.50 – 3.60 GPa and for EP-2
in the range of 1.60 – 1.90 GPa only [36].

Thus the technique suggested in [41] allows calculation of changes in the structure of epoxy polymers during their physical aging using ideas of the cluster model and the fractal analysis. The coefficient c in equation (6.7) may possibly be the function of aging temperature and, in principle, an analytical correlation for its determination can be deduced. Of much higher probability is that the increase of v (and, consequently, d_f) with temperature will take into account the aging intensification. These assumptions require to be proved experimentally. Of importance is that structural restructuring during physical aging involve both molecular and supermolecular levels [42].

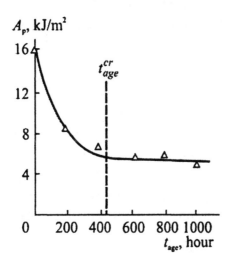

Figure 6.16. Dependence of blow viscosity, A_p, on aging time, t_{age}, for PC. Vertical dashed line shows t_{age}^{cr} value, calculated by equation (6.7) [48]

The above-considered concepts are used for description of PC blow viscosity, A_p, dependence on physical aging duration. Expectedly, due to general ideas [37], PC plasticity is decreased during physical aging, which is reflected by A_p decrease with aging time, t_{age} (Figure 6.16). Clearly the data in the Figure indicate A_p decrease during initial ~500 hours of aging, then it fits an asymptotic curve with low absolute values of A_p (of about 4.5 kJ/m^2). This run of $A_p(t_{age})$ dependence can be explained by reaching some border state of the structure, which is defined as the quasi-equilibrium state (refer to Chapter 5). For this state, D_f value $\left(D_f^{cr}\right)$ is determined by the equation [22]:

$$D_f^{cr} = \frac{4\pi T}{\ln\left(1/f_f\right)T_g}.$$ (6.8)

As shown in Chapter 5, polymer structure can be simulated as a thermal cluster, i.e. the cluster for which formation both geometrical and thermal interactions are important. The order parameter of such cluster, φ_{cl}, is determined from the comparison (5.74).

Figure 6.17 shows dependence $d_f(t_{age})$, calculated in accordance with equations (3.12) and (5.4), illustrating evolution of PC structure during physical aging. Clearly at $t_{age} > 500$ hours d_f reaches the asymptotic value equal ~2.46. Further on, D_f^{cr} was calculated [48] by equation (6.8), and then the appropriate d_f value $\left(d_f^{cr}\right)$ was estimated using the equation (5.43). In accordance with equations (6.8) and (5.43) the following values of fractal dimensionalities were obtained: $D_f^{cr} \approx 2.68$ and $d_f^{cr} \approx 2.41$. The latter value correlates well with the above-shown d_f for asymptotic branch of dependence $d_f(t_{age})$ (Figure 6.17). Thus A_p fitting the asymptotic branch, shown in Figure 6.16, can be explained by reaching quasi-equilibrium state by PC structure during physical aging.

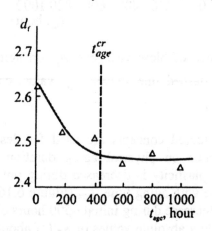

Figure 6.17. Dependence of the fractal dimensionality, d_f, of PC structure on aging time, t_{age}. Vertical dashed line shows t_{age}^{cr} value, calculated by equation (6.7) [48]

Then in the quasi-equilibrium state φ_{cl} can be estimated by the equation (5.74) via determination $\beta_T = \beta$ in accordance with the equation (5.75), i.e. under the assumption that indices of thermal and percolation clusters are equal. This calculation gives $\varphi_{cl}^{cr} = 0.619$. Subsequently, one can estimate d_f^{cr} value from the general correlation (5.7), which gives $d_f^{cr} = 2.45$. This value does also correlate well with the above-shown assessments. Thus thermal cluster model gives an adequate description of the polymer structure, the quasi-equilibrium state conditions, in particular. In accordance with equations (5.38), (6.8) and (5.43), D_f^{cr} was estimated as 2.85, 2.68 and 2.82, respectively, which also demonstrates conformity between the mentioned techniques [48].

Thereafter, physical aging duration, t_{age}^{cr}, required for reaching the quasi-equilibrium state was estimated [48] using the equation (6.7). For this purpose, parameters in the equation were estimated as follows. Values φ_{cl}^{cr} and φ_{cl}^0 are calculated at aging temperature, T_{age}, in accordance with equations (5.74) and (5.72), respectively. Then at T_{age} parameters d_f and d_f^{cr} are calculated by the equation (5.7) using appropriate values of φ_{cl}^0 and φ_{cl}^{cr}. Parameter t_{age}^{cr}, calculated by the equation (6.7) under the mentioned conditions, equals 438 hours. In Figures 6.16 and 6.17 it is presented by vertical dashed lines. Clearly t_{age}^{cr} conforms well to t_{age} value corresponded to the transition of $A_p(t_{age})$ and $d_f(t_{age})$ dependencies to the asymptotic branch. This allows determination of duration of transitional (non-stationary) period of the physical aging, t_{age}^n, during which structure and properties of the polymer are changed ($t_{age}^n \leq t_{age}^{cr}$).

As mentioned above, A_p increases with molecular mobility, the latter being associated with the mobility of chains in packless zones of the polymer structure [49]. In the framework of the cluster model, such zones are presented by the packless matrix surrounding the clusters. Figure 6.18 shows the dependence of A_p on the relative part of packless matrix, $\varphi_{p.m.}$. Expectedly [37], linear increase of A_p with $\varphi_{p.m.}$ is observed, and the condition $A_p = 0$ is realized at $\varphi_{p.m.} \approx 0.352$ instead of expected $\varphi_{p.m.} = 0$. In accordance with the rule $\varphi_{p.m.} = 1 -$

φ_{cl}, this value of $\varphi_{p.m.}$ conforms to $\varphi_{cl} \approx 0.648$, which is quite close to t_{age}^{cr} at T = 293 K (≈ 0.619). This correlates with the postulate of completely stretched chains in the quasi-equilibrium state limiting φ_{cl} increase and thus completely suppressed molecular mobility that leads to $A_p = 0$, e.g. completely friable polymer [48].

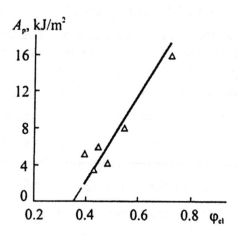

Figure 6.18. Dependence of blow viscosity A_p on relative part of the packless matrix $\varphi_{p.m.}$ for PC polymer [48]

As follows from the equation (6.8), parameter D_f^{cr} is the function of temperature (T) with regard to low logarithmic dependence of it on f_t, the value of which hereinafter is accepted constant and equal 0.075 [51]. In its turn, this means that the fractional dimensionality d_f^{cr} of the polymer structure in the quasi-equilibrium state will also be the function of T (refer to equation (5.43)). Experimentally, this statement can be proved by comparison of D_f and d_f parameters, calculated by the mentioned equations [50].

Thus in the used approximation, D_f^{cr} parameter is the function of T only, which is typical of the thermal cluster [52]. D_f^{cr} increases with temperature that conforms to behavior of the order parameter, described by the equation (5.74). It is illustrated in Figure 6.19 by $D_f^{cr}(T)$ dependence (curve 1).

Values of D_f^{cr} parameter, calculated as the function of T for the quasi-equilibrium state and primary and aged PC film samples by equations (6.8) and

(5.38), respectively, are also shown in the Figure. The comparison indicates higher D_f values for primary PC samples, rather than for the quasi-equilibrium state. However, heat aging also decreases D_f for aged PC samples, for which D_f^{cr} and D_f, calculated by equations (6.8) and (5.38), respectively, are in good conformity. All this proves that heat interactions during physical aging of PC induce structural changes in the polymer, and its quasi-equilibrium structure is determined by both configurative and heat interactions that gives an opportunity to consider it as the thermal cluster [52].

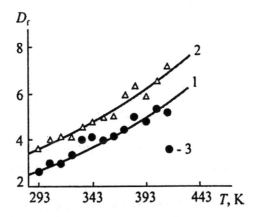

Figure 6.19. Temperature dependencies of dimensionalities of excessive energy localization zones, D_f, for PC polymer: 1 – calculated by equation (6.8); 2, 3 – calculated by equation (5.38) for primary (2) and aged (3) PC [50]

Figure 6.20 compares values of d_f parameter, calculated for the quasi-equilibrium state and primary and aged PC samples by equations (5.43) and (5.4), respectively. Clearly as before, heat interactions during physical aging induce structural changes in the polymer. Therefore, after aging d_f parameters of the PC polymer structure and the quasi-equilibrium state conform well.

Calculation performed by equation (6.8) using d_f value at $T = T_{age}$ ($d_f = 2.82$) has given $t_{age}^{cr} \approx 6.2$ hours for film PC samples that correlates well with the aging mode used [50]. Nevertheless, in accordance with the equation (6.8) the use experimental value $t_{age} = 72$ hours must give $\varphi_{cl} = 0.243$ or according to equation (5.7) $d_f = 2.655$. As follows from Figure 6.20, this value is not real that talk of physical reality of the polymer structure quasi-equilibrium [22].

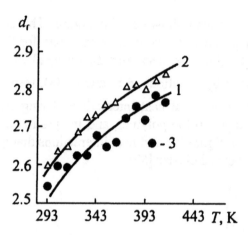

Figure 6.20. Temperature dependencies of fractal dimensionality, d_f, for PC polymer structure: 1 – calculated by equation (5.43); 2, 3 – calculated by equation (5.4) for primary (2) and aged (3) PC [50]

The techniques considered above in the framework of the thermal cluster concept were used for description of epoxy polymer physical aging [23]. Figure 6.21 shows dependencies of the order parameter φ_{cl} on K_{st} parameter, calculated by equations (6.7), (6.8) and (5.74), and experimental φ_{cl} values, obtained by the equation (5.7), for epoxy polymers: primary (EP-1) and aged under natural conditions during 3 years (EP-2). Clearly all methods mentioned above give well coinciding results with maximal deviation below 13%. There are two differences between assessments and approximation method [41], used in ref. [53]. Firstly, every cross-link frequency, ν_c (or every K_{st} value), of epoxy polymer displays self D_f^{cr} value of the quasi-equilibrium state and φ_{cl}^{cr}, respectively: the smaller ν_c is, the higher D_f^{cr} and the smaller φ_{cl}^{cr} are. For all epoxy polymers, $D_f^{cr} = 3$ was accepted [41]. Secondly, for more accurate methods, the difference in φ_{cl} and φ_{cl}^0 (or ν_{cl} and ν_{cl}^0) is lower rather than for approximated one [41].

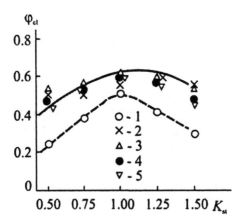

Figure 6.21. Dependencies of the relative part of clusters, φ_{cl}, on K_{st} for EP-1 (1) and EP-2 (2 – 6): 1, 2 – experimental data (equation (5.7)); calculation: 3 – equation (6.7); 4 – equation (6.8); 5 – equation (5.74); 6 – equation (6.10) [53]

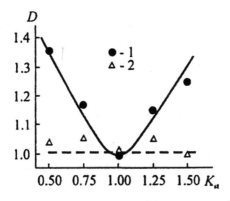

Figure 6.22. Dependencies of the fractal dimensionality, D, of the chain segment between clusters on K_{st} for EP-1 (1) and EP-2 (2) [53]

As mentioned above, reaching of the quasi-equilibrium state assumes straining of chains between clusters that prevents further increase of φ_{cl} and makes formation of an equilibrium structure possessing the parameter $\varphi_{cl} = 1.0$ impossible [22]. Molecular mobility of chains can be characterized by the fractal dimensionality, D, of the chain segment between clusters ($1 < D \leq 2$) [22, 54]. Parameter D can be calculated from the equation (5.78). Figure 6.22 shows dependencies $D(K_{age})$ for EP-1 and EP-2. It indicates an abrupt decrease of D during physical aging. For epoxy polymer EP-2, D values are close to the border (or expected) one $D = 1.0$ (marked by horizontal dashed line in Figure 6.22). Thus data in Figure 6.22 prove the concept of quasi-equilibrium state, induced by chain strain, under the condition $D = 1.0$ [22].

Moreover, data in Figure 6.22 demonstrate reasons for glassy-like polymer blistering during physical aging. As mentioned above, whenever the reason is, molecular mobility decreases with dissipation of externally supplied mechanical energy and, consequently, increases friability of the polymer [37]. Decrease of D parameter to ~1.0 means complete suppression of molecular mobility. That is why physical aging can make the polymer extremely friable. All this represents qualitative description, and quantitatively, the blistering degree can be expressed via the stretch rating (λ_s) decrease to the degradation point, which is presented as follows [55]:

$$\lambda_s = D_f^{D-1}. \tag{6.9}$$

Physical aging induces decrease of both D_f and D that, in accordance with equation (6.9), causes decrease of λ_s, i.e. makes the polymer friable. At $D = 1.0$, $\lambda_s = 1$ or the failure strain $\varepsilon_s = 0$, and the polymer becomes extremely friable [55].

Considered below is the increment introduced by approximation $d_f = 2.5$ for the quasi-equilibrium state [41]. Figure 6.23 compares experimentally obtained values of parameter d_f (equation (5.4)) with the ones, calculated by equations (6.8), (5.74), (5.7) and (5.43). Clearly there is a deviation between experimental d_f values and $d_f = 2.5$ (horizontal dashed line) that can be up to 7%. Though this value itself is quite low, the exponential dependence in the equation (6.7) can form a significant mistake for assessed φ_{cl} values. The deviation between experimental and theoretical values of d_f parameter are much lower and do not exceed ~3%. Therefore, though the approximation $d_f = 2.5$ can be the first approximation for the quasi-equilibrium state [41], more accurate result will be obtained under application of d_f value, calculated by one of the above-mentioned methods [56, 57].

To conclude this part, note also that the equation (5.78) can represent one of the methods for estimating φ_{cl} values in the quasi-equilibrium state. For this purpose, the condition $D = 1.0$ must be fulfilled. Thereafter, the equation (5.78) is reduced to the form as follows [56]:

$$\varphi_{cl}^{cr} = \frac{2}{C_\infty}. \tag{6.10}$$

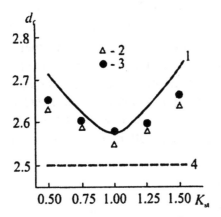

Figure 6.23. Dependencies of the fractal dimensionality, d_f, of the structure on K_{st} for EP-2: 1 - experimental data (equation (5.4)); calculation: 2 – equations (6.8) and (5.43); 3 – equations (5.74) and (5.7); 4approximation $d_f = 2.5$ [53]

Figure 6.21 shows the dependence $\varphi_{cl}(K_{st})$, plotted in accordance with the equation (6.10). Clearly it conforms well to both experimental data and differently obtained φ_{cl} values.

Thus the above-considered results indicate that structural changes in the polymer during physical aging are limited by reaching a quasi-equilibrium state, characterized by equilibrium between the local order increase (approaching a thermodynamically equilibrium structure) and entropic chain straining. This approach allows quantitative explanation of polymer blistering during aging and forecast of structural changes (and, consequently, properties) in polymers as the function of their initial structure, time and temperature [48, 50, 53, 56, 57].

6.5. ORIENTATION

Structural changes proceeding during solid-phase extrusion molding of ultrahigh-molecular weight polyethylene (UHMWPE) and polymerization-filled composite (componor) derived from it, UHMWPE-bauxite, were studied [58,

59]. Figure 6.24 shows dependence of relative part of the packless matrix, $\varphi_{p.m.}$, on extrusion stretch rating, λ, for the mentioned systems. Clearly variation of $\varphi_{p.m.}$ with λ is not monotonous reaching its maximum $\lambda = 5$ for UHMWPE and $\lambda = 3$ for UHMWPE-bauxite. Such dependencies assume rapid degradation of the local order zones (clusters) already at low λ values. Insertion in Figure 6.24 shows the relation between of the crystallinity increment, ΔK, determined by the X-ray method, and decrease of the local order level, $\Delta\varphi_{cl}$. Clearly on early stages of extrusion the former process proceeds slower than the latter one. To put it differently, no direct transformation of clusters into crystallites happen, and the latter are mainly formed from packless zones of the amorphous phase. Decline of $\varphi_{p.m.}(\lambda)$ dependence after the maximum reflects the mentioned process.

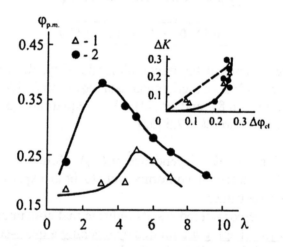

Figure 6.24. Dependencies of relative part of the packless matrix, $\varphi_{p.m.}$, on the extrusion stretch rating, λ, for UHMWPE (1) and componor UHMWPE-bauxite (2). Insertion: correlation between crystallinity increment, ΔK, and decrease of the relative part of clusters, $\Delta\varphi_{cl}$ [58]

In Figure 6.24 different locations of $\varphi_{p.m.}$ maximums by λ axis do not mean different rates of φ_{cl} decrease. Figure 6.25 shows dependence of φ_{cl} on the molecular stretch rating, λ_{mol}, that reflects orientation processes for the studied systems much more accurately than λ does [60, 61]. Data from the Figure indicate an abrupt decline of φ_{cl} in the range $\lambda_{mol} \approx 1 - 4$, after which φ_{cl} does not practically change and possesses low absolute value (~ 0.05). It can be said that at $\lambda_{mol} \approx 4$ the local order decay stipulated by orientation processes is practically complete. The value $\lambda_{mol} \approx 4$ is not random but is determined by the following factors. It is known [56] that the maximum of molecular stretch rating, λ_{max}, can be determined as follows:

$$\lambda_{max} = \frac{L_c}{R_c}, \tag{6.11}$$

where L_c and R_c are the length of the chain segment and the distance between molecular entanglements, respectively.

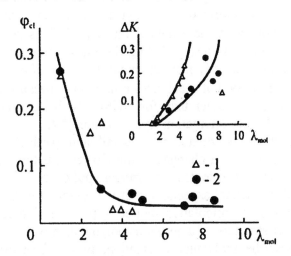

Figure 6.25. Dependencies of the relative part of clusters, φ_{cl}, on molecular stretch rating, λ_{mol}, for UHMWPE (1) and componor UHMWPE-bauxite (2). Insertion: dependencies $\Delta K(\lambda_{mol})$ for these materials [59]

Assessment by equation (6.11) gives $\lambda_{max} \approx 4.2$ [58]. This means that at $\lambda_{mol} = \lambda_{max}$ the chain is completely stretched between entanglements, which corresponds to the minimum of φ_{cl}. It should be noted that the assessment of D in accordance with the equation (5.42) at $d_f = 2.90$ corresponded to $\lambda_{mol} \approx 4$ gives $D \approx 1.06$, which is also corresponded to almost completely stretched chain [58, 59].

The above-shown assessments demonstrate that maximal (over 80%) decay of clusters corresponds to completely stretched macromolecules in the

UHMWPE amorphous phase. Note also that though the presence of a filler does not affect the local order decay, it hampers orientation crystallization that follows from the plot of $\Delta K(\lambda_{mol})$ dependence (Figure 6.25, the insertion).

Thus the model considered in works [58, 59] allows the following interpretation of structural changes in UHMWPE and componor UHMWPE-bauxite with solid-phase extrusion molding. Orientation of macromolecules increases with λ, which is reflected by λ_{mol} increase. This induces proceeding of two parallel processes: the local order zones (clusters) decay and orientation crystallization. However, rates of these processes are different. In the range λ_{mol} ≈ 1 - 4 clusters are abruptly decayed, after which their number remain practically constant. One may assume that clusters possessing orientation the most favorable for the extrusion direction do not decay. The crystallinity is increased monotonously in the whole λ range [63]. This suggests crystallite formation from packless zones, including the ones formed during the cluster decomposition, but not directly from the latter. Presence of a filler has no direct effect on the cluster decay, but accelerates it indirectly, because for equal λ, λ_{mol} for componors is higher rather than for UHMWPE [61]. Simultaneously, the filler presence decelerates orientation crystallization [59].

For amorphous polymers, more detail and particular description of orientation process effect on changes in the cluster structure can be obtained. Note that properties of oriented amorphous polymers were studied quite frequently. Besides practical aspects, one of the reasons for these studies is the possibility of obtaining 'orientation degree – properties' correlations on the example of simpler structure of amorphous polymers, in which crystalline phase is absent. To describe molecular orientation, two deformation schemes are usually used: the so-called "affine" and "pseudo-affine" ones, which themselves and possibilities of their application to real polymers are described in many works (for example, [64, 65]). The basic statement in all these modified and unmodified deformation schemes is the presence of macromolecular entanglement carcass [66, 67].

Since at present several alternatives of molecular entanglements in polymers do exist (refer to Chapter 1), three features of the cluster entanglement network (compared with the macromolecular hooking network), sufficient for future discussion, should be reminded (refer to Figure 1.2).

1) Cluster network cross-link points possess clearly determined finite size (the segment length in the cluster is accepted equal the length of statistical segment in the polymer [68]).

2) Cluster network frequency, v_{cl}, is the function of temperature and decreases as it increases; at $T = T_g$ the cluster network is completely decomposed. At $T \leq (T_g - 50\ K)$ v_{cl} increase with decreasing temperature is slowed down significantly.

3) Frequency v_{cl} is approximately by an order of magnitude higher than the appropriate parameter of macromolecular hooking network.

Note that the effect of local order zones, which are clusters, on orientation behavior of amorphous polymers has also been observed before [69, 70].

Thorough check of appropriate results [64 – 67, 69 – 71] has detected some features of behavior of oriented polymers, which do not fit the version of the carcass of macromolecular hookings. That is why direct application of two above-mentioned deformation schemes is prohibited. These features are [72, 73]:

1) Dependencies of molecular orientation on the stretch rating, λ, are qualitatively different for the areas above and below T_g.

2) An alternative of "temporary" molecular carcass with high frequency, stipulated by electrostatic interactions between chain units, have been suggested [66]. As shown for poly(methyl methacrylate) [66], parameter v_{temp} is practically constant below $T = 323\ K$.

3) Above T_g the shrinkage force variation assumes presence of a permanent "residual" carcass.

4) It is assumed that a single mechanism providing for entanglement network frequency variation with temperature and deformation is enough for description of the molecular orientation.

5) It is assumed [64] that in poly(ethylene terephthalate) orientation is associated with microcrystallites, which dimensions are equal or smaller than the wavelength of light in the Brillouin scattering.

6) Low n_{seg} (the number of statistical segments between entanglements) that determines high carcass frequency.

As mentioned above, all considered non-correspondences have caused occurrence of several modified deformational schemes, one of which can be described by the equation below [66]:

$$\Delta n = c v_0 \alpha \left(\lambda^2 - \lambda^{-1} \right) \exp\left(-k\lambda \right), \qquad (6.12)$$

where Δn is the path-length difference in birefringence measurements, which characterizes molecular orientation degree; v_0 is the frequency of the primary molecular carcass; α is the difference of polarizability values of the statistical unit parallel and transversal to its axis; λ is the stretch rating; k is an index characterizing cross-link point degradation in the macromolecular carcass during stretching. Constant c is determined as follows [67]:

$$c = \left(\frac{2\pi}{45}\right)\left(\frac{\left(\bar{n}^2 + 2\right)^2}{\bar{n}}\right), \tag{6.13}$$

where $\bar{n}$ is the mean refraction index.

Botto, Duckett and Ward [65] have noted that equation (6.12) describes well the experimental data at $T < T_g$, but gives poor correspondence at $T > T_g$. To replace the equation (6.12), they have suggested its modification as follows:

$$\Delta n = \alpha c \left(v_{const} + v_{temp} \exp(-k(\lambda - 1))\right)\left(\lambda^2 - \lambda^{-1}\right), \tag{6.14}$$

which assumes the presence of two carcasses of macromolecular entanglements – constant and 'temporary' ones, with frequencies v_{const} and v_{temp}, respectively.

Though application of the equation (6.14) to description of experimental data on PMMA was successful (both at $T < T_g$ and $T > T_g$) [65], two notes on this point should be made. Firstly, v_{const}, v_{temp} and k values were obtained by adapting to experimental data, and the absence of independent appraisal methods for these parameters drops the sufficiency of equation (6.14). Secondly, v_{temp} values were found by an order of magnitude exceeding the macromolecular hooking network frequency, v_h. As it is known, the alternative of the carcass stipulated by electrostatic interactions [66] has not been ever used anywhere, though the absence of influence of macromolecular carcass with such frequency on other properties of the polymer can be hardly assumed.

Table 6.1 shows comparison of macromolecular carcass frequencies obtained by superimposition on experimental data [65, 67] and independent methods [74]. Clearly v_0 (v_{temp}) values obtained in works [65, 67] are almost two-fold different that indicates their adapting origin. At the same time, good conformity is observed for the couples of v_{const} [65] and v_h and v_{temp} and v_{cl} values. Thus works [72, 73] basically use the model of two carcasses, suggested in [65], and the difference is in accurate physical identification and independent determination of frequencies of these carcasses in [72, 73].

Table 6.1

Structural parameters for PMMA used in calculations by equations (6.12) and (6.14) [72]

T, K	Ref. [67]		Ref. [65]			Ref. [74]		
	v_0, 10^{26} m^{-3}	k	v_{const}, 10^{26} m^{-3}	v_{temp}, 10^{26} m^{-3}	k	v_{cl}, 10^{26} m^{-3}	$(v_{cl}+v_h)$, 10^{26} m^{-3}	k
303÷323	15.1	1.42	-	-	-	9.23	9.7	2.0
363	8.4	1.22	0.38	4.70	0.89	3.83	4.3	2.0
373	6.4	1.18	-	-	-	1.33	1.8	1.3
389.5	2.4	0.58	-	-	-	-	0.47	-
408	-	-	0.31	0.31	0.61	-	0.47	-

Figure 6.26 compares Δn values as the function of λ for PMMA, calculated by the equation (6.14), and experimental ones [65]. Since the experimental data are obtained at 408 K, i.e. at $T > T_g$ ($T_g = 378$ K [75] for PMMA), the calculation was executed for $v_{const} = v_h$, $v_{temp} = v_{cl} = 0$ and $k = 0$. Clearly quite good correlation between theory and experiment is obtained. Balankin [76] has suggested the fractal concept of the rubber elasticity state, in which stress F depends upon elasticity modulus E as follows:

$$F = \frac{E}{4.5}\left(\lambda^2 - \lambda^{-2.5}\right). \tag{6.15}$$

Since F is proportional to Δn and E is proportional to the carcass frequency [67], the following simple formula can be used for assessment of $|\Delta n|$ (at $T \geq T_g$):

$$|\Delta n| = c v_h \alpha\left(\lambda^2 - \lambda^{-2.5}\right). \tag{6.16}$$

Application of this formula indicates (Figure 6.26) even better conformity to experiment at $\lambda < 3$ than the equation (6.14). Somewhat overestimated calculation results, obtained by both these equations at $\lambda \geq 3$, indicate the necessity of introducing a coefficient $k \neq 0$ for high deformations considering probable degradation of hooking points, especially for chains with low molecular mass [77].

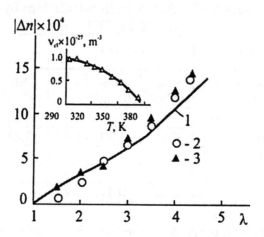

Figure 6.26. Dependencies of birefringence $|\Delta n|$ on macroscopic stretch rating λ at $T = 408$ K for PMMA: 1 – experimental data [65]; 2 – calculation by equation (6.14); 3 – calculation by equation (6.16)

Figure 6.27 compares experimental data [67] and calculation results by the equation (6.14) on dependencies $|\Delta n|(\lambda)$ for samples stretched at 303, 363 and 373 K. Since in this case orientation in PMMA proceeded at $T < T_g$, the calculation was executed for $v_{const} = v_h$, $v_{temp} = v_{cl}$ and $k = 2.0$. In this case, good correspondence between theory and experiment is also observed, and some decrease of calculated $|\Delta n|$ values in relation to experimental ones with temperature can be easily eliminated by k lowering. For example, for the curve $|\Delta n|(\lambda)$ at $T = 373$ K the coefficient selected was $k = 1.3$.

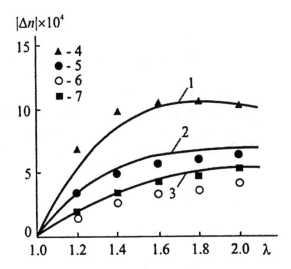

Figure 6.27. Dependencies of birefringence $|\Delta n|$ on macroscopic stretch rating λ at $T = 303$ K (1, 4), 363 K (2, 5) and 373 K (3, 6, 7) for PMMA: 1, 2, 3 – experimental data [67]; 4, 5, 6 – calculation by equation (6.14) for $k = 2.0$; 7 – calculation by equation (6.14) for $k = 1.3$ at $T = 373$ K [73]

Thus the above-mentioned result indicate that parameters of the entanglement cluster network, determined by an independent method, give quite good description of experimental data on molecular orientation of PMMA.

Solid-phase extrusion molding of amorphous glassy-like polymers allows obtaining of articles with high mechanical properties [78]. Hence, compared with extruded amorphous-crystalline polymers, elasticity modulus, E, increase is usually low, but a noticeable increase of the failure strain (or the border stretch ratio) is observed [79]. Structural changes proceeding during processing by this method are not yet studied. Nevertheless, two assumptions can be made on this point. The first [79] is that bulky side groups are "cut off" during solid-phase extrusion molding of PMMA that increases its deformability. Note that this assumption has not been proved experimentally, and it is unclear still which structural changes may proceed in amorphous polymers possessing no side groups. Milagin and Shishkin [69] have suggested proceeding of the following structural changes during orientation stretching of PMMA: transition to higher equilibrium structure due to improvement of the molecular packing and relaxation of internal stresses. These assumptions were checked [80] on the example of extruded amorphous polyarylates using the cluster model of the polymer amorphous state structure.

Figure 6.28 shows dependence of v_{cl} on extrusion stretch rating λ for polyarylates DV and DF-10, where v_{cl} for non-oriented DV ($\lambda = 1$) is accepted equal the appropriate value for the sample produced by press molding [16]. Clearly extrusion of polyarylates causes a significant (over 1.5-ford) increase of structure ordering, characterized by this v_{cl} value; moreover, this effect is reached already at $\lambda = 2$ and v_{cl} remains practically unchanged with future increase of λ [80].

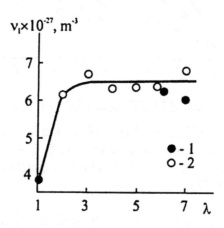

Figure 6.28. Dependence of entanglement cluster network frequency, v_{cl}, on extrusion stretch rating, λ, for polyarylates DV-1 (1) and DF-10 (2) [80]

As indicated [16], the value of E is determined by contributions of both clusters and the packless matrix. This is clearly displayed in Figure 6.29 by dependence $E(v_{cl})$, determined for DV film samples. However, though the dependence of relaxation modulus E_∞ on v_{cl} is presented by a straight line parallel to $E(v_{cl})$ dependence, it is extrapolated to $E_\infty = 0$ at $v_{cl} = 0$. This indicates that the stress relaxation is completely realized in the packless matrix of amorphous polymer. The dependence $E(v_{cl})$ for DV and DF-10 extrudates applied on the Figure 6.29 for DV film sample will fit $E_\infty(v_{cl})$ line, but not $E(v_{cl})$ one. The latter circumstance suggests stress relaxation in the packless matrix of extruded polyarylates. The above-shown results explain the reasons for obtaining lower elasticity modulus and higher yield stress (σ_y) for extrudates rather than for film samples of the same polymer. For extrudates, decrease of E

is stipulated by stress relaxation in the packless matrix and, consequently, by elimination of its contribution into E (Figure 6.29). Increase of σ_y is stipulated by a significant increase of v_{cl} for extrudates rather than for the film samples. An indirect proof of this conclusion was given by tests of polycarbonate samples [79], extruded at different temperatures, T_e. In these experiments, the lowest values of E were obtained for $T_e > T_g$ (433 K), e.g. under the same extrusion conditions as for polyarylates [80]. Moreover, in this case, E value for PC extrudate was found below the elasticity modulus of the initial sample. Thus the main structural changes proceeding in the solid-phase extrusion molding of polyarylates (at $T_e > T_g$) are increase of the structure ordering and stress relaxation in the packless matrix. It shall be noted that under current conditions of extrusion, the absence of the molecular orientation is absolute [81].

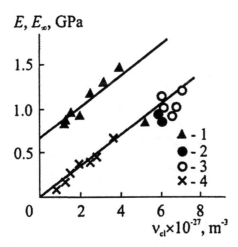

Figure 6.29. Dependencies of elasticity modulus E (1, 2, 3) and relaxation modulus E_∞ (4) on entanglement cluster network frequency v_{cl} for DV film samples (1, 4) and DV (2) and DF-10 (3) extrudates [80]

It has been shown [82] that at the solid-phase extrusion of polymer networks, orientation of intercross-link chains and mechanical degradation processes proceed in them simultaneously. The competition of these processes is the reason for observed behavior of the mentioned materials during mechanical tests. The more so, the type of dependencies of mechanical properties of the extrudates is assumed to be determined by the chemical cross-links network frequency [82]. At high v_c degradation effects dominate already at comparatively low values of the stretch rating [82] that deteriorates parameters of articles. At the same time, annealing of extrudates promotes restoration and even some increase of their physicomechanical parameters. In this connection, the effect of annealing on properties of epoxy polymer extrudates possessing low cross-linking frequency (infrequently cross-linked one), in which orientation effects dominate in broader deformation interval rather than in frequently cross-linked polymers [84] was studied [83]. Expectedly, structural changes in such polymers, induced by solid-phase extrusion and subsequent thermal processing, will cause more significant increase of deformation-strength characteristics, as it is observed for undercured epoxy polymer [84].

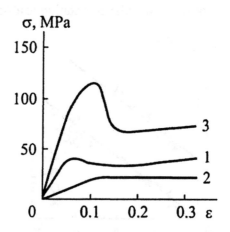

Figure 6.30. Diagrams stress-deformation (σ-ε) for infrequently cross-linked epoxy polymer (IEP): 1 – initial; 2 – extruded to $\lambda = 1.52$; 3 – annealed after extrusion [83]

Actually, for infrequently cross-linked epoxy polymer (IEP), solid-phase extrusion and subsequent annealing at 353 K during 2 hours induce significant changes in mechanical behavior of it [83]. They can be traced by appropriate stress-deformation (σ-ε) dependencies, shown in Figure 6.30. For the initial IEP ($v_c \approx 10^{26}$ m^{-3}), highly expectable behavior is observed and values E and σ_y for it are typical of polymers at the test temperature T, located approximately 40 K from T_g point [17]. A small decline of stress ($\Delta\sigma_y \sim 3$ MPa) after the yield stress (the yield drop) is observed, which is also typical of amorphous glassy-like polymers [85]. However, IEP extrusion before $\lambda = 1.52$ eliminated the yield drop and induces a significant decrease of E and σ_y. Moreover, the diagram σ-ε

is now more similar to the appropriate curve for rubber rather than for glassy-like polymer. Sample annealing at 353 K gives similar but opposite effect: sharp increase of elasticity modulus (approximately, two-fold compared with the extruded sample) and the yield stress. Moreover, clear yield drop is observed. Note also that sample shrinkage during annealing is low (~10%), which gives ~20% of extrusion deformation [83].

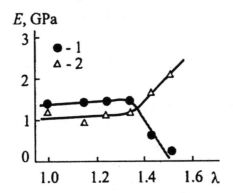

Figure 6.31. Dependencies of elasticity modulus E on extrusion stretch rating λ for IEP extrudates: 1 – initial; 2 – annealed [83]

General behavior of parameters E, σ_y and $\Delta\sigma_y$ as functions of λ is shown in Figures 6.31 – 6.33. Clearly the mentioned three parameters show general tendencies at λ variation: before $\lambda \approx 1.36$ a low increase of E, σ_y and $\Delta\sigma_y$ with λ is observed, their absolute values for extruded and annealed samples being close; at $\lambda > 1.36$ clear inverse relation in behavior of these parameters is observed. The effects observed can be explained in the framework of the cluster model [83].

Infrequently cross-linked epoxy polymer possesses low glass transition temperature, T_g, assessed by shrinkage measurements as ~333 K [83]. This means that the test temperature $T = 293$ K and T_g' are close, which is also indicated by low $\Delta\sigma_f$ for initial IEP. This assumes low segment density in unstable clusters v_{cl}^{unst} [85] and, because these clusters possess arbitrary orientation relating to the extrusion axis, increase of λ rapidly leads to their decomposition that induces mechanical devitrification of the packless matrix at $\lambda > 1.36$. Devitrified matrix gives an insignificant contribution into E, practically equal zero (refer to Figure 6.29), which induces an abrupt decrease of the elasticity modulus.

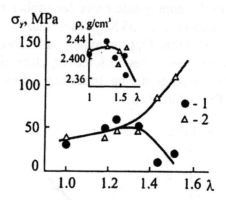

Figure 6.32. Dependencies of the yield stress, σ_y, on extrusion stretch rating, λ, for IEP extrudates: 1 – initial; 2 – annealed. Insertion: dependence of ρ on λ for IEP extrudates [83]

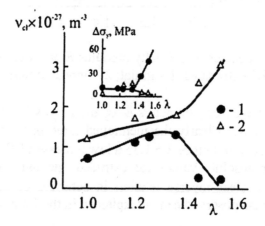

Figure 6.33. Dependencies of cluster network frequency, ν_{cl}, on extrusion stretch rating, λ, for IEP extrudates: 1 – initial; 2 – annealed. Insertion: dependence of ρ on λ for IEP extrudates [83]

Moreover, at $T > T_g'$ an abrupt decline of the cluster network frequency (ν_{cl}) is observed that means decrease of the number of segments both in stable and unstable clusters (refer to Figure 2.1). Since these very parameters (E and ν_{cl}, refer to equations (3.1) and (3.13)) regulate σ_y value, their decline induces an abrupt decrease of the yield stress. Thus essentially, extruded IEP represents a rubber with high cross-linking frequency, which is displayed by its $\sigma-\varepsilon$ diagram (Figure 6.30, curve 2).

As extruded IEP is annealed at temperatures approaching T_g, oriented polymer chains shrink. Since this process is realized in a narrow temperature range and during a short time, a great number of unstable clusters is formed. This effect is strengthened by existing molecular orientation (i.e. by preliminary favorable forming-up of segments) and is reflected by a rapid increase of $\Delta\sigma_y$ (Figure 6.33, insertion). Increase of v_{cl} induces E growth (Figure 6.31), and combined increase of v_{cl} and E sharply increases σ_f (Figure 6.32). A great number of newly formed unstable clusters also determine high $\Delta\sigma_y$ value (large yield drop) on $\sigma-\varepsilon$ diagram for annealed sample (Figure 6.30, curve 3).

Dependencies $v_{cl}(\lambda)$ shown in Figure 6.33 possess the type expected from the above description and represent its quantitative proof. Dependence of extruded IEP sample density, ρ, on λ is symbate to dependence $v_{cl}(\lambda)$ that also should be expected, because increase of closely packed segments' part must be shown up in increase of ρ.

An assumption has been made [84] that ρ change can be induced by formation of microcrack network in the sample that provides for ρ decrease at high λ (1.43 and 1.52) close to the border ones. Assessment of relative change, $\Delta\rho$, by the following equation:

$$\Delta\rho = \frac{\rho_{max} - \rho_{min}}{\rho_{max}} \tag{6.17}$$

gave $\Delta\rho \approx 0.033$. This value can be reasonable for free volume increment, necessary for packless matrix devitrification [33, 34], but it is too small for the case of microcrack formation. Experimental assessments indicate that IEP extrusion at $\lambda > 1.52$ is impossible due to sample cracking already during the extrusion. This allows a supposition that $\lambda = 1.52$ approaches the critical value. That is why one can estimate critical dilatation, $\Delta\sigma_{cr}$, required for formation of a microcrack cluster [86]:

$$\Delta\sigma_{cr} = \frac{2(1+v)(2-3v)}{11-19v}. \tag{6.18}$$

Accepting the average value $v \approx 0.35$ [46], $\Delta S_{cr} \approx 0.60$ is obtained, which is much higher than previously estimated $\Delta\rho$. These calculations assume that ρ decrease at $\lambda = 1.43$ and 1.52 is stipulated by decomposition of unstable clusters and corresponded loosening of the IEP structure [83].

Thus the above-shown results indicate that neither the cross-linking degree nor molecular orientation determined final properties of cross-linked polymers. The parameter controlling properties is the supersegmental (cluster) state of the polymer structure which, in its turn, can be purposefully controlled using molecular orientation and thermal processing. As a practical matter extrusion and subsequent annealing of infrequently cross-linked epoxy polymers allow obtaining of materials possessing properties analogous to frequently cross-linked EP.

6.6. FILLING

At present, it is common knowledge [87] that introduction of a filler into polymeric matrix increases inhomogeneity of the system due to agitation of a conformational state in the filler-polymeric matrix interface layers. The agitation degree was described quantitatively [88] in the framework of the cluster model on the example of two series of a graphite-filled polyhydroxyester: with the filler surface untreated (PHE-Gr-I) and treated (PHE-Gr-II) by a mixture of sulfuric and nitric acids.

Figure 6.34 shows dependencies of φ_{cl} on volumetric concentration of the filler, φ_n, for the mentioned composites. It is indicated that graphite injection into polymeric matrix of PHE reduces the local order in it, and φ_{cl} decline is specifically noticeable at low filler concentrations. Filler surface treatment suppresses agitating effect of graphite that supposes the local order basic decrease at the particle surface, i.e. in the interface filler-polymer layer.

This statement was proved by a simple experiment [88] as follows. As follows from previous data [87], annealing of a filled polymer below its T_g leads to structure leveling. At invariable average density of samples, density of packless zones increases approaching density of initial non-filled polymer. That is why PHE-Gr-II composite possessing $\varphi_n = 0.176$ was annealed at 353 K during 200 hours (for PHE, $T_g = 378$ K [89]). Blow tests of annealed PHE-Gr-II samples indicate an increase of their blow viscosity, A_p, by 1.5 times, approximately (from 10.5 to 15.2 kJ/m^2), accompanied by φ_{cl} increase from

0.361 to 0.445. This means that annealing under the above-mentioned conditions increases the local order degree in interface layers of graphite-filled polyhydroxyester [88].

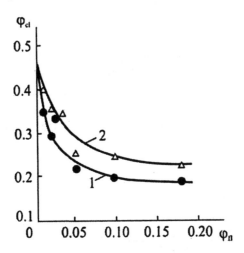

Figure 6.34. Dependencies of the relative part of clusters, φ_{cl}, on volumetric concentration of graphite, φ_n, for PHE-Gr-I (1) and PHE-Gr-II (2) composites [88]

Using ideas of the cluster model and fractal analysis, it has been shown [90] that the polymer matrix structure agitation under consideration is determined by a broad spectrum of changes at molecular, topological and supersegmental levels. In the framework of the fractal analysis, this effect is expressed by increase of fractal dimensionality, d_f, of its structure. It has also been demonstrated [90] that dispersed filler particles form a carcass in the polymeric matrix. This carcass possesses fractal (in the general case, multifractal [91]) properties and is characterized by the fractal (Hausdorff) dimensionality D_H. As a consequence, formation of the polymeric matrix structure of in dispersion-filled composite proceeds not in the Euclidean, but in the fractal space. Based on general physical principles in the framework of the fractal analysis, this fundamentally important circumstance explains the obligatory existence of structure agitation in the polymeric matrix as dispersed filler is injected.

To estimate d_f of the polymeric matrix structure [92], alterable in the space with dimensionality D_H, a model of fractal object formation on fractal lattices is used [93]:

$$\frac{1}{d_f} = \frac{2}{3D_H} \cdot \frac{3d_e - d_s}{2 + 2d_e - d_s},\tag{6.19}$$

where d_s is the spectral dimensionality of the object; d_e is the chemical or "extension" dimensionality, which characterizes the distance between two points of the fractal object not as geometrical distance (a straight line connecting points), but as a "chemical distance", i.e. the shortest path between these two points by particles of the objects [93]. Assessment methods for d_s and d_e are discussed in ref. [92].

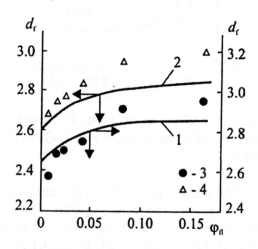

Figure 6.35. Dependencies of experimental, d_f^e (1, 2), and calculated by equation (6.19), d_f^t (3, 4), values of the fractal dimensionality of the polymeric matrix structure on volumetric concentration of the filler, φ_n, for composites PHE-Gr-I (1, 3) and PHE-Gr-II (2, 4) [92]

Figure 6.35 shows comparison of theoretical, d_f^t (equation (6.19)), and experimental, d_f^e (equation (5.4)), values of fractal dimensionalities of the polymeric matrix structure as the function of φ_n. Clearly good quantitative conformity (maximal deviation is below 6%) of d_f^t and d_f^e values. It should be noted that equation (6.19) represents an approximation, and its check on Serpinsky mats indicates deviation of d_f values, obtained by this method, from accurate values equals ~3.5 – 7.5% [93]. Such inaccuracy conforms to the above-mentioned one [92].

It is common knowledge [94 – 98] that the self-similarity of natural and many model structures cannot be described with the help of fractal dimensionality only. More full description of non-oriented structures requires calculation of a spectrum of different dimensionalities, i.e. multifractal formalism should be applied [95]. At present, there is a series of works displaying conformity between different multifractal characteristics and parameters of the real materials [90, 99 – 101]. One of the most wide-spread multifractal characteristics is "potential ordering" parameter, the physical meaning and the calculation method of which are given below. However, any fractal (multifractal) characteristics are extremely general mathematical notions not allowing structure specialization of one or another object. Obviously, this circumstance does not allow accurate and comprehensive characterization of real materials and limits application of the multifractal analysis by composing purely empirical correlations [97, 99, 101]. On this basis the physical meaning of "potential ordering" parameter for dispersion-filled composites was cleared out using ideas of the cluster model [102].

Application of the Williford modification [103] of Halsey et al. [95] multifractal formalism to description of the structure and properties of dispersion-filled polymeric composites was considered [100]. As the mathematical model, the Cantor set ("dust") was used. It is assumed that a segment l_1 long possessing the probability measure p_1 characterizes the filler, and the one with parameters l_2 and p_2 – polymeric matrix, respectively. This attribution is caused by belonging of segment 1 to the brittle failure branch in accordance with [103], and increase of the filler concentration embrittles the studied composites [90]. Values of l_1 and l_2 can be estimated from electron microphotographs of failure surfaces. If in the Cantor composition the middle part is removed from the primary segment of specific (unit) length, the length of two remaining parts equals ~0.667 [95]. Thereafter, this value can be divided proportionally to the size of filler particles (aggregates of particles) and the distance between surfaces of neighboring filler particles (using their average values), and this will be corresponded to l_1 and l_2 scales. Obviously, total failure probability of composites, p, equals 1. Then under the condition $p = p_1 + p_2$ this value is divided as follows. It is assumed that $p_1 = p_2 = 0.5$, if strength of the interface layer σ_a (destruction of filler particles is eliminated) equals the polymeric matrix shear strength τ_m, i.e. failure probabilities of the interface boundary, p_1, and polymeric matrix, p_2, are equal. For $\sigma_a = 0$, the condition $p_1 = 1$ is obvious. Intermediate values of p_2 are determined as follows [102]:

$$p_2 = \frac{0.5\sigma_a}{\tau_m}. \tag{6.20}$$

For PHE-Gr composites, σ_a and τ_m values are accepted in accordance with [104].

Further on, the multifractal composition scheme, suggested in [103], was used in the form of the following relations:

$$\ln(n/m)\ln(l_1/l_2) - \ln(n/m - 1)\ln(l_1) = q[\ln(p_1)\ln(p_2) - \ln(p_2)\ln(l_1)], \tag{6.21}$$

$$\alpha = \frac{\ln(p_1) + (n/m - 1)\ln(p_2)}{\ln(l_1) + (n/m - 1)\ln(l_2)}, \tag{6.22}$$

$$f = \frac{(n/m - 1)\ln(n/m - 1) - (n/m)\ln(n/m)}{\ln(l_1) + (n/m - 1)\ln(l_2)}, \tag{6.23}$$

$$(q - 1)D_q = q\alpha(q) - f(q), \tag{6.24}$$

where n and m are numbers of segments of every type in the Cantor set; q is an index; α is a scaling index characterizing concentration of singularities; f is the dimensionality of singularities α; D_q are generalized Renoui dimensionalities.

Multifractal diagram possesses several characteristic points. For example, at $q = \infty$:

$$D_\infty = \frac{\ln p_1}{\ln l_1}, \tag{6.25}$$

and at $q = -\infty$:

$$D_{-\infty} = \frac{\ln p_2}{\ln l_2}. \tag{6.26}$$

Applying parameters p_1, l_1, p_2 and l_2 estimated by the above-considered method, equations (6.25) and (6.26) allow calculation of dimensionalities D_{40}

and D_{-40}, because in common practice these Renoui dimensionalities are accepted for D_∞ and $D_{-\infty}$, respectively (at $q = 40$ and $q = -40$) [103].

Moreover, dimensionalities D_0 and D_1 (at $q = 0$ and $q = 1$, respectively) are named the Hausdorff and informative dimensionalities of the multifractal [95]. In accordance with [103], these dimensionalities characterize the entire multifractal and the greatest sub-fractal, respectively. For them the relation $D_1 \le D_0$ is valid, the equality being observed for regular fractals only [95, 105].

The Hausdorff dimensionality D_0 can be calculated as follows. First, the relation n/m at $q = 0$ is calculated by the equation (6.21). Subsequently, corresponded values of α and f are calculated by equations (6.22) and (6.23), and then D_0 is calculated by the equation (6.24). This method does not allow calculation of dimensionality D_1, because at $q = 1$ the parameter $(q - 1)$ at D_q in equation (6.24) becomes zero. That is why D_1 is estimated with the help of the technique as follows. As shown in ref. [106], for multifractals the following correlation is true:

$$D_q = D_0 - \frac{9q\varepsilon^2}{64},$$
(6.27)

where parameter ε is defined as follows:

$$\varepsilon = 4 - d.$$
(6.28)

In equation (6.28) d is the dimensionality of Euclidean space, to which the fractal is buried. For $d = 3$ it is obtained that

$$D_1 = D_0 - 0.1406.$$
(6.29)

The relation between D_0 and D_1 expressed analytically by equation (6.29) was experimentally deduced in the work [101].

Finally, the "potential ordering" parameter, Δ, is defined as follows [98]:

$$\Delta = D_1 - D_{40}.$$
(6.30)

The problem which parameter in purely "polymeric" models can be identified with Δ, i.e. can characterize the ordering degree in amorphous polymer matrix, should be considered. In the framework of the cluster model, relative part of clusters, φ_{cl}, is accepted for this parameter. It represents the parameter of the polymer amorphous state structure order in the strictest physical meaning of this term [107]. Figure 6.36 shows a correlation between parameters Δ and φ_{cl} indicating its approximation by a straight line passing through the origin of coordinates. To put it differently, for the polymer amorphous state (in the case under consideration, for polymers of dispersion-filled composite matrix), parameter Δ does not present any "potential ordering", but with an accuracy to a constant ($\Delta \cong 2.27\varphi_{cl}$) represents the structure order parameter of the mentioned state [102]. Complete conformity of physical meanings of Δ and φ_{cl} should be noted. For example, decrease of Δ means smoothing of the structure localization around defects [98]. In the present Chapter oriented zones in polymers (crystallites and clusters) are considered as defects of the amorphous state structure, and decrease of φ_{cl} means transition of a part of segments from clusters into packless matrix, e.g. the above-mentioned smoothing of the structure localization around defects [102].

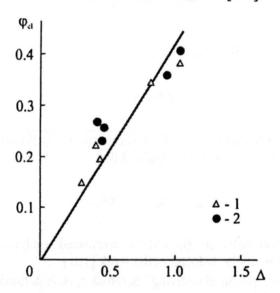

Figure 6.36. Dependence of the relative part of clusters, φ_{cl}, on "potential ordering" parameter, Δ, for PHE-Gr-I (1) and PHE-Gr-II (2) composites [102]

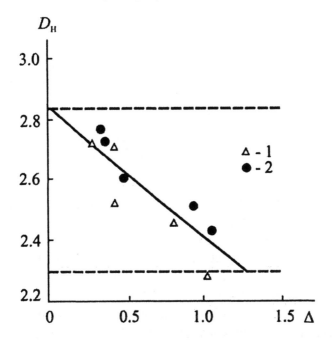

Figure 6.37. Dependence of the fractal dimensionality of carcass from filler particles, D_H, on "potential ordering" parameter, Δ, for PHE-Gr-I (1) and PHE-Gr-II (2) composites. Horizontal dashed lines show border D_H values [102]

It has been shown [90] that dispersion-filled polymer composite can be considered as a combination of two fractals (in the general case, multifractals): a carcass formed by filler particles and polymeric matrix possessing dimensionalities D_H and d_f, respectively. As mentioned above, D_H change induces variation of the polymeric matrix structure, namely, D_H increase determined φ_{cl} decrease. That is why of special interest is the relation between Δ and D_H, shown in Figure 6.37 for both series of PHE-Gr composites. Expectedly, D_H increase causes Δ decrease. Moreover, $D_H(\Delta)$ curve allows assessment of Δ border values. For this purpose, remind the technique for D_H assessment. This dimensionality was calculated using the following formula [108]:

$$D_H = \frac{\ln N}{\ln \rho},$$ (6.31)

where N is the number of filler particles (aggregates of particles) sized ρ.

Since N and ρ values are determined from electron microphotographs, i.e. in the plane ($d = 2$), recalculation for three-dimensional space has used correlation, obtained in the work [97]:

$$D3 = \frac{d + D2 \pm \left[(d - D2)^2 - 2\right]^{1/2}}{2}, \qquad (6.32)$$

where $D3$ and $D2$ are appropriate fractal dimensionalities in three- and two-dimensional Euclidean spaces, $d = 3$.

In accordance with the general regularities, $D2$ cannot be below zero, and this condition ($D2 = 0$) determines the maximum of $D3 = D_H \approx 2.83$. It follows from the expression in square brackets that $(d - D2)^2 \geq 2$ or $D2 \leq 1.59$ that gives the minimum of $D3 = D_H \approx 2.30$. These border values of D_H are presented in Figure 6.35 by horizontal dashed lines. Expectedly, maximal value $D_H \approx 2.83$ determines the maximal agitation of the structure: $\Delta = 0$ or $\varphi_{cl} = 0$ (refer to Figure 6.34). The minimum $D_H \approx 2.30$ gives the maximal value $\Delta \approx 1.30$ [102].

Subsequently, turning back to the curve in Figure 6.36 and using obtained values of Δ, border values of φ_{cl} can be estimated. At $D_H \approx 2.83$, $\Delta = 0$ and $\varphi_{cl} = 0$ that means polymer transition into the rubbery state [109]. However, for composites the effect indicated is not reached, because the filler "freezes" (i.e. suppresses) the molecular mobility. For example, for PHE-Gr composites the maximal fractal dimensionality D of the chain segment between clusters, which characterizes molecular mobility, equals ~1.60, whereas for devitrified polymer $D = 2.0$ [22]. As follows from the equation (5.7), at $\varphi_{cl} = 0$, $d_f = 3 = d$ and polymeric matrix loses its fractal properties. Moreover, as follows from the definition (6.30), $D_1 = D_{40} = D_0 = \ldots = D_{-\infty}$ at $\varphi_{cl} = 0$, i.e. the polymeric matrix loses its multifractal properties with no respect to the Renoui dimensionality [110]. To put it differently, the condition of polymer structure multifractality is existence of a local order ($\varphi_{cl} \neq 0$ and $\Delta \neq 0$) [102].

Extrapolation of the linear dependence $\varphi_{cl}(\Delta)$ to $\Delta = 1.30$ gives the border maximum $\varphi_{cl} = 0.57$. Theoretically, this value can be estimated in the framework of the thermal cluster model, according to which the maximum of the order parameter (in the present case, φ_{cl}) is given by the equation (5.72). For

PHE, $T_g \cong 378$ K and $T = 293$ K. Therefore, $\varphi_{cl} \cong 0.55$, which is close to extrapolated value of this parameter [102].

Another method for binding multifractal characteristics and parameters of the cluster structure of polymers was suggested [111]. It has also been mentioned [112] that generalized Renoui dimensionalities D_∞ and $D_{-\infty}$ (or D_{40} and D_{-40}) of the multifractal diagram characterize the loosest and the most concentrated sets in the system, respectively. For amorphous polymeric materials, the free volume characterized by the relative part (f_f) should be assumed to be the loosest set in the structure; closely packing zones are accepted for the most concentrated set, identified in the model [5] with the local order zones (clusters) and characterized by the relative part of clusters (φ_{cl}). Thereafter, dependencies $D_{40}(f_c)$ and $D_{-40}(\varphi_{cl})$ were plotted, and the following correlations were deduced:

$$D_{40} = \frac{8.4 \times 10^{-3}}{f_f}, \qquad (6.33)$$

$$D_{-40} = 1 + 4\varphi_{cl}. \qquad (6.34)$$

Equations (6.33) and (6.34) allow binding of extreme Renoui dimensionalities to structural characteristics of the polymer matrix and clearing out the physical meaning of the loosest and the most concentrated sets in the composite structure. In its turn, pair-wise comparison of equations (6.25) and (6.33), (6.26) and (6.34) allows estimation of changes in the failure mechanism characterized by p_1 and p_2 variation as the function of volumetric filler concentration, φ_n. Some data of this kind are shown in Table 6.2, from which the following conclusions can be made. Increase of l_1, i.e. φ_n increase at $f_f =$ const increases the failure probability for the interface, p_1. Similar effect is observed under the condition $l_1 =$ const (i.e. $\varphi_n =$ const) and increasing f_f. Note that in both cases, p_1 increase is rather low. Under the condition $\varphi_n =$ const the decrease of l_2 (φ_n increase, too) induces an abrupt decline the polymeric matrix failure probability, p_2, and vice versa, φ_n decrease under the condition $l_2 =$ const causes an insignificant increase of p_2. This approach, even its simplest form, allows general analysis of tendencies in changes of structure and properties of the dispersion-filled polymer composites by plotting schematic multifractal diagrams [111].

Table 6.2
Tendencies in changes of structural characteristics of dispersion-filled PHE-Gr composites [111]

l_1	f_f	p_1	l_2	φ_{cl}	p_2
0.17	0.042	0.70	0.49	0.35	0.182
0.47	0.042	0.86	0.20	0.35	0.021
0.17	0.120	0.88	0.49	0.16	0.312

Structural aspect of the interface adhesion in filled polymers, yet studied insufficiently, was considered [113]. The authors of ref. [114] have suggested characterization of the polymer-filler interaction level (the adhesion degree) using parameter A, determined from the following correlation:

$$A = \frac{1}{1-\varphi_{fl}} \cdot \frac{tg\delta_c}{tg\delta_m} - 1, \qquad (6.35)$$

where $tg\delta_c$ and $tg\delta_m$ are dissipation factors for filled polymer and pure polymeric matrix, respectively.

Strong interactions between the filler and the polymeric matrix at the interface tend to decline molecular mobility in the filler surface surrounding compared with the bulky polymeric matrix. This leads to a decrease of $tg\delta_c$ and, as a consequence, A. Thus low value of A indicates high degree of interaction or adhesion between phases of filled polymer [114].

Selection of parameter A as the measure of adhesion at the polymer-filler interface is induced by aggregation of filler particles. Traditional methods of adhesion measurement [115] are unable to reflect variations of this parameter, expected by virtue of the change in the surface structure of aggregates from particles of the initial filler due to aggregation of the latter [116].

However, the technique [114] does not consider the polymeric matrix structure near the filler surface, i.e. the interface surface. Changes of the mentioned structure may cause variation of molecular mobility in this layer and thus affect parameter A. As a consequence, parameter A will be affected by, at least, two factors in this interpretation: chemical and (or) physical interactions and the interphase layer structure. Quantitatively, the latter factor can be characterized in the framework of the cluster model by the relative part of the local order zones (clusters), φ_{cl}^0, directly at the filler surface, the values of which for PHE-Gr are accepted from data in the work [117]. Figure 6.38 shows dependencies of parameter A, calculated by equation (6.35), on φ_{cl}^0 for both series of PHE-Gr composites. Clearly φ_{cl}^0 increase induces A decrease or adhesion increase. This type of dependence $A(\varphi_{cl}^0)$ was expectable, because increase of the local order degree (φ_{cl}^0 increase) of structure of the interface leads to a decrease of molecular mobility in it [118, 119]. Of special attention is the fact that dependence $A(\varphi_{cl}^0)$ is split into two parallel lines for each series of studied composites. Lower values of A for PHE-Gr-II rather than for PHE-Gr-I at equal φ_{cl}^0 assume stronger physicochemical polymer-filler interactions for the former series rather than for the latter one. One of possible explanations of the effect observed can be as follows. As graphite is treated by a mixture of sulfuric and nitric acids, different chemically active oxygen-containing groups are formed on its surface. These groups irreversibly chemosorb bisphenol A by interacting with its hydroxyl groups, which increases the interface adhesion characterized by parameter A, for PHE-Gr-II rather than for PHE-Gr-I composites. Extrapolation of $A(\varphi_{cl}^0)$ lines to $\varphi_{cl}^0 = 0$ allows obtaining of A values, stipulated by physical and/or chemical polymer-filler interactions ($A_{ph\text{-}ch}$), which equal ~3.82 for PHE-Gr-I and ~2.60 for PHE-Gr-II [113].

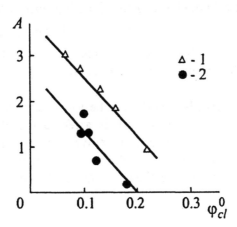

Figure 6.38. Dependence of parameter A on relative part of clusters in the interphase layer φ_{cl}^0 for PHE-Gr-I (1) and PHE-Gr-II (2) composites [113]

These results allow obtaining of an analytical relation between parameter A and two factors controlling it ($A_{ph\text{-}ch}$ and φ_{cl}^0) in the form as follows:

$$A = A_{ph\text{-}ch} - 12.5\,\varphi_{cl}^0.\qquad(6.36)$$

Critical value of φ_{cl}^0 $\left(\varphi_{cl}^{cr}\right)$ can be easily obtained from equation (6.36). At this value molecular mobility at the interface (with regard to the filler concentration, φ_{fl}) equals this index for the bulky matrix [113]:

$$\varphi_{cl}^{cr} = \frac{A_{ph-ch}}{12.5}.\qquad(6.37)$$

Calculation by the equation (6.37) gives φ_{cl}^{cr} = 0.306 for PHE-Gr-I and 0.208 for PHE-Gr-II. Lower φ_{cl}^{cr} value for PHE-Gr-II rather than for PHE-Gr-I means higher contribution of physical and/or chemical polymer-filler interactions into suppression of molecular mobility in the interface that determines lower critical packing density of the interface structure.

Under the condition $\varphi_{cl}^0 > \varphi_{cl}^{cr}$, $A < 0$ is obtained from the equation (6.36), i.e. molecular mobility in the interphase layer is lower than in bulky polymeric matrix. This result is obtained at relatively smooth surface of the filler particles [104].

Thus the polymer-filler interface adhesion, characterized by parameter A in accordance with the equation (6.35), is affected by, at least, two factors: intensity of physical and/or chemical polymer-filler interactions and the interface structure. Analytically, the effect of the mentioned factors can be described in the framework of the cluster model of the polymer amorphous state structure [113].

6.7. THERMOOXIDATIVE DEGRADATION

It has been shown [120, 121] that the dependence of relative blow viscosity (where A_p and A_p^0 are blow viscosities of the sample after an arbitrary period of heat aging and the initial non-aged one, respectively) on heat aging period, t_{age}, at $T < T_{melt}$ for poly(butylene terephthalate) (PBTP), modified by a high-dispersion Fe/FeO mixture (Z), splits into three areas (for initial PBTP, into two areas), which are shown in Figure 6.39 for PBTP and two PDTP + Z composites. Application of the fractal analysis allows clearing out physical reasons for such behavior of A_p/A_p^0 (t_{age}) function and determines chemical mechanisms of thermooxidative degradation for every part of it. However, plots in Figure 6.37 show that for PBTP and PBTP + Z composites with different Z concentrations, the part with the most rapid A_p/A_p^0 decline with t_{age} (and consequently, the most unfavorable for practical application of PBTP) ends at different A_p/A_p^0 values, followed by the part of much slower A_p/A_p^0 decline, i.e. more favorable from positions of application. An attempt to clear out such behavior of A_p/A_p^0 (t_{age}) dependencies for PBTP and PBTP + Z composites was undertaken in [122].

Let us consider A_p/A_p^0 (t_{age}) dependencies, shown in Figure 6.39, in more detail. In the first part of these dependencies, characterized by the condition $A_p/A_p^0 \cong$ const, inhibited thermooxidative degradation of PBTP + Z composites, described in the framework of the task on traps [123, 124], proceeds. The absence of thermooxidative degradation inhibitor in the initial PBTP causes the absence of the mentioned part on the curve for it. The mixture Fe/FeO is also a strong modifier of the structure (refer to Chapter 2 for details) that induces the extreme increase of φ_{cl} at Z concentration equal $C_Z \cong 0.05 - 0.10$ wt.% and corresponded increase of the first part duration (the induction period), t_{in}. Moreover, it is found that pseudo-monomolecular reaction (the reaction of intermediate products with the polymer macromolecule) is realized at the second part of A_p/A_p^0 (t_{age}) curve, and bimolecular reaction (the reaction between oxygen and the macromolecule) is realized on the third part of it (the second part for PBTP) [120].

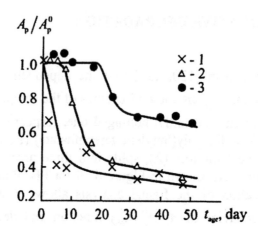

Figure 6.39. Dependencies of relative blow viscosity (A_p/A_p^0) on aging time (t_{age}) for initial PBTP (1) and PBTP + Z composites with Z concentrations as follows: 0.01 wt.% (2) and 0.05 wt.% (3) [121]

More thorough consideration of plots in Figure 6.39 indicates that t_{in} increase is accompanied by termination of pseudo-monomolecular reaction at higher A_p/A_p^0 values, i.e. by lowering the effect of this reaction. As mentioned above, t_{in} increase is stipulated by the extreme growth of φ_{cl} in the range $C_Z = 0.05 - 0.10$ wt.% and corresponded decrease of oxygen diffusion to non-crystalline zones of the polymer. Since the modifying effect of Z is preserved after exhaustion of its inhibition effect, the above-mentioned correspondence between t_{in} and A_p/A_p^0 value, at which transition from pseudo-mono- to bimolecular reaction happens, is stipulated by the same effects of structural stabilization. Reduction of the pseudo-monomolecular reaction part with t_{in} increase suggests proceeding of the mentioned reaction in the less closely packed zones of the polymer. In the framework of the cluster model these zones are identified as the packless matrix, the reduced relative part of which $\left(\varphi_{p.m.}^r\right)$ can be determined from the equation (5.80).

Figure 6.40 shows dependence of A_p/A_p^0 values, corresponded to transition from pseudo-monomolecular to bimolecular reaction of thermooxidative degradation, on $\varphi_{p.m.}^r$. Data in the Figure indicate the result confirming the above-mentioned suggestion that $\varphi_{p.m.}^r$ increase leads to a decrease of the border relative blow viscosity $(A_p/A_p^0)^b$ and an increase of pseudo-monomolecular reaction activity zone, which can be written down as $(1 - (A_p/A_p^0)^b$, where b is border). In the absence of packless zones, i.e. at $\varphi_{cl} + K = 1$, $(A_p/A_p^0)^b = 1$ and the mentioned reaction cannot proceeds (e.g. its duration equals zero). At $\varphi_{p.m.}^r = 1.0$ the reaction is the only one proceeding during thermooxidative degradation and lasts until the condition $A_p = 0$ is fulfilled [122].

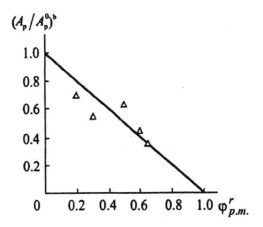

Figure 6.40. Dependence of the border relative blow viscosity $(A_p/A_p^0)^b$ on the reduced part of the packless matrix $\varphi_{p.m.}^r$ for PBTP and PBTP + Z composites [122]

From the above-said one more assumption can be made that $\varphi_{p.m.}^r$ increase must cause shortening of the pseudo-monomolecular reaction duration. Actually, the dependence of this reaction duration (t_{age}^{mm} on $\varphi_{p.m.}^r$), shown in Figure 6.41, proves this assumption. At $\varphi_{p.m.}^r = 0$, $t_{age}^{mm} \cong 8.7$ days, i.e. for the initial PBTP this time is the induction period, and then transition to bimolecular reaction happens. Expectedly, extrapolation of $t_{age}^{mm}(\varphi_{p.m.}^r)$ curve to $\varphi_{p.m.}^r = 1.0$ gives $t_{age}^{mm} = 0$. This means that under the mentioned conditions the monomolecular reaction proceeds at extremely high rate. This conclusion can be proved by results of PBTP thermal oxidation at temperatures above the melting point, i.e. thermal aging of its melt, where $K = 0$ and $\varphi_{cl} = 0$, and, consequently, in accordance with the equation (5.80) $\varphi_{p.m.} = \varphi_{p.m.}^r = 1.0$. In this case, thermooxidative degradation is completed (the polymer becomes practically non-viscous) during less than 1 hour, which in the time scale of the Figure 6.39 means $t_{age}^{mm} \cong 0$.

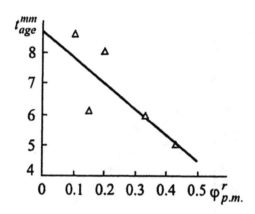

Figure 6.41. Dependence of pseudo-monomolecular reaction duration, t_{age}^{mm}, on reduced part of the packless matrix, $\varphi_{p.m.}^r$, for initial PBTP and PBTP + Z composites [122]

The following reason for proceeding of the pseudo-monomolecular reaction directly in the packless zones of initial PBTP and PBTP + Z composite can be suggested. Broadly speaking, this reaction suggests interaction of two polymer molecules during thermal degradation. Thus it requires a definite molecular mobility providing for the above-mentioned interaction. It is known [125] that non-zero molecular mobility can be realized in packless zones of amorphous-crystalline polymers exclusively.

The logical consequence of the above-considered structural mechanism of pseudo-monomolecular reaction is the assumption that bimolecular reaction represents thermooxidative degradation of closely packed zones (crystallites and clusters). This very circumstance determines an abrupt decrease of the rate of its proceeding compared with the rate of pseudo-monomolecular reaction. To prove this statement, duration of the bimolecular reaction, t_{age}^{bm}, is determined by extrapolation of the dependence $A_p/A_p^0 (t_{age})$ to $A_p = 0$ [122]. Thus the dependence of t_{age}^{bm} on total relative part of closely packed zones (crystallites and clusters) $(K + \varphi_{cl})$ is shown in Figure 6.42. The data shown indicate the linear increase of t_{age}^{bm} with $(K + \varphi_{cl})$, and at $(K + \varphi_{cl}) = 0$ $t_{age}^{bm} = 0$, too. This means that in the absence of closely packed zones bimolecular reaction does not take place. This conclusion is proved by the results of thermal aging of PBTP melt, where thermooxidative degradation is fully realized as the pseudo-monomolecular reaction.

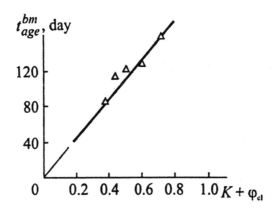

Figure 6.42. Dependence of bimolecular reaction duration t_{age}^{bm} on total part of closely packed zones $(K + \varphi_{cl})$ for initial PBTP and PBTP + Z composites [122]

As a consequence, the above-shown results indicate determination of chemical mechanisms of thermooxidative degradation of initial PBTP and PBTP + Z composites by structural mechanisms. Pseudo-monomolecular reaction proceeds in the packless zones of these amorphous-crystalline polymers and bimolecular reaction proceeds in closely packed zones exclusively. This indicates the importance of the structure stabilization effect at any stage of the thermooxidative degradation. There are two methods suppressing the most applicably unfavorable impact of the pseudo-monomolecular reaction. The first method (the physical one) requires formation of a closely packed polymer structure, which could be hardly provided [126]. The second method (the chemical one) requires creation of a complex stabilizer, the injection of which mixed with Fe/FeO into PBTP would neutralize the pseudo-monomolecular reaction [122].

Additional information on the processes proceeding during PBTP thermooxidative degradation can be obtained in the framework of the fractal kinetics of chemical reactions [127, 128]. Figure 6.43 shows dependence of the mean molecular weight, $\overline{M}_w$, on thermal aging duration, t_{age}, at 393 K for PBTP. The data indicate an abrupt decline of $\overline{M}_w$ during initial ten days, approximately, after which it is strongly slowed down. Thermooxidative degradation rate, k, can be described by the following correlation [127]:

$$k \sim t_{age}^{-h}, \qquad (6.38)$$

where h is the index of inhomogeneities ($0 \le h \le 1$).

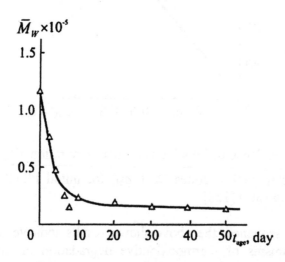

Figure 6.43. Dependence of the mean molecular weight, $\overline{M}_w$, on the aging duration, t_{age}, for PBTP [129]

If the reaction proceeds in fractal media, $h > 0$; in the case of the Euclidean medium (classical behavior), $h = 0$ and $k = \text{const}$ [127, 128]. Parameter k can be calculated as follows [130]:

$$k = \frac{\overline{M}_{w_i+1} - \overline{M}_{w_i}}{t_{age_i+1} - t_{age_i}}, \tag{6.39}$$

where $\overline{M}_{w_i+1}$ and $\overline{M}_{w_i}$ are molecular weights of the polymer at arbitrary aging times t_{age_i+1} and t_{age_i}, respectively.

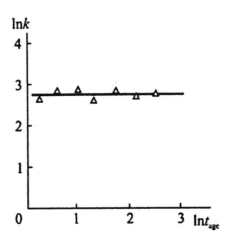

Figure 6.44. Dependence of the pseudo-monomolecular reaction rate k on the thermal aging duration t_{age} presented in double logarithmic coordinates for solid PBTP [129]

The dependence $k(t_{age})$ presented in double logarithmic coordinates in accordance with the correlation (6.38) allows determination of the inhomogeneity parameter, h. Figures 6.44 and 6.45 present these dependencies for the zones of pseudo-monomolecular and bimolecular reactions of PBTP thermooxidative degradation, respectively. In the first case (Figure 6.44), the reaction rate k is independent of t_{age}, i.e. the classical behavior is realized (the reaction proceeds in the Euclidean space, where k = const and $h = 0$). Because in polymers the Euclidean object can be represented by devitrified (rubbery) polymer only [22], the results obtained prove the above conclusion about PBTP pseudo-monomolecular degradation proceeding in the devitrified zones of the amorphous phase. For the purpose of comparison Figure 6.46 presents the dependence $k(t_{age})$ for PBTP thermal aging above the melting point $T_{melt} \cong 500$ K [75], e.g. for its melt. In this case, the entire sample represents the Euclidean object [22] and, expectedly, it also displays k = const and $h = 0$ (classical behavior). Thus no principal differences in proceeding of the pseudo-monomolecular reaction for devitrified zones of solid PBTP amorphous phase (at $T_g < T_{age} < T_{melt}$, where T_{age} is the thermal aging temperature) and completely devitrified PBTP are observed, because the structural zones mentioned are the Euclidean objects. However, pseudo-monomolecular reaction rate in the melt is about 20 times higher than in the solid polymer [129].

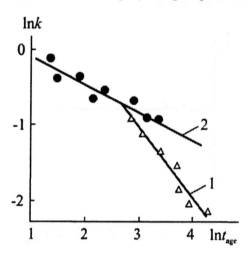

Figure 6.45. Dependencies of the reaction rate k on thermal aging duration t_{age} in double logarithmic coordinates for bimolecular reaction of PBTP (1) and thermooxidative degradation of PAr (2) [129]

Figure 6.45 shows the dependence $k(t_{age})$ in double logarithmic coordinates for bimolecular reaction zone of PBTP thermooxidative degradation (the second part of dependence $\overline{M}_w (t_{age})$ in Figure 6.43). Clearly in this case, k decline with t_{age} increase is observed that represents the proof of bimolecular reaction proceeding in the fractal space [127, 128]. Linearity of dependence $\ln k(\ln t_{age})$ allows determination of the inhomogeneity parameter, h, from its slope in accordance with the correlation (6.38) equal 1.0. It is shown [127, 128] that effective spectral dimensionality $d_s' = 0$, i.e. bimolecular reaction proceeds in the zero-dimensional space or, as defined by the authors of the work [127], in miniclusters. For amorphous-crystalline polymers such miniclusters are closely packed zones (clusters and crystallites), which also confirms the above conclusion about bimolecular reaction proceeding in closely packes PBTP zones. Clearly pseudo-monomolecular and bimolecular reactions represent border cases of thermooxidative degradation kinetics with the border conditions

$h = 0$ and $h = 1$ [127, 128]. On this basis it should be assumed that k values obtained for the mentioned reactions give border (maximal and minimal, respectively) thermooxidative degradation rates. Figure 6.43 also shows the dependence $\ln k(\ln t_{age})$ for polyarylate (PAr) derived from phenolphthalein, diane and a mixture of iso- and terephthalic anhydrides [131]. Data from this Figure indicate linearity of the mentioned dependence for PAr, and $h = 0.385$ can be determined from its slope. Expectedly, this means that thermooxidative degradation of PAr proceeds in the fractal space, because thermal aging of PAr proceeded at $T_{age} < T_g$ [131]. For PAr, parameter h gives an intermediate value from the range $0 \leq h \leq 1$. This circumstance assumes simultaneous proceeding of thermooxidative degradation both in packless and closely packed (cluster) zones of the PAr structure. Relatively low values of h for PAr allow an assumption that basic degradation processes proceed in packless zones of this polymer structure.

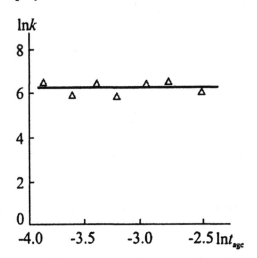

Figure 4.46. Dependence of pseudo-monomolecular reaction rate k on thermal aging duration t_{age} in double logarithmic coordinates for PBTP melt [129]

Thus in the framework of the fractal model, kinetics of PBTP thermooxidative degradation allows identification of structural zones, in which pseudo-monomolecular and bimolecular reactions proceed. The type of space is the main parameter for structural localization of the mentioned reactions: pseudo-monomolecular reaction proceeds in the Euclidean space, and bimolecular reaction – in the fractal one.

REFERENCES

1. Aloev V.Z., Burya A.I., and Kozlov G.V., *Doklady NAN Ukrainy*, 2003, No. 6, pp. 123 – 126. (Rus)
2. Borokhovsky V.A., Gasparyan K.A., Mirzoev R.G., and Baranov V.G., *Vysokomol. Soedin.*, 1976, vol. A18(11), pp. 2406 – 2411. (Rus)
3. Yekh G.S., *Vysokomol. Soedin.*, 1979, vol. A21(11), pp. 2433 – 2446. (Rus)
4. Vladovskaya S.G. and Baranov V.G., *Vysokomol. Soedin.*, 1983, vol. A25(2), pp. 258 – 264. (Rus)
5. Kozlov G.V. and Novikov V.U., *Uspekhi Fizicheskikh Nauk*, 2001, vol. 171(7), pp. 717 - 764. (Rus)
6. Vladovskaya S. G. and Baranov V. G., *Acta Polymerica*, 1982, vol. 33(2), pp. 125 – 130.
7. Aloev V.Z., Kozlov G.V., and Afaunova Z.I., *Manuscript deposited to VINITI RAS*, Moscow, December 30, 1999, No. 3931-V99. (Rus)
8. Frenkel S.Ya., *Khimicheskie Volokna*, 1977, No. 3, pp. 11 - 14. (Rus)
9. Baranov V.G., Martirosyan A.A., and Frenkel S.Ya., *Vysokomol. Soedin.*, 1975, vol. B17(4), pp. 261 – 262. (Rus)
10. Belyaev O.F., Aloev V.Z., and Zelenev Yu.V., *Vysokomol. Soedin.*, 1986, vol. A28(2), pp. 260 – 265. (Rus)
11. Barnstein R.S., Kirilovich V.I., and Nosovsky Yu.E., *Plasticizers For Polymers*, 1982, Moscow, Khimia, 200 p. (Rus)
12. Perepechko I.I. and Yakovenko S.S., *Vysokomol. Soedin.*, 1981, vol. A23(5), pp. 1166 – 1170. (Rus)
13. Zvonkova E.M., Mar'in A.P., Mikheev Yu.A., Kerber M.L., and Toptuigin D.Ya., *Vysokomol. Soedin.*, 1984, vol. A26(6), pp. 1228 – 1235. (Rus)
14. Zvonkovz E.M., Zvonkov V.V., and Kerber M.L., *Vysokomol. Soedin.*, 1985, vol. A27(3), pp. 595 – 603. (Rus)
15. Kozlov G.V., Sanditov D.S., and Lipatov Yu.S., In Coll.: *Fractals And Local Order In Polymeric Materials*, Ed. G.V. Kozlov and G.E. Zaikov, 2001, New York, Nova Science Publishers, Inc., pp. 65 - 82.
16. Shogenov V.N., Belousov V.N., Potapov V.V., Kozlov G.V., and Prut E.V., *Vysokomol. Soedin.*, 1991, vol. A33(1), pp. 155 – 160. (Rus)
17. Olkhovnik O.E. and Baranov V.G., *Vysokomol. Soedin.*, 1994, vol. A26(4), pp. 822 – 828. (Rus)
18. Deng Q., Sunder C.S., and Yean Y.C., *J. Phys. Chem.*, 1992, vol. 96(1), pp. 492 – 495.

19. Borek J. and Osoba W., *J. Polymer Sci.: Part B: Polymer Phys.*, 1996, vol. **34**(11), pp. 1903 – 1906.
20. Kozlov G.V., Sanditov D.S., Ovcharenko E.N., and Mikitaev A.K., *Fizika I Khimia Stekla*, 1997, vol. **23**(4), pp. 369 – 373. (Rus)
21. Sanditov B.D., Kozlov G.V., Serdyuk V.D., and Sanditov D.S., *Proc. All-Russian Scientific Conf. "Mathematical Modeling In Synergic Systems"*, 1999, Ulan-Ude – Tomsk, pp. 317 - 319. (Rus)
22. Kozlov G.V. and Novikov V.U., *Synergism And Fractal Analysis Of Polymer Networks*, 1998, Moscow, Klassika, 112 p. (Rus)
23. Budtov V.P., *Physical Chemistry Of Polymer Solutions*, 1992, Saint-Petersburg, Khimia, 384 p. (Rus)
24. Berens A.R. and Hodhe I.M., *Macromolecules*, 1982, vol. **15**(4), pp. 756 – 761.
25. Petrie S.E.B., *J. Macromol. Sci.-Phys.*, 1976, vol. **B12**(2), pp. 225 – 247.
26. Flick J.R. and Petrie S.E.B., In Coll.: *Second Symp. On Macromolecules: Structure And Properties Of Amorphous Polymers*, Amsterdam, 1980, vol. **10**, pp. 145 – 163.
27. Chan A.H. and Paul D.R., *J. Appl. Polymer Sci.*, 1979, vol. **24**(6), pp. 1538 - 1550.
28. Morgan R.J. and O'Neal J.E., *J. Polymer Sci.: Polymer Phys. Ed.*, 1976, vol. **14**(6), pp. 1053 - 1076.
29. Wyzgoski M.G., *J. Appl. Polymer Sci.*, 1980, vol. **25**(7), pp. 1455 - 1467.
30. Bubeck R.A. and Yasar H.Y., *Polymer Commun.*, 1989, vol. **30**(1), pp. 25 - 27.
31. Kozlov G.V., Burya A.I., Sanditov D.S., Serdyuk V.D., and Lipatov Yu.S., *Materialy, Tekhnologii, Instrumenty*, 1999, vol. **4**(2), pp. 51 - 54. (Rus)
32. Aleksanyan G.G., Berlin Al.Al., Gol'dansky A.V., Grineva N.S., Onishchuk V.A., Shantarovich V.P., and Safonov G.P., *Khimicheskaya Fizika*, 1986, vol. **5**(9), pp. 1225 - 1234. (Rus)
33. Boyer R.F., *J. Macromol. Sci.-Phys.*, 1973, vol. **B7**(3), pp. 487 - 501.
34. Sanchez I.C., *J. Appl. Phys.*, 1974, vol. **45**(10), pp. 4204 - 4215.
35. Song H.-H. and Roe R., *J. Macromolecules*, 1987, vol. **20**(11), pp. 2723 - 2732.
36. Kozlov G.V., Beloshenko V.A., Gazaev M.A., and Lipatov Yu.S., *Vysokomol. Soedin.*, 1996, vol. **B38**(8), pp. 1423 - 1426. (Rus)
37. Kausch H.H., *Polymer Fracture*, Berlin, Heidelberg, New York, Springer Verlag, 1978, 440 p.

38. Belousov V.N., Kotsev B.Kh., and Mikitaev A.K., *Doklady AN SSSR*, 1983, vol. 270(5), pp. 1145 - 1147. (Rus)
39. Kozlov G.V., Beloshenko V.A., Stroganov I.V., and Lipatov Yu.S., *Doklady NAN Ukrainy*, 1995, No. 10, pp. 117 - 118. (Rus)
40. Blueman A., Clufter J., and Cumofen G., In Coll.: *Fractals in Physics*, Ed. L. Pietronero and E. Tosatti, Amsterdam – Oxford – New York – Tokyo, Horth-Holland, 1986, 670 p.
41. Kozlov G.V., Beloshenko V.A., and Lipatov Yu.S., *Ukrainsky Khimichesky Zhurnal*, 1998, vol. 64(3), pp. 56 - 59. (Rus)
42. Kozlov G.V., Beloshenko V.A., and Lipskaya V.A., *Ukrainsky Khimichesky Zhurnal*, 1996, vol. 41(2), pp. 222 - 225. (Rus)
43. Kozlov G.V., Shustov G.B., and Zaikov G.E., *Zhurnal Prikladnoi Khimii*, 2002, vol. 75(3), pp. 485 - 487. (Rus)
44. Nigmatullin R.R., *Teoreticheskaya I Matematicheskaya Fizika*, 1992, vol. 90(3), pp. 354 - 367. (Rus)
45. Kozlov G.V., Novikov V.U., Sanditov D.S., Gazaev M.A., and Mikitaev A.K., *Manuscript deposited to VINITI RAS*, Moscow, November 21, 1995, No. 3037-V95. (Rus)
46. Filyanov E.M., *Vysokomol. Soedin.*, 1987, vol. 29(5), pp. 975 - 981. (Rus)
47. Kozlov G.V., Beloshenko V.A., Varyukhin V.N., and Lipatov Yu.S., *Polymer*, 1999, vol. 40(4), pp. 1045 - 1051.
48. Kozlov G.V., Dolbin I.V., and Zaikov G.E., *Zh. Prikl. Khim.*, 2003, vol. 76 (in press). (Rus)
49. Bartenev G.M. and Frenkel S.Ya., *Physics Of Polymers*, 1990, Leningrad, Khimia, 432 p. (Rus)
50. Kozlov G.V., Dolbin I.V., Novikov V.U., and Zaikov G.E., In Coll.: *Proc. Intern. Conf. "Baikal Readings-II on Process Simulation in Synergic Systems"*, Ulan-Ude – Tomsk, TGU, 2002, pp. 212 – 214. (Rus)
51. Belousov V.N., Beloshenko V.A., Kozlov G.V., and Lipatov Yu.S., *Ukrainsky Khimichesky Zhurnal*, 1996, vol. 62(1), pp. 62 - 65. (Rus)
52. Family F., *J. Stat. Phys.*, 1984, vol. 36(5/6), pp. 881 - 896.
53. Kozlov G.V. and Zaikov G.E., *Materialovedenie*, 2002, No. 12, pp. 13 – 17. (Rus)
54. Kozlov G.V., Beloshenko V.A., Gazaev M.A., and Varyukhin V.N., *High Pressure Physics And Technology*, 1995, vol. 5(1), pp. 74 - 80. (Rus)
55. Kozlov G.V., Serdyuk V.D., and Dolbin I.V., *Materialovedenie*, 2002, No. 12, pp. 2 - 5. (Rus)

56. Kozlov G.V. and Zaikov G.E., In Coll.: *Studies in Chemistry and Biochemistry*, Ed. Zaikov G. and Lobo V., New York, Nova Science Publishers, Inc., 2003, p. 206 - 212.

57. Kozlov G.V. and Zaikov G.E., In Coll.: *Chemical And Biochemical Physics. Problems And Solutions*, New York, Nova Science Publishers, 2003 (in press).

58. Kozlov G.V., Beloshenko V.A., Varyukhin V.N., and Novikov V.U., *Zhurnal Fizicheskikh Issledovanii*, 1997, vol. 1(2), pp. 204 - 207. (Rus)

59. Aloev V.Z., Kozlov G.V., and Beloshenko V.A., *Izv. KBNTs RAN*, 2000, No. 1(4), pp. 108 - 113. (Rus)

60. Watts M.P.C., Zachariades A.E., and Porter R.S., *J. Mater. Sci.*, 1980, vol. **15**(2), pp. 426 - 430.

61. Beloshenko V.A., Kozlov G.V., Slobodina V.G., Prut E.V., and Grinev V.G., *Vysokomol. Soedin.*, 1995, vol. **B37**(6), pp. 1089 - 1092. (Rus)

62. Kramer E.J., *Polymer Eng. Sci.*, 1984, vol. **24**(10), pp. 741 - 769.

63. Kozlov G.V., Beloshenko V.A., Slobodina V.G., and Prut E.V., *Vysokomol. Soedin.*, 1996, vol. **B36**(6), pp. 1056 - 1060. (Rus)

64. Brown D.Y., *Polymer Commun.*, 1985, vol. **26**(2), pp. 42 - 45.

65. Botto P.A., Duckett R.A., and Ward I.M., *Polymer*, 1987, vol. **28**(2), pp. 257 - 262.

66. Raha S. and Bowden P.B., *Polymer*, 1972, vol. **13**(4), pp. 174 - 184.

67. Kahar N., Duckett R.A., and Ward I.M., *Polymer*, 1978, vol. **19**(2), pp. 136 - 144.

68. Belousov V.N., Kozlov G.V., Mikitaev A.K., and Lipatov Yu.S., *Doklady AN SSSR*, 1990, vol. **313**(3), pp. 630 - 633. (Rus)

69. Milagin M.F. and Shishkin N.I., *Vysokomol. Soedin.*, 1972, vol. **A14**(2), pp. 357 - 362. (Rus)

70. Lomonosova N.V., *Vysokomol. Soedin.*, 1978, vol. **A20**(10), pp. 2270 - 2277. (Rus)

71. Milagin M.F. and Shishkin N.I., *Vysokomol. Soedin.*, 1988, vol. **A30**(11), pp. 2249 - 2254. (Rus)

72. Aloev V.Z., Kozlov G.V., and Afaunova Z.I., *Izv. KBNTs RAN*, 2000, No. 2(5), pp. 95 - 98. (Rus)

73. Aloev V.Z., Kozlov G.V., and Zaikov G.E., *Polymer Yearbook*, vol. **18**, Ed. R. Pethrick and G. Zaikov, Rapra Technology, Shrewbury, Shropshire, UK, 2003 (in press).

74. Sanditov D.S. and Kozlov G.V., *Fizika I Khimia Stekla*, 1993, vol. **19**(4), pp. 593 - 601. (Rus)

75. Kalinichev E.L. and Sakovtseva M.B., *Properties And Processing Of Thermoplasts*, 1983, Leningrad, Khimia, 288 p. (Rus)
76. Balankin A.S., *Doklady AN*, 1992, vol. 325(3), pp. 465 - 471. (Rus)
77. Shishkin N.I., Milagin M.F., and Gabaraeva A.D., *Fizika Tverdogo Tela*, 1963, vol. 5(12), pp. 3453 - 3462. (Rus)
78. Preston J., In Coll.: *Ultrahigh Modular Polymers*, Ed. A. Chieferri and I.M. Ward, London, Applied Science publishers, 1977, 271 p.
79. Inoue N., Nokayama T., and Ariyama T., *J. Macromol. Sci.-Phys.*, 1981, vol. B19(3), pp. 543 - 567.
80. Kozlov G.V., Temiraev K.B., and Shustov G.B., *Plast. Massy*, 1998, No. 9, pp. 27 - 29. (Rus)
81. Aloev V.Z., Kozlov G.V., Dolbin I.V., and Beloshenko V.A., *Izv. KBNTs RAN*, 2001, No. 1(6), pp. 70 - 78. (Rus)
82. Askadsky A.A., Beloshenko V.A., and Pakter M.K., *Vysokomol. Soedin.*, 1991, vol. A33(10), pp. 2206 - 2211. (Rus)
83. Kozlov G.V., Beloshenko V.A., Varyukhin V.N., and Novikov V.U., *High Pressure Physics And Technology*, 1997, vol. 7(1), pp. 112 - 116. (Rus)
84. Beloshenko V.A., Pakter M.K., Beresnev B.I., Zaika T.P., Slobodina V.G., and Shepel' V.M., *Mekhanika Kompozitnyikh Materialov*, 1990, vol. 25(2), pp. 195 - 199. (Rus)
85. Kozlov G.V., Belousov V.N., Serdyuk V.D., and Kuznetsov E.N., *High Pressure Physics and Technology*, 1995, vol. 5(3), pp. 59 - 64. (Rus)
86. Balankin A.S., *Synergism Of Deformable Body*, 1991, Moscow, MO SSSR, 404 p. (Rus)
87. Moicya E.G., Semenovich G.M., and Lipatov Yu.S., *Vysokomol. Soedin.*, 1973, vol. A15(6), pp. 1337 - 1342. (Rus)
88. Kozlov G.V., Abaev A.M., Serdyuk V.D., Nagaeva D.A., Sanditov D.S., Beev A.A., and Oshpoeva R.Z., *Plast. Massy*, 1999, No. 11, pp. 27 - 28. (Rus)
89. Kozlov V.G., Abaev A.M., Novikov V.U., and Komalov A.S., *Materialovedenie*, 1997, No. 5, pp. 31 - 35. (Rus)
90. Kozlov V.G. and Mikitaev A.K., *Mekhanika Kompozitnyikh Materialov I Konstruktsii*, 1996, vol. 2(3-4), pp. 144 - 157. (Rus)
91. Kozlov G.V., Shustov G.B., and Dolbin I.V., *Eur. Conf. On Computational Physics*, Aachen, Germany, September 5-8, 2001, Abstr., vol. 8, p. B62.
92. Kozlov G.V., Yanovsky Yu.G., and Lipatov Yu.S., *Mekhanika Kompozitnyikh Materialov I Konstruktsii*, 2002. (Rus)
93. Vannimenus J., *Physica D.*, 1989, vol. 38(2), pp. 352 - 355.

94. Hentschel H.G.E. and Procaccia I., *Physica D.*, 1983, vol. **8**(3), pp. 435 - 445.

95. Halsey T.C., Jensen M.H., Kadanoff L.P., Procaccia I., and Shraiman B.I., *Phys. Rev. A.*, 1986, vol. **33**(2), pp. 1141 - 1151.

96. McCauley J.L., *Int. J. Modern Phys. B.*, 1989, vol. **3**(6), pp. 821 - 852.

97. Vstovsky G.V., Kolmakov A.G., and Terentiev V.F., *Metally*, 1993, No. 4, pp. 164 - 178. (Rus)

98. Vstovsky G.V., Bunin I.Zh., Kolmakov A.G., and Tanitovsky I.Yu., *Doklady AN*, 1995, vol. **343**(5), pp. 613 - 615. (Rus)

99. Kolmakov A.G., *Metally*, 1996, No. 6, pp. 37 - 43. (Rus)

100. Novikov V.U., Kozlov G.V., and Bilibin A.V., *Materialovedenie*, 1998, No. 10, pp. 14 - 19. (Rus)

101. Semenov B.I., Agibalov S.N., and Kolmakov A.G/. *Materialovedenie*, 1999, No. 5, pp. 25 - 31. (Rus)

102. Kozlov G.V., Dolbin I.V., and Lipatov Yu.S., In Coll.: *Proc. Internat. Symp. "Order, Disorder And Properties Of Oxides" ODRO-2001*, Sochi, pp. 174 - 180. (Rus)

103. Williford R.E.. *Scripta Metal.*, 1988, vol. **22**(11), pp. 1749 - 1754.

104. Novikov V.U., Kozlov G.V., and Bur'yan O.Yu., *Mekhanika Kompozitnyikh Materialov*, 2000, vol. **36**(1), pp. 3 - 32. (Rus)

105. Balankin A.S., Izotov A.D., and Lazarev I.B., *Neorganicheskie Materialy*,1993, vol. **29**(4), pp. 451 - 457. (Rus)

106. Cates M.E. and Witten T.A., *Phys. Rev. A.*, 1987, vol. **35**(4), pp. 1809 - 1824.

107. Kozlov G.V., Gazaev M.A., Novikov V.U., and Mikitaev A.K., *Pis'ma V ZhTF*, 1996, vol. **22**(16), pp. 31 - 38. (Rus)

108. Ishikawa K., *J. Mater. Sci. Lett.*, 1990, vol. **9**(4), pp. 400 - 402.

109. Beloshenko V.A., Kozlov G.V., and Lipatov Yu.S., *Fizika Tverdogo Tela*, 1994, vol. **36**(10), pp. 2903 - 2906. (Rus)

110. Hayakawa Y., Sato S., and Matsushita M., *Phys. Rev. A.*, 1987, vol. **36**(4), pp. 1963 - 1966.

111. Kozlov G.V. and Ovcharenko E.N., *Izv. KBNTs RAN*, 2001, No. 2(7), pp. 93 - 101. (Rus)

112. Novikov V.U., Kozitsky D.V., and Ivanova D.V., *Materialovedenie*, 1999, No. 8, pp. 12 - 16. (Rus)

113. Kozlov G.V. and Lipatov Yu.S., *Voprosy Khimii I Khimicheskoi Tekhnologii*, 2002,No. 3, pp. 65 - 67. (Rus)

114. Kubat J., Rigdahl M., and Welander M., *J. Appl. Polymer Sci.*, 1990, vol. **39**(5), pp. 1527 - 1539.

115. Schnell R., Stamm M., and Creton C., *Macromolecules*, 1998, vol. **31**(7), pp. 2284 - 2292.
116. Kozlov G.V., Ovcharenko E.N., and Lipatov Yu.S., *Doklady NAN Ukrainy*, 1999, No. 11, pp. 128 - 132. (Rus)
117. Mashukov Kh.Kh., Novikov V.U., Kozlov G.V., and Bur'yan O.Yu., *Materialovedenie*, 2000, No. 3, pp. 35 - 37. (Rus)
118. Kozlov G.V., Beloshenko V.A., Novikov V.U., and Lipatov Yu.S., *Ukrainsky Khimichesky Zhurnal*, 2001, vol. **67**(3), pp. 57 - 60. (Rus)
119. Kozlov G.V., Yanovsky Yu.G., and Lipatov Yu.S., *Mekhanika Kompozitnyikh Materialov I Konstruktsii*, 2002, vol. **8**. (Rus)
120. Borukaev T.A., Mashukov N.I., Kozlov G.V., Mikitaev A.K., and Zaikov G.E., *Vestnik KBGU, Ser. Khimicheskie Nauki*, 2001, Iss. 4, pp. 101 - 104. (Rus)
121. Kozlov G.V., Mashukov N.I., Zaikov G.E., Mikitaev A.K., and Borukaev T.A., In Coll.: *Fractals And Local Order In Polymeric Materials*, Ed. G. Kozlov and G. Zaikov, New York, Nova Science Publishers, Inc., 2001, pp. 43 - 53.
122. Kozlov G.V., Dolbin I.V., Shustov G.B., and Zaikov G.E., *Proc. Intern. Conf. "Natural Science At The Turn Of The Century"*, vol. **1**, Dagomys, 2001, pp. 36 – 37. (Rus)
123. Sahimi M., McKarnin M., Nordahl T., and Tirrell M., *Phys. Rev. A.*, 1985, vol. **32**(1), pp. 590 - 595.
124. Afaunov V.V., Kozlov G.V., Mashukov N.I., and Zaikov G.E., *Zhurnal Prikladnoi Khimii*, 2000, vol. **73**(1), pp. 136 – 140. (Rus)
125. Boyd R.H., *Polymer*, 1985, vol. **26**(3), pp. 323 - 347.
126. Bernstein V.A., Egorov V.M., Marikhin V.A., and Myaskikova L.P., *Vysokomol. Soedin.*, 1990, vol. **A32**(12), pp. 2380 - 2387. (Rus)
127. Klymko P.W. and Kopelman R., *J. Phys. Chem.*, 1983, vol. **87**(23), pp. 4565 - 4567.
128. Kopelman R., In Coll.: *Fractals In Physics*, Ed. L. Pietronero and E. Tozatti, 1988, Moscow, Mir, pp. 524 - 527. (Rus)
129. Kozlov G.V., Dolbin I.V., and Zaikov G.E., In Coll.: *Aging Of Polymers, Polymer Blends And Polymer Composites*, vol. **1**, Ed. G. Zaikov, A. Bouchachenko, and V. Ivanov, New York, Nova Science Publishers, Inc., 2002, pp. 145 - 150.
130. Afaunov V.V., Shogenov V.N., Mashukov N.I., and Kozlov G.V., *Doklady Adyigeiskoi (Cherkesskoi) Mezhdunarodnoi AN*, 2000, vol. **5**(1), pp. 100 - 104. (Rus)

131.   Shogenov V.N., Gazaev M.A., and Kardanova M.Sh., In Coll.: *Polycondensation Processes And Polymers*, Ed. V.V. Korshak, 1987, Nalchik, KBGU, pp. 8 - 12. (Rus)

Chapter 7.

Description of polymer properties in the framework of the cluster model

---

7.1. MOLECULAR-WEIGHT CHARACTERISTICS

Molecular-weight characteristics are specific parameters of polymers and significantly affect practically all properties of them [1]. As shown in Chapter 2, injection of highly dispersed Fe/FeO mixture into HDPE induces an extreme change in properties of the initial polymer (for example, blow viscosity, melt viscosity, etc.). The same changes are observed for HDPE molecular weight increase. In accordance with the data by Union Garbide Company (USA), the mean molecular weight of polyethylenes, $\overline{M}_w$, can be estimated by the following empirical correlation [2]:

$$\lg \overline{M}_w = \lg(129{,}000) - 0.263 \cdot \lg MFI_{21.6}^{190},\qquad (7.1)$$

where $MFI_{21.6}^{190}$ is the melt flow index, measured at 190°C and load of 21.6 N.

For HDPE, dependence of $\overline{M}_w$ estimated by the equation (7.1) on Fe/FeO mixture concentration (C_Z) possesses a maximum at $C_Z = 0.05$ wt.%, when $\overline{M}_w$ is greater than the appropriate value for the initial HDPE by 25%, approximately (Table 7.1). In accordance with these data, injection of Fe/FeO mixture into HDPE causes an extreme increase of the polymer molecular weight. Hence, mechanical characteristics change in the same manner, as they would do with $\overline{M}_w$ increase [3]. For example, Table 7.1 shows dependence of blow viscosity A_p on C_Z, which possesses a maximum completely corresponded to the maximum of $\overline{M}_w(C_Z)$ dependence. The equation shown below presents correlation between viscosity at blow stretch of HDPE, a_p, and MFI:

$$\lg a_p = 1.492 - 0.22 \cdot \lg MFI_{21.6}^{190}.$$ (7.2)

Table 7.1

Molecular and mechanical characteristics of HDPE + Z composites [3]

C_Z, wt.%	$\overline{M}_w \times 10^{-5}$	A_p, kJ/m²	A_p, kJ/m²
0	1.56	19.5	34.5
0.01	1.48	15.8	34.1
0.03	1.70	23.5	39.5
0.05	1.93	37.2	44.8
0.07	1.76	29.8	40.5
0.10	1.45	11.7	34.0
0.15	1.44	12.5	33.9
0.20	1.40	14.8	33.5
0.50	1.47	13.4	34.0
1.0	1.42	19.2	33.5

Dependence of a_p on C_Z calculated by the equation (7.2) (refer to Table 7.1) is also analogous to $\overline{M}_w(C_Z)$ one [3].

However, the data obtained do not allow an assumption that injection of the Fe/FeO mixture (which is Z) into HDPE does really cause $\overline{M}_w$ increase. This is clearly demonstrated by the experiment as follows [4]. After treatment of HDPE + Z composite possessing C_Z = 0.01 – 0.15 wt.% in boiling p-xylene during 5 hours, MFI and consequently $\overline{M}_w$ were reduced to practically initial values. The processing modes (temperature and duration) of HDPE composites in p-xylene were selected so that van-der-Waals bonds were broken only, and covalent ones were preserved. This is confirmed by equality of MFI values for HDPE composites, initial and processed in p-xylene. As a consequence, one can suggest that injection of Fe/FeO mixture into HDPE does not change its molecular weight. The extreme increase of $\overline{M}_w$ can be explained by an increase of the entanglement cluster network frequency, which is initiated by local magnetic field of Z particles. As shown [5], polymer melt viscosity at zero shear η_0 is determined from the equation (2.61). Quite good approximation of η_0 can be obtained using MFI, because MFI measuring technique is not associated with high shear stresses. For this purpose, the reverse MFI value is more suitable, because its increase is corresponded to η_0 raise. Hence, parameter MFI^{-1} reflects the integral change of both polymer molecular weight and macromolecular entanglement network frequency [3].

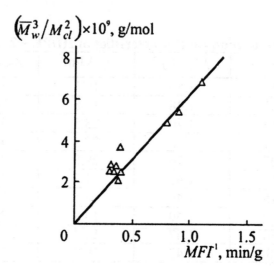

Figure 7.1. Correlation between reverse MFI and $\overline{M}_w^3/M_{cl}^2$ ratio for HDPE + Z composites [3]

Data in Figure 7.1 show approximation of the correlation between MFI^1 and $\overline{M}_w^3/M_{cl}^2$ by a straight line passing through the origin of coordinates that proves the above assumption of $\overline{M}_w$ constancy and M_{cl} influence on the melt viscosity of HDPE + Z composites. Bsed on these data, the following conclusions can be made. Firstly, assessment of $\overline{M}_w$ using equation (7.1), as well as any other empirical correlation, shall be made very carefully. As shown by the data obtained for practical purposes, $\overline{M}_w$ can be calculated by the equation (7.1) for assessment of the polymer quality (refer to Table 7.1), because MFI measurement is simple and can be executed in a short time. Secondly, the dependence $\overline{M}_w^3/M_{cl}^2 (MFI^1)$ allows determination of M_{cl} by MFI measurement results only. Such technique is the most simple and reliable for unmodified HDPE, because in this case equation (7.1) gives quite correct $\overline{M}_w$ values. For modified HDPE, $\overline{M}_w$ equal that of the initial polymer is accepted and, vice versa, the correlation $\overline{M}_w^3 / M_{cl}^2$ (MFI^1) can be used for $\overline{M}_w$ estimation, if M_{cl} determined by an independent method is known [3].

7.2. DISSOLUTION HEAT

Studies on dissolution heats of amorphous polymers were performed quite frequently [5, 6]. Presumably, the dominating contribution into the dissolution heat, ΔH_d, is made by the heat effect of system transition from a metastable (non-equilibrium) glassy-like state to a quasi-equilibrium solution [6]. Hence, substantial contribution can be made by a component, determined by the polymer structure, by the presence of oriented zones [5], in particular. Such interpretation allows application of the cluster model ideas [7] to description of polymer dilution processes.

Of special attention are several general features of ΔH_d behavior [5]. Firstly, above T_g, ΔH_d is independent of the dissolution temperature, T_d. Presumably in the framework of the cluster model, local order zones (clusters) are thermofluctuationally decomposed at T_g, i.e. it should be expected that variation in ΔH_d behavior at T_g is regulated directly by this effect. Secondly, T_g increase induces ΔH_d increase at fixed T_d, and in the framework of the cluster model polymers with higher T_g possess higher ν_{cl} values at fixed test temperature (refer to Figure 5.29). At third, dissolution temperature rise leads to ΔH_d decrease and, as suggested in the cluster model, ν_{cl} decreases simultaneously. Of the greatest interest is linear dependence of ΔH_d on the temperature difference $(T_g - T_d)$. For different polymers, such dependence [5] suggests ΔH_d independence of the structural features of amorphous polymers and its determination by approach to T_g only. For three amorphous polymers studied (PMMA, PC and PAr) [7], ν_{cl} dependencies on $(T_g - T_d)$ displayed similar shapes, but differed by absolute ν_{cl} values. Much more general characteristic of the local ordering degree of the amorphous polymer structure is the relative part of clusters, φ_{cl}, determined by equation (2.11). Assessment of φ_{cl} for PMMA shall be discussed in more detail. Owing to the presence of bulky side groups this polymer possesses high cross-section of the macromolecule ($S \approx 64$ Å^2 [8]), which gives unreal values $\varphi_{cl} > 1$ determined by the equation (2.11) for this polymer. Obviously, side groups cannot enter closely packed clusters: this is possible for backbone segments only. Low values of ν_{cl} for

PMMA rather than for PC and PAr are stipulated by steric hindrances to cluster formation, caused by the presence of bulky side groups. That is why S value for PMMA backbone was used, estimated from dependence of the Bragg interval (refer to equation (4.1)) on the size of side groups, and then dependence of S on the value of this interval [7].

For studied polymers, all data in $\varphi_{cl} - (T_g - T)$ coordinates fit the same line (Figure 5.29). As a consequence, similar to ΔH_d, the local ordering degree in polymers is determined by T closeness to T_g. This allows an assumption of correlation between ΔH_d and v_{cl} or φ_{cl}. Note that similarly shaped dependencies of polymer properties on $(T_g - T)$ are frequently met in the literature. The example is presented by dependencies of a series of mechanical properties on $(T_g - T)$ for PMMA, PC and PAr [9].

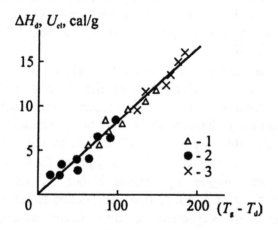

Figure 7.2. Dependencies of dissolution heat, ΔH_d, and cluster dissociation energy, U_{cl}, on temperature difference $(T_g - T_d)$ for PMMA (1), PC (2) and PAr (3) [7]. Straight line presents dependence ΔH_d (T_g and T_d) in accordance with the data from [5]

If segments in the cluster are interpreted as linear defects (the analogue of dislocations), one can apply mathematical apparatus of the dislocation theory to description of amorphous polymers' structure and properties (refer to Chapter 3). For example, the dissociation energy per segment, U_{cl}, is given by the equation (3.19), and dissociation energy of closely packed segments, U_{dis}, per specific polymer volume can be presented as follows [7]:

$$U_{dis} = U_{cl}v_{cl}. \tag{7.3}$$

For three studied polymers, comparison of U_{dis} and ΔH_d values as functions of $(T_g - T_d)$ shows their close quantitative correspondence (Figure 7.2). This allows the statement that ΔH_d represents the energy required for dissociation (decomposition) of clusters during polymer dissolution. The latter event means transition of the polymer from metastable non-equilibrium state to a quasi-equilibrium solution [7].

7.3. HEAT EXPANSION

Besides obtaining necessary technical characteristics, study of the thermal expansion of polymers gives interesting physical information [10]. This is associated with the chain structure of polymer molecules, which lead to a strong difference between intra- and intermolecular interaction that induces strong anisotropy of heat dynamics of the polymer macromolecule relating to its axis [11, 12]. The features of polymer crystallite lattice expansion (strong transversal expansion at longitudinal contraction) are quite thoroughly studied by X-ray diffraction methods [10]. For amorphous polymers, due to absence of sharp X-ray reflexes in them, detection of temperature behavior of molecular aggregations in the polymer volume is more complicated problem [13].

The difference between micro- and macro-expansion in amorphous polymers was shown and explained by definite ordering of the chain macromolecules [13]. To put it differently, for a series of amorphous polymers, an interconnection between thermal expansion and supermolecular structure was observed in this work [13]. This allows application of the cluster model to description of the mentioned interconnection [14].

The study of wide-angle X-ray diffraction patterns for amine (EP-1) and anhydride (EP-2) cross-linked epoxy polymers has indicated systematic shift of the center of gravity of amorphous halo along the angle θ with K_{st} variation that means corresponded change in the Bragg interval, d_B, varied within the range of $4.48 - 4.82$ Å [15]. From both physical meaning of halo for amorphous polymer [16] and d_B values a conclusion can be made that distances between axial lines of neighbor approximately parallel segments of macromolecules are registered, i.e. interchain distances D_{ic}, which can be estimated by the known Keezom formula (equation (4.1)). In accordance with the technique [13] micro-expansion values, $\alpha_{\perp}^m$, in transversal direction to the macromolecule axis, for

EP-1 system this value being counted off the state at K_{st} = 1.0, and for EP-2 system – from the state at K_{st} = 1.25, where d_B values were minimal.

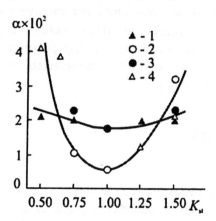

Figure 7.3. Dependencies of linear macro-, α, and micro-expansion, $\alpha_\perp^m$, determined by dilatation (1, 2) and X-ray diffraction (3, 4) methods, respectively, on K_{st} for epoxy –olymers EP-1 (1, 3) and EP-2 (2, 4) [14]

Figure 7.3 shows dependencies of linear macro-expansion α and $\alpha_\perp^m$ on K_{st} for both epoxy systems studied. Similar to amorphous polymers from the work [13], much stronger variation of $\alpha_\perp^m$ rather than α is observed, though tendencies of dependencies shown in Figure 7.3 are analogous. This difference can be explained in the framework of the cluster model, which is analogous to the one used by the authors in [13], except for the type of local order zones. Slutsker and Filippov [13] have assumed that these zones represent the analogue of crystallite with folded chains, whereas clusters represent amorphous analogue of CEC (Figure 5.3). Following the authors' reasoning [13] one can conclude that the cluster thermal expansion will be strongly anisotropic, and it can be described using the following equation [13]:

$$3\alpha = 2\alpha_\perp^m + 2\alpha_\parallel^m,$$ (7.4)

where $\alpha_\parallel^m$ is the micro-expansion parallel to the macromolecule axis.

Table 7.2

Structural and heat-physical characteristics of epoxy polymers [14]

Epoxy polymer	K_{st}	$\alpha_\perp^m \times 10^2$	$\alpha_\parallel^m \times 10^2$	$\alpha \times 10^2$
EP-1	0.50	4.15	− 2.03	2.09
	0.75	3.94	− 2.09	1.93
	1.25	1.04	3.83	1.97
	1.50	2.07	1.83	1.99
EP-2	0.75	1.08	2.81	2.23
	1.0	0.43	4.30	1.72
	1.50	3.03	0.24	2.10

Equation (7.4) allows calculation of $\alpha_\parallel^m$ and all obtained values of characteristics of macro- and micro-expansion of epoxy polymers are shown in Table 7.2. The data in the Table indicate that absolute values of α, $\alpha_\perp^m$ and $\alpha_\parallel^m$ are close, and tendencies of variation – to analogous values for amorphous linear polymers, shown in the work [13]. However, a significant difference, concluded in the possibility for $\alpha_\parallel^m$ to be both positive and negative, is also observed. One of the factors determining negative values of $\alpha_\parallel^m$, indicated in the work [13], is the possibility of conformational *trans-gauche*-transitions, additionally strengthening longitudinal contraction of the macromolecule. Obviously for studied epoxy polymers, positive sign and high absolute values of $\alpha_\parallel^m$ suppose high probability of reverse (*gauche-trans*) transitions at K_{st} change [14].

Previously, it has been shown [17] that the entanglement cluster network frequency increases with the part of stretched *trans*-conformations. On this basis one may assume that $\alpha_\parallel^m$ increase that means *gauche-trans* transition probability rise must also increase v_{cl}. Figure 7.4 shows the dependence $v_{cl}(\alpha_\parallel^m)$ for studied epoxy systems, which despite a definite dispersion of the results, indicates v_{cl} increase with $\alpha_\parallel^m$. This proves the above-made assumption.

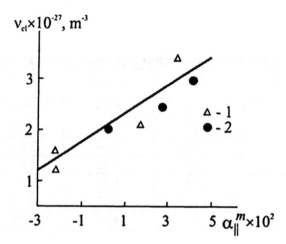

Figure 7.4. Dependence of the entanglement cluster network frequency, v_{cl}, on micro-expansion, $\alpha_{\parallel}^m$, for EP-1 (1) and EP-2 (2) [14]

For inextensible thread, which simulates a part of a macromolecule, under the assumption of sinusoidal oscillations, wavelength λ_k can be presented as follows [13]:

$$\lambda_k \cong \pi d_B \sqrt{\frac{\alpha_{\perp}^m}{|\alpha^m|}},$$ (7.5)

where α^m is total micro-expansion of the polymer.

Wavelengths of oscillations obtained can be approximately compared with the longitudinal length of the local order zone [13], which in the framework of the cluster model equals the length of the statistical segment, l_{st} (equation (2.8)). Figure 7.5 compares λ_k and l_{st}, which indicates both identical tendencies of their variation and satisfactory conformity of absolute values. A deviation between λ_k and l_{st} in the area of their high values is probably caused by approximate type of executed calculations [13]. Nevertheless, data in Figures 7.4 and 7.5 assume that quantitative parameters of the cluster model (cluster network frequency, v_{cl}, and cluster longitudinal size, l_{st}) can be used for assessment of the heat expansion factor, which is one of the most important characteristic of polymer.

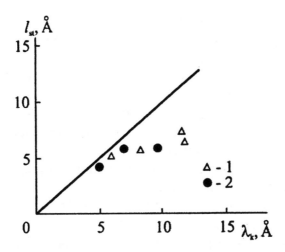

Figure 7.5. Correlation between the wavelength, λ_k, and the statistical segment length, l_{st}, for epoxy polymers EP-1 (1) and EP-2 (2). Straight line presents 1:1 ratio [14]

Thus the results discussed in this Section confirm the interconnection between heat expansion and supermolecular structure of the polymers. Cluster model can be used for quantitative estimation of micro- and macro-expansion of amorphous polymers [14].

7.4. GLASS TRANSITION TEMPERATURE

Highly demonstrative notion of the connection between the structure and the glass transition temperature of polymers can be obtained on the example of aged epoxy systems. At present, it is accepted [18] that for polymer networks, the increase of chemical cross-link network frequency (v_c) induces glass transition temperature rise. The presence of such dependence is associated with the molecular chain mobility limitation by chemical cross-link points [18, 19]. This conclusion is supported by a series of experimental results displaying symbate change of T_g and v_c [19]. At the same time, of special attention are the following facts. Data from the work [20] show that at $T < T_g$ heat aging of epoxy polymers increases T_g remaining v_c unchanged. The mentioned circumstances allow an assumption that there is no direct connection between T_g and v_c, but an indirect one may be realized via supermolecular structural organization of polymer networks [21].

Two series of anhydride cross-linked epoxy polymer (EP-2) were studied [21]: one directly after preparation and another after natural aging under atmospheric conditions during three years. Table 7.3 shows T_g, v_c for these epoxy polymers. Clearly for the majority of samples, v_c decreases and v_{cl} increases with aging, especially in the case of high deviation of the epoxy system composition from stoichiometric relation (refer to Figure 6.15). Thus as shown in Section 6.4, the observed variation of the polymer structure in the framework of the cluster model proceeding during heat aging can be interpreted as an increase of the local order, which is characterized by v_{cl}.

For aged epoxy polymer, T_g is higher than for the initial one (refer to Table 7.3). Analysis of data shown in the Table indicates symbate variation of v_{cl} and T_g during aging, whereas such relation between v_c and T_g is absent. As a consequence, T_g does not display direct dependence on v_c, but is determined by supersegmental structure conditions, characterized by v_{cl} value.

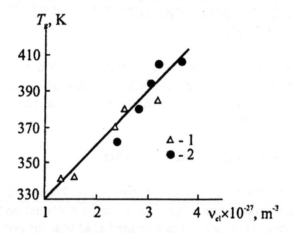

Figure 7.6. Dependence of the glass transition temperature, T_g, on the entanglement cluster network frequency, v_{cl}, for epoxy polymers of anhydride cross-linking type: 1 – initial samples; 2 – aged samples [21]

Figure 7.6 shows dependencies $T_g(v_{cl})$ for both series of EP-2. They are linear and fit the same line that allows consideration of the polymer network devitrification as thermofluctuational decay of the local order "frozen" below T_g [22]. As a consequence, the results obtained [21] testify about determination of behavior of the polymer network glass transition temperature by frequency variation of macromolecular entanglement cluster network [21].

Table 7.3

Comparison of EP-2 epoxy polymer characteristics before and after aging [21]

K_{st}	T_g, K	$v_c \times 10^{-26}$, m^{-3}	$v_{cl} \times 10^{-27}$, m^{-3}
0.50	342/363	4/3	1.3/2.4
0.75	372/383	10/5.3	2.3/2.8
1.0	399/408	11/7.4	3.3/3.5
1.25	378/398	10/11.7	2.4/3.0
1.50	343/408	8/10.4	1.6/3.2

Note: First values correspond to initial and second values to aged samples, respectively.

Presently it is found for amorphous polymers [23] that elasticity modulus E and glass transition temperature T_g vary symbately, because they are determined by symbately changing parameters: cohesion energy and chain rigidity, respectively [24]. However, polymer networks, epoxy ones in particular, may display different (including antibate) type of E and T_g behavior. For example, it has been shown [25 – 27] that as concentration of cross-link points in chemical network is changed due to K_{st} variation, the elasticity modulus at compression and longitudinal ultrasonic wave rate change antibately to T_g.

This effect is explained [24] using the frequency fluctuation density, $\psi(\infty)$. For a series of polymers, assessment of $\psi(\infty)$ from the existing data from the literature [28] testifies about its approximate constancy at both T_{ll} and T_g expectedly due to the correlation (2.26). This allows transformation of the equation (2.37) with respect to $\chi^{-1} = K_t$ (where χ is isothermal compressibility; K_t is isothermal bulk compression (dilatation) modulus) to the following form [24]:

$$\psi_{cr} = \frac{\rho k T_g}{K_t}, \qquad (7.6)$$

where ψ_{cr} is the critical value of $\psi(\infty)$ at T_g.

Parameters K_t and E are associated with one another by the well-known correlation [23] as follows:

$$K_t = \frac{E}{3(1-2v)}. \tag{7.7}$$

Since for $v \cong 0.33$ typical of glassy-like polymers, an approximate equality $K_t \cong E$ is true [23], it follows from the equation (7.6) that the criterion of the constant ψ_{cr} (refer to Section 2.4 for detail) cannot be applied to the glass transition of epoxy polymers studied before [25 – 27]: E decrease shall induce T_g decrease which is inconsistent with the experimental data. To explain this seeming contradiction an assumption was made [24] that chemical cross-link points of the network limit fluctuations of the cluster segments thus decreasing the frequency of fluctuations but preserving the basic postulate of the cluster model telling that polymer devitrification is determined by decomposition of "frozen" local order. Similar ideas have been developed by Flory [29] suggesting that physical entanglements in rubbers, considered by many authors as a "short living" local order, limit fluctuations of chemical cross-link points. As mentioned above, direct connection between v_c and T_g seems low probable, because glass transition is the critical event, at which all properties of the polymer suffer serious changes, whereas v_c parameter at overcoming T_g does not change. As before, this is the reason for the assuming the chemical cross-linking effects on formation of the physical supersegmental structure characterized by v_{cl} in polymers and, subsequently, T_g value [24].

Analytically, the effect of v_c on ψ_{cr} can be hardly expressed. However, one can assume that limitations of ψ_{cr} become stricter with v_c increase, i.e. ψ_{cr} decreases. These limitations are considered empirically [24] (in analogue to considerations by Flory [29]), namely, by introduction of Cv_c^n factor to the right part of the equation (7.6), where C and n are constants. In this case, with respect to correlation (7.7) the equation (7.6) for cross-linked systems is reduced to the form as follows [24]:

$$\psi_{cr} = \frac{3\rho k T_g(1-2v)}{ECv_c^n}. \tag{7.8}$$

Applying several simple assumptions, discussed in [24], equation (7.8) can be transformed as follows:

$$T_g \cong 4.45 \times 10^{-21} K_t \nu_c^{1/2}, \qquad (7.9)$$

where T_g is given in Kelvin degrees, K_t in Pa, and ν_c in m^{-3}.

Figure 7.7 shows $T_g(K_{st})$ dependencies, obtained experimentally [26] and calculated by equation (7.9), for epoxy polymers of amine and anhydride curing type (EP-1 and EP-2 systems, respectively). Generally, good conformity was observed for these dependencies. Deviations observed for EP-2 system are probably associated with approximations applied.

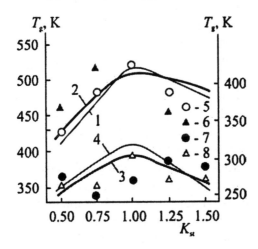

Figure 7.7. Dependencies of the glass transition temperature, T_g, on the curing agent:oligomer ratio, K_{st}: experimental (1 – 4) and calculated by equation (7.9) (5 – 8) for EP-1 (1, 2, 5, 6) and EP-2 (3, 4, 7, 8), cured under hydrostatic pressure of 0.1 MPa (1, 3, 5, 7) and 200 MPa (2, 4, 6, 8) [24]

As a consequence, the presence of chemical cross-link network both affects the local order degree (as mentioned above [30], ν_c parameter) and limits heat fluctuations of segments in the clusters. This effect is sufficient for formation of the cross-link system properties, in particular, allows explanation of observed antibate variation of E and T_g [24].

Let us now consider interpretation of characteristic temperatures for amorphous-crystalline polymers in the framework of the cluster model. For recent 40 years, glass transition (and the glass transition temperature T_g) in amorphous-crystalline polymers is the problem under discussion. This problem is not clear still, though a great number of works devoted to it has been performed. This can be easily shown on the example of well-studied polyethylenes (PE), which are typical illustration of the prevailing situation. In the reference literature T_g of polyethylenes usually equals ~250 K (for example [28]), whereas in the scientific literature T_g varies significantly. For example, Bartenev, Shut and Kaspersky [31] consider that correct T_g value for PE falls within the range of 183 – 193 K. For explanation of this problem, the authors in [32] have used the Brillouin spectroscopy method. The spectrum was obtained for the temperature range of PE possessing density 0.919 g/cm^3, crystallinity degree 0.26, and the melting temperature T_{melt} = 379 K. Temperature dependencies of the Brillouin frequency shear and the Brillouin band width possess a knee corresponded to 201 – 204 K. On this basis, the most probable value of T_g suggested for PE [1] equals 201 K. Stehling and Mandelkern [33] have studied relaxation transitions in linear PE by heat expansion, calorimetry and dynamic mechanical spectroscopy methods. The assume that for PE, T_g equals 143 K. This schedule can be significantly broadened. Brief review of the problem prehistory has been given by Popli *et al.* [34]. Note also that the most modern experimental techniques [31 – 38] were applied to solve this problem. That is why the mentioned readings are hardly subjective.

Briefly, the essence of the problem is as follows. Besides the melting transition, polyethylenes (both linear and branched) display several more relaxation transitions. Usually, three main temperature zones of such transitions are observed, shown below in accordance with the common classification [34]: γ-transition in the temperature range of 123 - 173 K, β-transition in the temperature range of 243 - 283 K, and α-transition in the temperature range of 303 - 393 K. Considering the origin of the latter transition, one can observe the following similarity in the points of view of different authors: it is stipulated by mobility of chain units present in crystalline zones of polyethylenes. Positively, the basic reason for the benefit of such conclusion is a correlation between α-transition intensity and crystallinity degree of the polymer [34]. Moreover, the dependence between α-transition temperature and thickness of crystallites has been determined [34].

Transitions of γ and β types are much more ambiguous in the scientific literature. There appears a question, to which transition the glass transition temperature must be belonged. Different points of view have defined the above-mentioned broad dispersion of T_g values. Leaving aside the discussion details, note that sufficient proofs of T_g belonging to β-transition are shown in the work [35], and the work [34] presents reasons for its belonging to γ-transition. Each of these points of view possesses many advantages and disadvantages. In this connection, note that the basic (and the most objective) factor determining accurate identification of T_g for amorphous-crystalline polymers is complexity of their entire composition and amorphous state structure, in particular. Essentially, the glass transition temperature belongs to some hypothetic, completely amorphous polymer, which is not yet obtained and the structure of which is quite different from the amorphous phase of real amorphous-crystalline polymer [39]. The existence of amorphous and crystalline phases combined with transitional (interphase) zones indicates the multiplicity of transitions, which makes the problem much more complicated. That is why proofs of one or another point of view are usually used as indirect arguments [40].

Since the traditional way of solving this problem reached a deadlock, there appeared a necessity in alternative interpretation taking into account specificity of the structure of amorphous-crystalline polymers, more particularly, the presence of the crystalline phase in them and their effect on the amorphous zone structure [39]. Perhaps, Boyer [41] was the first who paid attention to this point. He has indicated that amorphous-crystalline polymers do not possess clearly determined T_g and has shown two glass transition temperatures (243 and 193 K) for crystalline and amorphous polyethylenes, respectively. Much more clear opinion was presented by Kargin, Andrianova and Kardash [42]. They have indicated for amorphous-crystalline oriented polymers extremely broad (compared with amorphous polymers) distribution of glass transition temperatures embracing tens of degrees, which is probably associated with their structure. Specification and concretization of structural factors participating in this distribution were given in the works [38, 43].

Egorov and Zhizhenkov [43] have studied features of the glass transition of amorphous interlayers in oriented flexible-chain polymers, mainly polycaproamide (PCA), basing on the well-known structural model of amorphous-crystalline polymers [38, 44] that suggests the presence of chains with different strain in amorphous interlayers. Figure 7.8 shows plots of the mobile fraction part, C_m (the part of devitrified zones), with temperature for PCA fibers, unloaded and externally quasi-elastically stretched by 15% (the stretch rating $\lambda = 5.2$). For unloaded sample at temperature below 350 K, the narrow component in the NMR spectrum is absent: all amorphous zones are glass transited. At 350 K, the narrow component appears in the spectrum. It indicates the beginning of micro-Brownian motion. For amorphous-crystalline polymers, this temperature with respect to the time factor of the method is usually accepted for the glass transition temperature, T_g. (Note that for non-oriented polyethylene, this temperature falls within the range of 150 – 170 K

[38]). But as follows from data in Figure 7.8, at T_g just a small part of amorphous zones is devitrified. Devitrification proceeds with temperature increase, and all amorphous zones transit into the rubbery state and the condition $C_m = \alpha_{am}$ (α_{am} is the part of the amorphous phase) is fulfilled at 420 K only. Growth of C_m observed above 445 K is associated with the beginning of crystallites' destruction. As a consequence, the real local glass transition temperature for PCA falls within the range of 350 – 420 K. Similar phenomenon can also be observed for other polymers [43]. Thus devitrification, i.e. transition of zones from glassy to rubbery state, shall be characterized by two temperature parameters: macroscopic glass transition temperature, T_g, and true local glass transition temperature, determined from the condition as follows: if at temperature increase by ΔT (Figure 7.8) mobile fraction is increased by ΔC_m, this means that for the part of amorphous zones equal C_m local glass transition temperature falls within the range of $T - (T + \Delta T)$ [43].

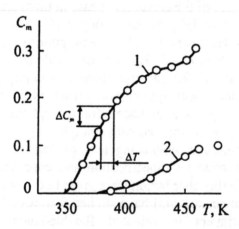

Figure 7.8. Temperature dependence of the part of non-glass transited zones, C_m, for PC fibers: 1 – initial sample; 2 – sample quasi-elastically stretched by 15% [43]

Stretching of samples (characterized by quasi-elastic deformation, ε) also causes the glass transition effect. Macroscopic glass transition temperature increases with ε (Figure 7.9, curve 2). At fixed temperature above T_g the part of devitrified amorphous zones is deceased with deformation growth and mechanical glass transition occurs (Figure 7.9, curve 1). Mechanical effects are reversible: after load removal and sample contraction molecular mobility is restores.

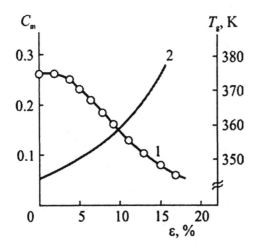

Figure 7.9. Mechanical glass transition of PCA: 1 – decrease of the part of non-vitrified zones C_m at sample stretching; 2 – increase of macroscopic glass transition temperature T_g at stretching (ε is quasi-elastic deformation) [43]

Obviously at temperatures above T_g the polymer can be partly glassy still. The value of C_m characterizes the aggregate state of amorphous zones more selectively rather than macroscopic glass transition temperature T_g [43].

Analogous interpretation of the glass transition temperature for amorphous-crystalline polymers can be obtained in the framework of the cluster model [40]. In amorphous glassy-like polymers the "frozen" local order below T_g is stipulated by a sharp viscosity increase and, as a consequence, corresponded decrease of molecular mobility, and for polymers of the polyethylene type at about room temperature existence of the local order is stipulated by absolutely different reasons (refer to Section 2.5 for detail). It has been assumed before that at low deformations (of about 1%) much less rigid devitrified amorphous zones will be deformed first (and consequently, determine elasticity modulus E). It was also considered obvious due to the rubbery state of amorphous zones that E value can be calculated in the framework of rubbery concepts. However, assessments have indicated that E values calculated in this manner are by two orders of magnitude smaller than experimentally determined ones. Krigbaum *et al.* [45] have shown that this deviation is stipulated by chain strain in amorphous interlayers during crystallization. In its turn, this strain induces formation of local order zones (clusters) by analogy with formation of the liquid-crystal order [46]. Thus the local order formation is also stipulated by "freezing up" of the molecular mobility due to mechanical chain strain (refer to Figure 2.32).

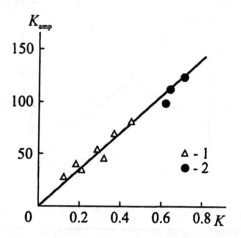

Figure 7.10. Dependence of amplification coefficient K_{amp} on crystallinity degree K for HDPE (1) and LDPE (2) [40]

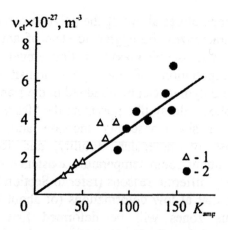

Figure 7.11. Dependence of entanglement cluster network frequency v_{cl} on amplification coefficient K_{amp} for HDPE (1) and LDPE (2) [40]

If the coefficient K_{amp} is defined as the relation of elasticity modules, experimentally measured and calculated by equations of the rubber elastisity, and then its dependence on the crystallinity degree K is plotted (Figure 7.10), one can observe that $K_{amp}(K)$ correlation is linear and passes through the origin of coordinates [17]. To put it differently, the chain strain and, as a consequence, the amorphous phase local ordering are fully determined by the presence of the crystalline phase. Temperature increase induces surface melting of crystallites [47, 48] and decreases the local ordering degree (Figure 7.11).

Obviously, in such interpretation the part of mobile fraction C_m is completely identical to the relative part of packless matrix, $\varphi_{p.m.}$ Figure 7.12 compares C_m values, accepted from [49], and $\varphi_{p.m.}$ for two polyethylenes and indicates their good conformity.

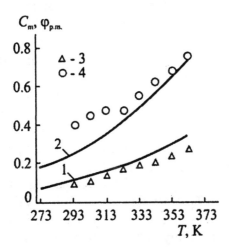

Figure 7.12. Temperature dependencies of relative parts of "mobile fraction" C_m (by data from [49]) (1, 2) and packless matrix $\varphi_{p.m.}$ (3, 4) for HDPE (1, 3) and LDPE (2, 4) [40]

Summing up the above-discussed data one may suggest the following model of the amorphous phase devitrification in amorphous-crystalline polymers. At some temperature T_g (as mentioned above, for polyethylenes T_g = 150 – 170 K [38]) devitrification of amorphous zones is initiated, expressed by occurrence of relatively small part of the amorphous phase with macromolecules capable of micro-Brownian motion (quantitatively determined by C_m or $\varphi_{p.m.}$ values). However, a great part of this phase consists of local order zones (clusters), the presence of which is stipulated by straining intermediate chains during crystallization. As temperature increases the chain strain decreases that induces a decline of the local order and appropriate increase of C_m or $\varphi_{p.m.}$ (Figure 7.12). As T_{melt} is approached, decrease of crystallinity degree due to surface melting of crystallites and relieve of chain sliding through them lead to the cluster decomposition, and the amorphous phase devitrifies (rubberizes). Termination temperature of this process coincides with the melting point of the polymer. Thus similar to amorphous glassy-like polymers, devitrification in amorphous-crystalline ones is associated with decomposition of the "frozen" local order (cluster decaying). The entire devitrification process is spread upon a very broad temperature interval: for example, from 150 – 170 K to 400 – 410 K for polyethylenes (~250 K).

It should be noted that behavior of the glass transition process in amorphous-crystalline polymers is in many ways analogous to melting behavior in amorphous polymers, where the unique T_{melt} is absent, and melting itself is spread upon a definite temperature interval. It is suggested [40] that this difference is fundamental and stipulated by strictly antipodal "ideal" structures of the mentioned types of polymers (Chapter 3). For crystallizing polymers the "ideal" structure is a defect-free monocrystal, and for amorphous polymers this role is played by chaotic structure of interpenetrating macromolecular coils (the "felt" model), in which the short-range (local) order and *a fortiori* the long-range one are absent. For the former "ideal" structure characteristic temperature is T_{melt}, and for the latter – T_g. The deviation degree of a real structure from the "ideal" one determines the width of the temperature interval for processes characterized by T_{melt} or T_g. Note also that such intervals do exist in any case: for both T_{melt} in amorphous-crystalline polymers and T_g in amorphous ones by virtue of practical inaccessibility of "ideal" structures [40].

7.5. HEAT CAPACITY JUMP AT T_g

Heat capacity jump, ΔC_p, at the glass transition temperature, T_g, was estimated [50] using the Wunderlich expression [51]:

$$\Delta C_p = R \frac{\upsilon_0}{\upsilon_h} \left(\frac{\varepsilon_h}{RT_g} \right)^2 \exp\left(-\frac{\varepsilon_h}{RT_g} \right), \tag{7.10}$$

where R is the universal gas constant; υ_0 is the molar volume occupied by macromolecules; υ_h is the molar free volume; ε_h is the energy of free volume microcavity formation.

Table 7.4

Comparison of heat capacity jumps at T_g, ΔC_p, estimated by equations (7.10) and (7.11) for epoxy polymers EP-1 and EP-2 [50]

Epoxy polymer	K_{st}	ΔC_p, J/(g·deg)	
		Equation (7.10)	Equation (7.11)
	0.50	32.6	32.0
	0.75	26.6	28.5
EP-1	1.0	21.8	24.6
	1.25	18.8	25.7
	1.50	24.8	26.7
	0.50	29.3	30.4
	0.75	24.4	28.0
EP-2	1.0	22.9	26.1
	1.25	25.2	27.5
	1.50	31.0	30.2

Assessments can easily be performed by reduced formula (7.10) under the conditions $\upsilon_0/\upsilon_h \cong 5$ [51], $\varepsilon_h/RT_g = \ln(1/f_f)$ (refer to equation (5.5)) and $\exp(-\varepsilon_h/RT_g) = f_f$ [23]. Table 7.4 presents ΔC_p values, obtained in this manner. Correctness of this calculation can be proved by an independent ΔC_p estimation using empirical Boyer expression as follows [52]:

$$\Delta C_p \approx (104.25 \text{ J/g})/T_g. \qquad (7.11)$$

Clearly data from Table 7.4 indicate that ΔC_p values, calculated by equations (7.10) and (7.11), are quite close. It is also clear that equation (7.10) suggests ΔC_p calculation basing on the single parameter only, f_f. Taking into account linear correlation $f_f(\varphi_{p.m.})$ (refer to Figures 5.21 and 5.22), one can state that ΔC_p calculation in the framework of the cluster model will give similar result [50, 53].

7.6. MOLECULAR MOBILITY AND STRUCTURAL RELAXATION

Problems of molecular mobility in polymers are of special attention [23, 54, 55]. The reasons for this are obvious: polymers represent thermodynamically non-equilibrium solids, and their physical properties are determined by molecular relaxation processes proceeding in them [54]. In its turn, molecular mobility (relaxation) processes depend upon structural features of molecular chains and structural organization of polymers. However, there is no common point of view on parameters describing these factors. For example, it is assumed [54] that fast relaxation processes are determined by mobility of free chains, located between closely packed zones, which simultaneously represent cross-link points of physical entanglements of macromolecules. Though such interpretation is related to elastomers, it displays full conformity to the basic statements of the cluster model, the advantage of which is the possibility of quantitative description of the mentioned structural elements. Application of these notions to amorphous glassy-like polymers means "freezing" of closely packed zones, e.g. a sharp increase of their lifetime. This suggests that in the glassy state the main factor determining molecular mobility is mobility of free chains, located between clusters [56].

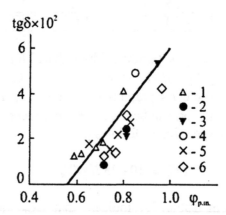

Figure 7.13. Dependence of the loss tangent, tgδ, on the relative part of packless matrix, $\varphi_{p.m.}$, for SP-OFD-10/OSF-10 diblock-copolymers with the formal block content: 0 (1), 5 (2), 10 (3), 30 (4), 50 (5)m and 70 (6) mol% [56]

It is common knowledge [23] that in the general case, increase of the part of packless zones in the polymer structure leads to intensification of the molecular mobility, which is most clearly demonstrated by amorphous-crystalline polymers [57]. That is why one may expect that an increase of this structural parameter (relative part of packless matrix, $\varphi_{p.m.}$) will induce increase of dielectric loss factor, tgδ, which characterizes the molecular mobility. Figure 7.13 shows dependence tg$\delta(\varphi_{p.m.})$ for a series of diblock-copolymers of oligoformal 2,2-di-(4-oxyphenyl)-propane-oligosulfone phenolphthalein (SP-OFD-10/OSF-10), which proves the above assumption. Extrapolation of linear dependence tg$\delta(\varphi_{p.m.})$ to $\varphi_{p.m.}$ = 1.0 gives tg$\delta \cong 0.062$, which is typical for polymers under consideration at T_g [56]. Such result was expectable, because in the framework of the cluster model devitrification of polymer is accepted as decomposition of "frozen" local order, i.e. φ_{cl} = 0 or $\varphi_{p.m.}$ = 1.0 [24]. Of much higher interest is extrapolation of tg$\delta(\varphi_{p.m.})$ dependence to tgδ = 0, for which the finite non-zero value $\varphi_{p.m.}$ = 0.56 is obtained, though it would be expected that suppression of molecular mobility must occur at φ_{cl} = 1.0 or $\varphi_{p.m.}$ = 0. This non-zero $\varphi_{p.m.}$ value corresponds to the quasi-equilibrium state of the copolymer structure (refer to Section 5.5), when φ_{cl} increase causes strain of free chains between clusters, which is the effect suppressing the molecular mobility. More accurate structural interpretation of molecular mobility in polymers is given in the framework of the fractal analysis [56].

The method of photochromic labels is quite widely spread for the studies of free volume parameters in polymers [58, 59]. This method allowed obtaining of new quality information about the free volume in polymers and gave an opportunity to compare it with different theoretical simulations (for example, [60]). Experimental results, obtained by this method, were compared with parameters, obtained for the cluster model, on the example of epoxy polymers of amine (EP-1) and anhydride (EP-2) cross-link type, and aged epoxy polymer EP-2 (EP-3) [61].

The method of photochromic labels determines the value and distribution of the free volume for different structural zones of polymer networks: for free (trailing) chains, between neighbor chains of the carcass, and at curing agent fragments (cross-link points), for which different labels are used [58, 59]. For instance, the label for the first structural area under consideration [59] in the epoxy oligomer analogous to bisphenol A diepoxide diglycidyl ether cross-linked by p,p'-diaminodiphenylsulfone, studied in the work [61], was the reactive p,p'-aminoazobenzene (AA) monodiamine. For the second area, p,p'-diamine azobenzene derivative (DAA), in which four hydrogen atoms in amine were replaced by ethyl groups (*tt* – DAA) and for the third – DAA included to the carcass network as the cross-linking agents, are labels. For epoxy carcasses [59], after the gel formation point it was found that kinetics of photoisomerization of the areas under consideration can be characterized in the framework of two processes with significantly different rates (by two orders of magnitude, approximately). The part of "fast" process, α_s, is interpreted [59] as integral square for the free volume distribution curve at a separate label. It is implied that for label isomerization, a critical volume of free volume microcavity around it is required. That is why the method under consideration estimates conditions of different parts of epoxy polymer structure surrounding the label – its neighborhood [59].

In accordance with the data shown in [59], the radius of the free volume microcavity in epoxy polymers equals 6.5 Å (positron spectroscopy gives somewhat smaller values [62, 63]). The assessments performed [61] indicate that for epoxy polymers under consideration the distance between clusters, R_{cl}, shown in the equation (2.14), varies within the range of ~19 – 31 Å, and the size of the packless matrix zone between two neighbor clusters is the more smaller. Thus the free volume microcavity dimensions and the distance to the closest cluster are comparable. Under this circumstance it can be accepted (in terms of [59]) that the packless matrix zone is the closest neighborhood of the free volume microcavity present in it and, consequently, identify the part of the "fast" process, α_s, proceeding in epoxy polymers as the part of the packless matrix, $\varphi_{p.m.}$ [61]. One more reason for such binding is proportionality between $\varphi_{p.m.}$ and relative part of the fluctuation free volume in polymers (refer to Figure 5.22). To complete the point, one more circumstance shall be noted that proves identification of α_s as $\varphi_{p.m.}$. It has been indicated [58, 59] that impacts on polymer as temperature increase, plasticization and deformation increase α_s, whereas physical aging decreases this parameter. As shown in Chapters 2 and 5, behavior of $\varphi_{p.m.}$ parameter under the effect of the same factor is identical.

Figure 7.14 shows dependencies of α_s for AA (line 1) and tt – DAA (line 2) cross-linked carcasses in accordance with the data from [59] and $\varphi_{p.m.}$ on the glass transition temperature T_g of epoxy polymers. Clearly data on $\varphi_{p.m.}$ fall between lines 1 and 2, displaced from the line 1 towards the line 2 with the increase of cross-linking frequency (or T_g). As mentioned above, label AA (line 1) sounds free volume in the zones of trailing chains and label tt – DAA (line 2) in the areas between chain fragments participating in the carcass. Obviously, with increase of T_g (and v_c) the amount of trailing chains will be decreased due to their cross-linking and inclusion into the carcass and, consequently, the part of free volume, induced by them, is decreased, too. Then the basic role for quite frequently cross-linked carcass (in which molecular mass can reach 10^4 [64]) will be played by the free volume between chain fragments, fixed at both ends by chemical cross-link points. In Figure 7.14 this process suggested is presented by curve 5.

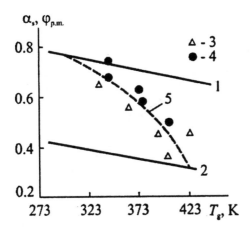

Figure 7.14. Dependencies of the "fast" fraction part, α_s, for labels AA (1) and
tt – DAA (2) and the relative part of packless matrix, $\varphi_{p.m.}$, (3 – 5)
on the glass transition temperature T_g for EP-1 (3) and EP-2 (4).
Curve 5 shows the suggested transition of places with dominant
content of free volume microcavities [61]

The dependence of α_s on radius of the free volume microcavity, r_h,
possessing the physical meaning of the free volume distribution was obtained in
[59]. Similar curve $\varphi_{p.m.}(r_h)$ is shown in Figure 7.15. Contrary to the graph
$\alpha_s(r_h)$ [59], it possesses somewhat different physical meaning. Since the latter
correlation includes data for different epoxy polymers, it indicates the regulated
change of free volume macrocavity size at variation of thermodynamic non-
equilibrium degree of the structure of these polymers. For instance, the local
order increase ($\varphi_{p.m.}$ decrease) induces r_h growth. Figure 7.15 also shows data
on epoxy polymer of the anhydride cross-linking type (EP-3), aged under
natural conditions. Expectedly, compared with the initial EP-2 polymer $\varphi_{p.m.}$ is
decreased and r_h increased simultaneously. As described in Section 6.4, this
indicates strengthening of thermodynamic balance of epoxy polymer structure
during physical aging.

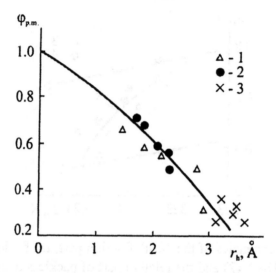

Figure 7.15. Correlation between the relative part of packless matrix, $\varphi_{p.m.}$, and the mean radius of free volume microcavities, r_h, for EP-1 (1), EP-2 (2), and EP-3 (3) [61]

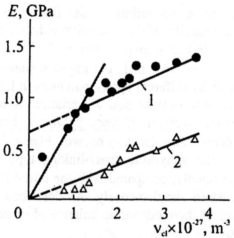

Figure 7.16. Dependencies of elasticity modulus E (1) and balance modulus E_∞ (2) on entanglement cluster network frequency (ν_{cl}) for PC [66]

Thus the results of the work [61] suggest a possibility of "fast" fraction α_s identification using the method of photochrolic labels as the relative part of the packless matrix, $\varphi_{p.m.}$, in the framework of the cluster model, which conforms completely to the physical meaning of α_s. Data from Figure 7.15 indicate the measure of thermodynamic balance of the polymer network structure, which is not only relative fluctuation free volume f_f, interpreted as the disorder parameter [65], but also the average size of free volume microcavities (r_h or V_h). Note that the suggested interpretation of parameter α_s was also accepted in [59], the authors of which suggested is bound to high segmental mobility associated with less dense zones in a glassy polymer.

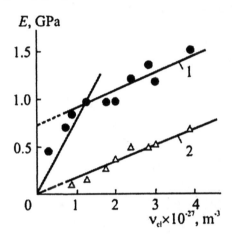

Figure 7.17. Dependencies of elasticity modulus E (1) and balance modulus E_∞ (2) on entanglement cluster network frequency (v_{cl}) for PAr [66]

In discussion below structural changes accompanying relaxation processes in amorphous glassy-like and amorphous-crystalline polymers, i.e. structural relaxation [66], will be considered. Figures 7.16 and 7.17 show dependencies of elasticity modulus E and balance modulus E_∞ on v_{cl} for PC and PAr, respectively. The graphs mentioned illustrate structural relaxation principles for amorphous polymers (refer to Figure 6.29 also). Dependencies $E(v_{cl})$ are separated into two linear segments, bordered by glass transition temperature of the packless matrix, T_g'. Thus below T_g' parameter E is determined by total contribution of both clusters and packless matrix, and above it by the contribution of clusters only [67]. This is quite clear, because above T_g' elasticity modulus of devitrified packless matrix equals about 2 MPa [68], which in the scale of Figures 7.16 and 7.17 is practically negligible value. Of special interest is the fact that packless matrix contribution to E, determined by graph $E(v_{cl})$ extrapolation to $v_{cl} = 0$, is independent of the test temperature. This is not random case but the one for which fundamental reasons are, which will be discussed below.

Dependencies $E_\infty(v_{cl})$ for PC and PAr are also shown in Figures 7.16 and 7.17. They differ from $E(v_{cl})$ dependencies at $T < T_g'$ by the negligible contribution of the packless matrix only. Essentially, a parallel transfer of $E(v_{cl})$ graph is observed that indicates full realization of relaxation processes in the packless matrix of amorphous polymers at $T < T_g'$.

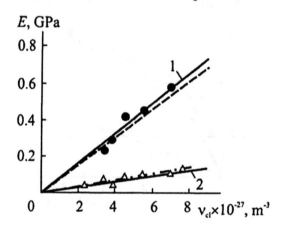

Figure 7.18. Dependencies of elasticity modulus E (1) and balance modulus E_∞ (2) on the entanglement cluster network frequency v_{cl} for HDPE. Dashed lines show appropriate data for PC from Figure 7.16 [66]

Figure 7.18 shows dependencies $E(v_{cl})$ and $E_\infty(v_{cl})$ for HDPE, obtained in quasi-static stretch tests. Compared with PC and PAr behavior at $T < T_g'$ (Figures 7.16 and 7.17), principally different situation is observed. Both mentioned graphs are linear and extrapolated to zero at $v_{cl} = 0$. Proceeding of relaxation processes is illustrated by the change of graph $E_\infty(v_{cl})$ inclination compared with $E(v_{cl})$. As equality to zero of E and E_∞ at $v_{cl} = 0$ is determined by low elasticity modulus (~2.1 MPa [68]) of the packless matrix (devitrified [38] under test temperatures equal 293 – 363 K, applied in [66]), then inclination decrease indicates a change in the structural relaxation mechanism, which is now determined by the second structural component, represented by clusters.

The reasons for presumed change in the structural relaxation mechanism shall be discussed. As shown before [69, 70], two mechanisms of amorphous polymer deformation are possible: occurrence and "freezing up" of non-equilibrium conformations of macromolecules (mechanism I) and mutual displacement of supermolecular structure elements bonded by tie chains (mechanism II). Essentially, processes proceeding during relaxation are opposite to those proceeding in deformation. That is why the above mentioned interpretation can be used for explaining the data obtained.

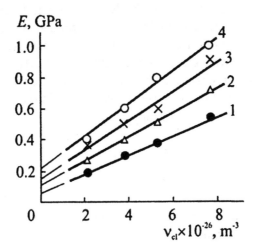

Figure 7.19. Dependencies of elasticity modulus E on entanglement cluster network frequency v_{cl} for HDPE, obtained from blow tests for samples with sharp undercuts 0.5 (1), 0.9 (2), 1.2 (3), and 1.5 mm (4) long [71]

Obviously in the case of glassified packless matrix, structural relaxation by means of clusters displacement is hindered (or anyway requires extremely long time intervals). That is why relaxation is realized by chain returning from the imbalanced conformations to balance ones (the mechanism I). As a stress is applied, this process is quite simply realized in the packless matrix, for which f_f is high. In the case of devitrified packless matrix (PC and PAr at $T > T_g'$, and HDPE), relaxation processes in it proceed at very high rate. Thus its viscosity is decreased significantly [23] and a possibility of cluster displacement occurs (the mechanism II) that now determines the process of structural relaxation. Dashed lines in Figure 7.18 show $E(v_{cl})$ and $E_\infty(v_{cl})$ plots for PC at $T > T_g'$. Clearly they almost completely coincide with the corresponded graphs for HDPE that proves identity of structural relaxation mechanisms. It should be noted that indicated coincidence of the graphs for PC at $T > T_g'$ and HDPE represents an indirect proof of binding temperature transitions for HDPE, suggested in Section 7.4, where the temperature interval $T \approx 140 - 170$ K is associated with T_g', and T_{melt} is associated with T_g [40].

One more proof for presumable identity can be obtained from the data, shown in Figure 7.19. They represent $E(\nu_{cl})$ dependencies for HDPE, obtained in blow tests. Undercut elongation induces shortens time to sample destruction and, as a consequence, increases E due to incompleteness of relaxation processes [72]. Clearly the data in Figure 7.19 indicate simultaneous induction of mechanical glass transition of a part of packless matrix, the contribution of which into E value is now high (compare with Figures 7.16 and 7.17).

For the parameter characterizing temperature dependence of relaxation properties, some authors [73, 74] use the reverse relative stress decline during some definite duration of its relaxation, β, as follows:

$$\frac{1}{\beta} = \frac{\sigma_l}{\sigma_l - \sigma_t},$$

(7.12)

where σ_l is the stress at the moment of loading; σ_t is the stress after a definite time of relaxation proceeding.

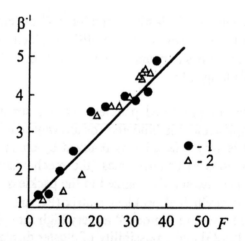

Figure 7.20. Correlation between relaxation index, β^{-1}, and cluster functionality, F, for PC (1) and PAr (2) [66]

In the framework of suggested mechanisms it is expected that decrease of cluster size with temperature must simplify the process proceeding by the mechanism II and, consequently, correlation between β^{-1} and F (remind that the number of segments in the cluster equals $F/2$ [53]). The graph in Figure 7.20 proves this assumption, and β^{-1} value approaches the unit at complete thermofluctuational decomposition of clusters, i.e. at $F \rightarrow 2$ and $T \rightarrow T_g$ [75].

Finally, the depth of relaxation proceeding can be expressed by the ratio E/E_∞: the lower is E_∞, the more complete these processes are and the greater E/E_∞ ratio is. Analogous interpretation was suggested by Boyd [57], which indicates secondary relaxation peaks determination by the difference $(E - E_\infty)$. Figure 7.21 shows dependence E/E_∞ on the temperature difference $(T_g - T)$ (for HDPE, on the difference $(T_{melt} - T)$), i.e. on the approach of T to T_g (T_{melt}). For amorphous-crystalline polymers, the analogy is discussed in Section 7.4. As follows from Figure 7.21, completeness of relaxation processes increases for three mentioned polymers as T_g (T_{melt}) is approached. There can be two reasons for that: firstly, viscosity decrease of devitrified packless matrix and, secondly, decrease of the cluster volume (F decrease, refer to Figure 2.2) with T increase. Both mentioned reasons promote for simplifying clusters' displacement in accordance with the mechanism II.

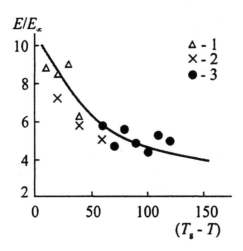

Figure 7.21. Dependence of E/E_∞ ratio on temperature difference $(T_g - T)$ for PC (1) and PAr (2) and $(T_{melt} - T)$ for HDPE (3) [66]

As a consequence, the cluster model allows identification of structural relaxation mechanisms for polymers. In the case of glassy packless matrix, relaxation is realized in conformational restructurings in this structural component (mechanism I), and in the case of devitrification by mutual displacements of clusters (mechanism II) [66].

7.7. ELASTICITY MODULUS

For the first time, behavior of elasticity modulus E was described in the framework of the cluster model in the work [67]. Figure 7.22 show dependencies $E(v_{cl})$ for film samples of polyarylate sulfone (PAS). Such correlation is easily explained in the framework of the model under discussion: v_{cl} increase means increase of the amount of closely packed segments and corresponded intensification of intermolecular interaction between them that determines E increase [76].

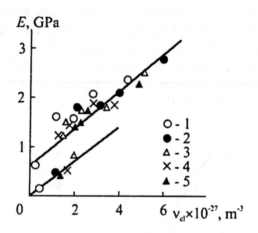

Figure 7.22. Dependencies of elasticity modulus E on entanglement cluster network frequency v_{cl} for PAS films, obtained from solutions in tetrachloroethane (1), tetrahydrofurane (2), chloroform (3), N,N-dimethylformamide (4), and methylene chloride (5) [67]

As in the case of PC and PAr (Figures 7.16 and 7.17), correlation $E(v_{cl})$ for PAS shown in Figure 7.22 is separated into two parallel lines, limited by T_g' ≈ 425 K for PAS ($T_g \approx 425$ K [77]). Reasons for this separation into two parts are described in the previous Section. It should be noted that in the framework of the rubbery theory the elasticity modulus is proportional to the entanglement ("hooking") network frequency [78], which in accuracy to a constant is also true for PAS (with respect to v_{eng} substitution by v_{cl}) in the whole interval of test temperatures (Figure 7.22).

As is known [38], deformation of amorphous glassy-like polymers, which occupy the intermediate position between liquids and solids by structure, are described satisfactorily by both molecular-kinetic (liquid) and solid models. They can be quantitatively connected on the example of kinetic theory of the fluctuation free volume [23], related to the first type of models, and the cluster model representing the second type. Interconnection of these concentrations is illustrated in Section 5.3, and on their basis temperature dependence of shear modulus G is analytically described on the example of two amorphous glassy-like polymers (PC and PAr) [79]. The mentioned dependence can be easily obtained by combining equations (2.8), (2.33), (3.12), (3.19) and (5.59) as follows [79]:

$$G = \frac{4\pi(1-v)kT_g \ln(1/f_f)}{b^2 l_0 C_\infty \ln(r/r_0)}. \tag{7.13}$$

For one and the same polymer, parameters k, T_g, b, l_0 and C_∞ are constants and, consequently, experimentally observed change of G with temperature is stipulated by temperature dependencies of v and f_f. Equation (7.13) illustrates the necessity of v and f_f variation with temperature (these parameters are bound with one another by the approximate correlation (2.33)) and indicates that constancy of these parameters below T_g, often postulated in connection with "freezing" of the polymer structure, is too rough approximation [80].

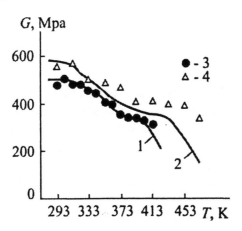

Figure 7.23. Dependencies of shear modulus G on test temperature T for PC (1, 3) and PAr (2, 4): experimental (solid curves 1 and 2) and calculated by equation (7.13) (3, 4) [79]

Assessment results of shear modulus G temperature dependencies in accordance with equation (7.13) for PC and PAr are compared with the experimental data in Figure 7.23 indicating their good conformity. As a consequence, low absolute variations of ν and f_f with temperature can significantly effect on mechanical behavior of polymers [79].

Specificity of the structure – elasticity modulus relation for polymer networks is discussed with respect to separation of the fluctuation free volume of epoxy polymers into two components, described in Section 5.3 [26]: f_f^d stipulated by the presence of chemical cross-link points and f_f' stipulated by decay of clusters, i.e. properly fluctuation component ($f_f = f_f^d + f_f'$, where $f_f^d \approx 0.024$, refer to Figure 5.22). The following interconnection between microhardness H_v and E is suggested [23]:

$$\frac{H_v}{E} = \frac{(1-2\nu)}{6(1+\nu)}. \qquad (7.14)$$

Though formulae (3.12) and (7.14) are extrinsically identical, they give different values of H_v and σ, because parameter ν values in them are different for different zones of polymer deformation. As follows from equations (2.33) and (7.14) [26]:

$$E \approx 35.3 f_f H_v. \qquad (7.15)$$

The authors of [26] use this correlation for obtaining theoretical dependencies of E on parameter K_{st}. For epoxy polymer of amine cross-linking type cured under different hydrostatic pressures, calculated and experimental dependencies $E(K_{st})$ are symbate (Figure 7.24). Hence, the use of f_f' values for calculation of E in accordance with the equation (7.15) gives better conformity rather than f_f. Similar results were obtained for epoxy polymers of the anhydride cross-linking type [26].

Data in Figure 7.24 help in understanding why in the work [26] better analogy between E and ν_{cl} rather than E and ν_c for epoxy polymers was observed. Apparently, in the area of elastic deformation mechanical properties of epoxy polymers are determined not by entire free volume, f_f, but by its part only, f_f', caused by association (dissociation) of segments into (from) clusters. It should be noted that at different stages of deformation properties of polymers are regulated by different structural zones. For example, in the ranges of elasticity (E) and local plasticity (H_v) the packless matrix (f_f') is mainly effective, and at macroscopic yielding (σ_y) chemical cross-link zones are also active $(f_f = f_f' + f_f^d)$. This is the explanation for different values of σ_y and H_v, obtained from equations (3.12) and (7.14), despite the extrinsic similarity of the latter [26].

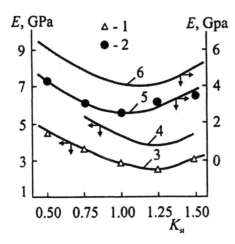

Figure 7.24. Experimental (1, 2) and theoretical (3 – 6) dependencies of elasticity modulus E on the curing agent:oligomer ratio, K_{st}, for epoxy polymer of the amine cross-linking type under pressures as follows: 0.1 MPa (3, 4) and 200 MPa (5, 6); 3, 5 – calculation by equation (7.15) using f_f'; 4, 6 – calculation by equation (7.15) using f_f [26]

Let us now discuss elastic properties of amorphous-crystalline polymers on the example of two polyethylenes (LDPE and HDPE). Presently, there are several concepts which tend to explain correlation between the structure and elasticity modulus for mentioned polymers. One of them [82] assumes that amorphous-crystalline polymer is a two-phase composite consisting of alternating crystalline and amorphous zones. That is why E value is mainly determined by high elasticity modulus of the crystalline phase. Since crystalline zones are much more rigid, greater deformation of the bulky sample shall be realized in amorphous interlayers. Moreover, in the general case, polymeric macromolecules will pass through both crystalline and amorphous zones, due to which crystallites act similar to cross-links preventing sliding of macromolecules in relation to one another. Thus guides in structural organization of amorphous-crystalline polymers and elastomeric carcasses are easily detected. A strict argument against this concept is the fact that calculated modules are much lower than experimentally observed values [45]. As mentioned above, Krigbaum *et al.* [45] have modified this approach. They have suggested that formation of crystallites leads to significant strain of remaining amorphous chains. Thus the great increase of the modulus, observed during crystallization, is stipulated by the fact that chains in the carcass are almost completely stretched even in the absence of external load [45]. Mandelkern *et al.* [83, 84] have studied different polyethylenes and shown that crystallinity cannot be adequately explained elasticity modulus of amorphous-crystalline polymers. The concepts [45, 83, 84] suggest that the factor determining E value is molecular structure of amorphous zones. As a consequence, crystallinity changes conditions of amorphous interlayers, which, in its turn, causes variation of E.

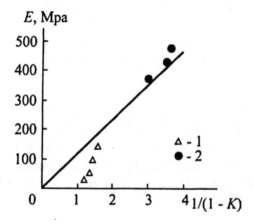

Figure 7.25. Dependence of elasticity modulus E on parameter $1/(1-K)$ for LDPE (1) and HDPE (2) [17]

Pakhomov *et al.* [85] have suggested an approach, according to which the value of parameter E of amorphous-crystalline polymers is completely determined by their conformational state, namely, by concentration of *trans*-conformers. The mentioned concentrations were compared with the cluster model on the example of LDPE and HDPE [17].

Figure 7.25 shows variation of elasticity modulus of LDPE and HDPE as a function of the parameter $1/(1 - K)$ (where K is the crystallinity degree). This dependence correlates well with the theoretical results of Krigbaum *et al.* [45]. For LDPE, some deviations between calculated and experimental data are observed, which is probably stipulated [17] by non-consideration of variation of crystallite sizes with temperature [82]. Nevertheless, the observed correspondence proves correctness of the model [45].

In Section 7.4 the notion of amplification coefficient, K_{amp}, was introduced. It indicates by how many times experimental value of E is greater than that calculated in the framework of the rubber state concepts. It should be noted that temperature dependencies of K_{amp} (Figure 7.26) and bulky crystallinity K [86] are similar. Expectedly, as follows from the above ideas, quite good conformity between these parameters is observed (Figure 7.10). Tanabe, Strobl and Fischer [47] have published analysis of small-angle X-ray diffraction patterns, which indicates high changes in thicknesses of crystalline and amorphous zones in polyethylene with temperature. They have observed that thickness of amorphous interlayers changes from 30 Å at room temperature to 150 – 200 Å near the melting point. In its turn, this means a decrease of amorphous chain strain with temperature, which is displayed by data in Figure 7.10.

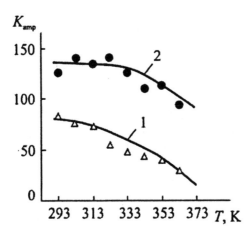

Figure 7.26. Temperature dependencies of elasticity modulus amplification coefficient, K_{amp}, for LDPE (1) and HDPE (2) [17]

As mentioned above, the relative part of clusters, φ_{cl}, represents characteristic of the local order degree of the polymer amorphous state structure. Figure 7.27 shows correlation $E(\varphi_{cl})$ for LDPE and HDPE. Clearly these dependencies were found linear and, despite the presence of the crystalline phase, are extrapolated to $E = 0$ at $\varphi_{cl} = 0$. Maximal values of E for non-oriented polyethylenes as the function of their density at the test frequency of 1.45 MHz, i.e. for the case, when viscoelastic contribution to E can be neglected, are reported [87]. These E values correlate well with the ones extrapolated at $\varphi_{cl} = 0$, obtained from the graph in Figure 7.27 ($E = 1.3$ and 1.1 GPa for PNP and 2.8 and 2.6 GPa for HDPE). The above-shown results indicate determination of E value for non-oriented polyethylenes by the state of crystalline zones, which in its turn, is regulated by crystalline morphology [45, 83, 84].

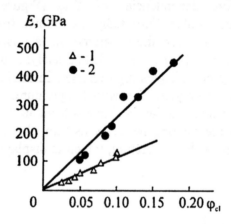

Figure 7.27. Dependencies of elasticity modulus E on relative part of clusters φ_{cl} for LDPE (1) and HDPE (2) [17]

Concentration of *trans*-conformers, c_t, in LDPE and HDPE using the technique [85] was calculated [17]. Expectedly, c_t is decreased with temperature increase. Figure 7.28 shows correlation between the relative part of clusters and concentration of *trans*-conformers for two polyethylenes, wherefrom it follows the c_t increase rises the local order degree of non-crystalline zones and leads to increase of elasticity modulus (Figure 7.27).

Let us now turn back to data in Figure 7.19, which indicate the increase of mechanically glass transited part of packless matrix with sharp cutoff length increase (or increase of the local deformation rate [88]) for HDPE during blow tests [71]. This effect is also stipulated by thermodynamic non-equilibrium of the structure of polymers and, as a consequence, φ_{cl} depends upon the time scale of tests. Note that in the present case, φ_{cl} corresponds to generalized definition of this parameter, i.e. φ_{cl} the time-dependent order parameter that can be presented as follows [89]:

$$\varphi_{cl} = \varphi_{cl}^{0} \exp\left(-\frac{t_d}{\tau_0}\right), \qquad (7.16)$$

where φ_{cl}^{0} is the value of φ_{cl} at $t_d = 0$ (maximum theoretically accessible); t_d is time to sample destruction; τ_0 is the relaxation time.

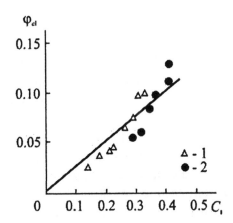

Figure 7.28. Correlation between the relative part of clusters, φ_{cl}, and concentration of *trans*-conformers, c_t, for LDPE (1) and HDPE (2) [17]

Figure 7.29 shows comparison of φ_{cl} values, experimental and calculated by equation (7.16), which shows their good conformity. This means that at blow loading of HDPE with devitrified amorphous phase, mechanical glass transition of a definite part of it proceeds increasing with reduction of the test time [90]. This effect determines the increase of E with the sharp cutoff length (Figure 7.19) [71].

Quantitatively, this increase of elasticity modulus can be described in the framework of the carcass connectivity model (Section 5.5), where theoretical value of elasticity modulus (E^t) is determined by the equation (5.69).

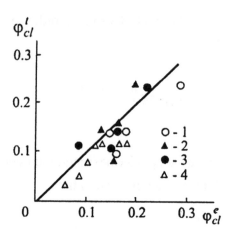

Figure 7.29. Correlation between relative parts of clusters, experimental and calculated by equation (7.16) (φ_{cl}^{e} and φ_{cl}^{t}, respectively), for HDPE at 293 K (1), 313 K (2), 333 K (3), and 353 K (4) [90]

As shown by estimations performed by this equation, E^{t} values were found somewhat lower than corresponded experiments ones, E^{e} [90]. This assumes the presence of one more structural factor affecting the value of elasticity modulus. This factor can be macromolecular "hookings" network, which can be taken into account by additional probability of bonds' formation, $\eta_{x_{0}}$ = const, determined by the equation (5.70). The value $\eta_{x_{0}}$ gives density of the "hooking" carcass per cross-section of the sample, but not the volume of it, as the "hooking" network frequency does. That is why there is an approximate correlation between mentioned parameters as follows [90]:

$$\eta_{x_{0}} \cong v_{l}^{2/3}. \tag{7.17}$$

Comparison of values E^{t}, calculated by the equation (5.69) under the condition $\eta_{x} = \varphi_{cl} + \eta_{x_{0}}$, and E^{e} indicated their good conformity (discrepancy < 15%). At the highest test temperature (353 K) E^{t} values were found higher than E^{e} by 40%, approximately. It is assumed [90] that this effect is stipulated by sharp increase of chain sliding through macromolecular "hookings" at increased temperatures [91], therefore they are unable to transmit load and affect E value.

Calculation by the equation (5.69) at $\eta_x = \varphi_{cl}$ gave the proper conformity between E^t and E^e at $T = 353$ K [90].

Finally, let us consider description of polymers' behavior in the elasticity zone in the framework of percolation models [92 – 94]. Inhomogeneus statistical mixture of a solid and a liquid displays the properties of the former (for example, non-zero shear modulus G) only when the component of it forms a percolation cluster (for instance, as at gel formation in polymeric solutions). If the liquid component is replaced by vacuum, the bulky modulus B below the percolation threshold is also equal zero [95]. Such model gives the following correlation for elastic constants [95, 96]:

$$B, G \sim (p - p_c)^{\eta}, \qquad (7.18)$$

where p is the volumetric part of the solid component; p_c is the percolation threshold; η is an index.

In the simulation of fractal structures of the "Serpinsky carpet" type, for the index η the following equation is deduced [95]:

$$\left(\eta / \nu_p\right) = d - 1, \qquad (7.19)$$

where ν_p is the correlation length index in the percolation theory; d is the Euclidean space dimensionality, in which the fractal is considered.

Dependencies of the shear modulus G and equilibrium shear modulus G_∞ on φ_{cl} will display the shape similar to dependencies $E(\nu_{cl})$ and $E_\infty(\nu_{cl})$, for PAr with regard to equations (2.11) and (5.41) shown in Figure 7.17. Comparison of three lines, presented in this Figure, indicates for the case of thermodynamically quasi-equilibrium state of the packless matrix the description of G value by the unique order parameter (φ_{cl}). Thus the percolation threshold will be $p_c = \varphi_{cl}^0 = 0$. Thermodynamic quasi-equilibrium is reached for either devitrivied or relaxed packless matrix (Figure 7.17). For PAr samples possessing glassy-like thermodynamically non-equilibrium packless matrix, the unique order parameter is insufficient for description of the shear modulus behavior. It is known [97, 98] that for glassy polymers the Prigozhin-Defue relation is not the equal one, which in turn means impossibility of describing structure of the mentioned polymers by the unique order parameter. Data in Figure 7.17 indicate the general reason for this event, which is the non-equilibrium state of the packless matrix, but in particular cases (when dependence $G(\varphi_{cl})$ passes through the origin of coordinates) the polymer structure can be described using the unique order parameter (in the case under consideration, φ_{cl}) [94].

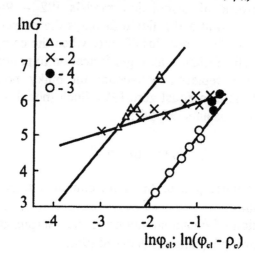

Figure 7.30. Dependencies of shear modulus G (1 – 3) and equilibrium shear modulus G_∞ (4) on parameters φ_{cl} (1, 2, 4) and $(\varphi_{cl} - \rho_c)$ (3) for PAr samples. Stretch tests were performed at $T > T_g'$ (2, 3) [93]

Figure 7.30 presents dependencies $G(\varphi_{cl})$ for PAr samples in double logarithmic coordinates, corresponded to the correlation (7.18) under the conditions $p = \varphi_{cl}$ and $p_c = \varphi_{cl}^0 = 0$. In both cases, linear dependencies with significantly different inclinations were obtained. For PAr samples, tested at $T > T_g'$, the inclination is $A = 1.5$, and if $v_p = 0.88$ is accepted [99], $d = 2.70$ is obtained that correlates well to dimensionality d_f of this polymer structure [53]. As PAr samples are tested at $T \leq T_g'$, the following parameters are correspondingly obtained: $A = 0.48$ and $d = 1.55$, the latter being physically meaningless, because $2 \leq d_f < 3$ [100]. The reasons for this nonconformity are obvious and considered above. For the last series of PAr samples, one more order parameter, associated with the packless matrix and displaying its thermodynamic non-equilibrium level, should be used. The parameters, which can be selected for this purpose, are quite numerous [97]. Some of them will be more precise from theoretical positions and others will be more suitable for practical application. The authors of the works [92 – 94] have chosen an intermediate alternative and used the value of relative fluctuation free volume,

f_f, for this purpose [23]. More precisely, deviation of this parameter from the quasi-equilibrium value, f_f^{cl}, was used. This choice is stipulated by the following reasons: on the one hand, parameter f_f characterizes the non-equilibrium degree of the polymer structure quire accurately [65]; on the other hand, it can be simply calculated by equation (2.33).

In accordance with equations (2.33) and (2.34), the value selected for f_f^{cl} equaled 0.060, which corresponded to the Poisson coefficient $v \cong 0.32$ [23]. Further on, the correlation (7.18) can be used, but the variable percolation threshold p_c shall be applied, expressed as follows [93]:

$$p_c = \frac{f_f^{cl} - f_f}{f_f}. \tag{7.20}$$

Essentially, this parameter determines the non-equilibrium degree of the packless matrix as the relative part of the fluctuation free volume exceeding f_f^{cl} [93].

Figure 7.30 shows dependencies $G(\varphi_{cl} - p_c)$ and $G_\infty(\varphi_{cl})$ for PAr samples tested at $T < T_g'$. It indicates data for mentioned series fitting the single line with the inclination $A = 1.5$ that gives meaningful d_f values in accordance with the equation (7.29).

As a consequence, the above-shown results indicated the possibility of elastic modules description for amorphous glassy-like polymers in the framework of the percolation theory. For the purpose of identification of the order parameters, the cluster model was used. In the general case, description of the structure and properties of polymer of this type require application of two order parameters, the necessity of the second one being determined by thermodynamic non-equilibrium of the packless matrix. Application of the percolation correlations (7.18) and (7.19) to amorphous glassy-like polymers requires modification of these expressions: a variable percolation threshold shall be introduced and dimensionality d_f instead of d shall be used. In this case, these correlations allow obtaining of proper structure – shear modulus ratio for polymers under consideration [94].

7.8. YIELDING CHARACTERISTICS

As shown using positron spectroscopy methods [100], the yielding in polymers is realized in closely packed zones. Theoretical analysis in the framework of the plasticity fractal concept [102] illustrates that the Poisson coefficient value in the yielding point, v_y, can be estimates as follows:

$$v_y = v\chi + 0.5(1 - \chi), \tag{7.21}$$

where v is the Poisson coefficient in the elastic deformation zone; χ is the relative part of elastically deformed polymer.

Figure 7.31 shows comparison of values χ and $\varphi_{p.m.}$ for PC and PAr indicating their good conformity. Data in this Figure assume that the packless matrix can be identified as the realization zone of elastic deformation process, and clusters are identified as the zone of elastic deformations [103]. These results prove the conclusion made in the work [101] about proceeding of non-elastic deformation processes in the close packing zones of amorphous glassy-like polymers and indicate correctness of the plasticity fractal theory at their description.

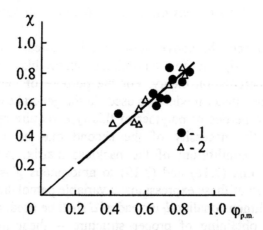

Figure 7.31. Correlation between relative part of the packless matrix $\varphi_{p.m.}$ and probability of elastic state realization for PC (1) and PAr (2) [103]

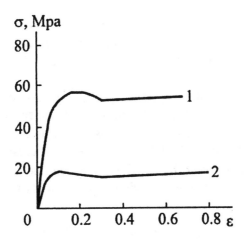

Figure 7.32. Stress – deformation ($\sigma - \varepsilon$) curves for PAr at test temperatures: 293 K (1) and 453 K (2) [53]

The yielding of amorphous glassy polymers is frequently considered as their mechanical devitrification [104]. However, if typical stress – deformation ($\sigma - \varepsilon$) graph for such polymers is considered (Figure 7.32), clearly outside the yield stress σ_y the stimulated elasticity (cold flow) σ_B is practically equal σ_y, i.e. equals several ten MPa, whereas for devitrified polymer this value is, at least, by the order of magnitude lower. Furthermore, σ_B is the function of the test temperature T, whereas for devitrified polymer such dependence must be much weaker and, which is the most important, possess the opposite tendency (σ_B increase with temperature). This non-correspondence is simply eventuated in the framework of the cluster model, where cold flow of polymers is associated with deformation processes in packless matrix, in which clusters are "floating". However, thermal devitrification of the packless matrix happens at temperature T_g', which is approximately 50 K lower than T_g. That is why it should be expected that amorphous polymer in the temperature range $T_g' - T_g$ will be subject to yielding under application of even extremely low stress (of about 1 MPa). Nevertheless, as shown by the $\sigma - \varepsilon$ graphs in Figure 7.32, this does not happen and $\sigma - \varepsilon$ curve for PAr in $T_g' - T_g$ range is qualitatively analogous to the graph $\sigma - \varepsilon$ at $T < T_g'$ (curve 1). Thus is shall be suggested that devitrification of the packless matrix is the consequence of the yielding, but not its criterion. Taking into account realization of non-elastic deformation process in the clusters (Figure 7.31) one can suggest that the sufficient condition of yield in the polymer is the loss of stability by the local order zones in the field of external mechanical stress, after which the deformation process proceeds without increasing the stress σ (at least, nominal), contrary to deformation below the yield stress, where a monotonous increase of σ is observed (Figure 7.32).

Now using the model suggested [105] one can demonstrate that the clusters lose their stability, when stress in the polymer reaches the macroscopic yield stress, σ_y. Since the clusters were postulated as the selection of closely packed collinear segments, and arbitrary orientation of cluster axes in relation to applied tensile stress σ should be expected, they can be simulated as "inclined plates" (IP) [105], for which the following expression is true:

$$\tau_y < \tau_{IP} = 24G_{cl}\varepsilon_0(1 + \nu_2)/(2 - \nu_2), \tag{7.22}$$

where τ_y is the shear stress in the yielding point; τ_{IP} is the shear stress in IP (cluster); G_{cl} is the shear modulus induced by the presence of clusters and determined from the graphs similar to the ones from Figures 7.16 and 7.17, i.e. $G(\nu_{cl})$; ε_0 is the intrinsic IP deformation; ν_2 is the Poisson coefficient for clusters.

Since the equation (7.22) characterizes inelastic deformations of clusters, the following can be accepted: $\nu_2 = 0.5$. Further on, under assumption that $\tau_y = \tau_{IP}$, the expression for the minimal (with regard to inequality in the left part of the equation (7.22)) intrinsic deformation ε_0^{min} is obtained [53]:

$$\varepsilon_0^{min} = \frac{\tau_y}{\sqrt{24G_{cl}}}. \tag{7.23}$$

The condition for clusters (IP) stability looks as follows [105]:

$$q = \sqrt{\frac{3}{2}} \cdot \frac{\varepsilon_0}{\tau_y} \left\{ \left| 1 + \frac{\tilde{\varepsilon}_0}{\varepsilon_0} \right| - \sqrt{\frac{3}{8} \frac{\tau_y}{G_{cl}\varepsilon_0(1+v_2)}} \right\}, \tag{7.24}$$

where q is the parameter characterizing plastic deformation; $\tilde{\varepsilon}_0$ is the intrinsic deformation of the packless matrix.

The cluster stability distortion condition is fulfillment of the following inequality [105]:

$$q \leq 0. \tag{7.25}$$

Comparison of the expressions (7.24) and (7.25) gives the following criterion of stability loss for IP (clusters) [53]:

$$\left|1+\frac{\tilde{\varepsilon}_0}{\varepsilon_0}\right| = \sqrt{\frac{3}{8}} \frac{\tau_y^t}{G_{cl}\varepsilon_0(1+v_2)}, \tag{7.26}$$

from which theoretical stress $\tau_{s(}\tau_y^t)$ can be determined, after reaching which the criterion (7.25) is fulfilled.

To perform quantitative estimations, one shall make two reducing assumptions [53]. Firstly, for IP the following condition is fulfilled [105]:

$$0 \leq \sin^2 \theta_{IP}(\tilde{\varepsilon}_0/\varepsilon_0) \leq 1, \tag{7.27}$$

where θ_{IP} is the angle between the normal line to IP and the main axis of intrinsic deformation.

Since for arbitrarily oriented IP (clusters) $\sin^2\theta_{IP} = 0.5$, for fulfillment of the condition (7.27) the assumption $(\tilde{\varepsilon}_0/\varepsilon_0)=1$ is enough. Secondly, the equation (7.23) gives the minimal value of ε_0, and for the sake of convenience of calculations parameters τ_y and G_{cl} were replaced by σ_y and E, respectively. Parameter E value is greater than the elasticity modulus E_{cl} induced by the presence of clusters (refer to graphs in Figure 7.16 or 7.17). That is why to compensate two mentioned effects the deformation ε_0, estimated by the equation (7.23), was twice increased. The final equation is as follows [53]:

$$\varepsilon_0 \cong 0.64\frac{\sigma_y}{E} = 0.64\varepsilon_{el}, \tag{7.28}$$

where ε_{el} is the elastic component of yielding macroscopic deformation [106], which is physically corresponded to deformations ε_0 and $\widetilde{\varepsilon}_0$ [105].

Combination of equations (7.26) and (7.28) with the graphs similar to the ones shown in Figures 7.16 and 7.17, wherefrom $E_{cl}(G_{cl})$ can be determined, allows an assessment of theoretical yield stress σ_y^t $\left(\sigma_y^t = \sqrt{3}\tau_y^t [71]\right)$ and compare it with experimental values σ_y. Such comparison is shown in Figure 7.33 demonstrating satisfactory conformity between σ_y^t and σ_y that proves the suggestion made in the work [53] and justifies the above assumptions.

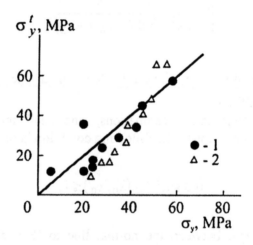

Figure 7.33. Correlation between experimental, σ_y, and calculated by equation (7.26), σ_y^t, yield stress values for PAr (1) and PC (2) [53]

Thus realization of the yielding in amorphous glassy-like polymers requires cluster stiffness (stability) loss in the mechanical stress field, after which mechanical devitrification of the packless matrix proceeds. Similar criterion was obtained for amorphous-crystalline polymers [53].

As indicated by the results obtained [27, 107], behavior of polymer networks at yielding is just slightly different from the above-described one of PC and PAr. However, further progress for polymer networks in this branch is quite difficult due to, at least, two reasons: extreme overestimation of the role of chemical cross-links and the absence of quantitative structural model. The yielding mechanism for polymer network has been suggested [107], based on the application of the cluster model and the latest achievements in the branch of synergism of deformable body [108] on the example of two already known epoxy polymers of amine (EP-1) and anhydride (EP-2) cross-linking type.

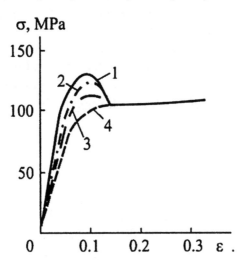

Figure 7.34. Stress – deformation (σ - ε) dependencies at loading up to fracture (1) and at periodical load application (2 – 4) for EP-2 ($K_{st} = 1.0$): 2 - the first loading cycle; 3 - the second loading cycle; 4 - the third loading cycle [107]

Figure 7.34 shows σ - ε graphs for EP-2 under axial compression of the sample up to destruction (curve 1) and at consecutive loading up to deformation ε exceeding the yield deformation ε_y (curves 2 – 4). Comparison of these graphs indicates consecutive lowering of the "yield peak" under constant cold flow stress, σ_B. High values of σ_B assume corresponded values of stable cluster network frequency v_{cl}^{st} (refer to Chapter 3), which is much higher than the frequency of chemical cross-link network v_c [27]. Thus though behavior of a polymer network at the cold flow plateau is described in the framework of the rubber state theory, the stable cluster network in this part of σ - ε graphs is preserved. The only process proceeding is decay of unstable clusters determining devitrification of the packless matrix. This process is initiated at the stress equal the proportionality limit that correlates with the data from [74], where the effect of this stress and temperature $T_2 = T_g'$ is assumed analogous. The analogy between cold flow and glass transition processes is just partial: the only one component, the packless matrix, is devitrified. Moreover, complete decay of unstable clusters proceeds not in the point of reaching the yield at σ_y, but at the start of cold flow plateau at σ_B. This can be observed from σ - ε graphs shown in Figure 7.34. As a consequence, the yielding is regulated not by devitrification of the packless matrix, but by a different mechanism. As shown above, for such mechanism the loss of stiffness (stability) by clusters in the mechanical stress field can be assumed, which also follows from the common fact of derivative $d\sigma/d\varepsilon$ turning to zero in the yield point [109]. In accordance with [108], critical shear deformation γ_* leading to the loss of shear stability by a solid equals:

$$\gamma_* = \frac{1}{mn},\qquad(7.29)$$

where m and n are indices in the Mie equation [23] setting the interconnection between the interaction energy and distance between particles. The value of parameter $1/mn$ can be expressed via the Poisson coefficient, v [23]:

$$\frac{1}{mn} = \frac{1-2v}{6(1+v)}.\qquad(7.30)$$

As follows from equations (3.12) and (7.30):

$$\frac{1}{mn} = \gamma_* = \frac{\sigma_y}{E}.\qquad(7.31)$$

Equation (7.31) gives the value of deformation with no regard to viscoelastic effects, i.e. deviation of σ - ε graph from linearity outside the proportionality. Taking into account that tensile deformation is twice greater, approximately, than the corresponded shear deformation [109], theoretical yield deformation (ε_y'), corresponded to the stability loss by a solid, can be calculated. Figure 7.35 shows comparison of ε_y' with experimental ε_y values. Approximate equality of these parameters is observed that suggests association of the yielding with the stability loss by polymers. More precisely, we are dealing with the stability loss by clusters, because parameter v depends upon the cluster network frequency v_{cl} (equation (3.14)), and ε_y is proportional to v_{cl} [67].

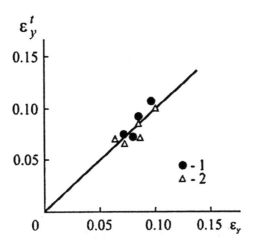

Figure 7.35. Correlation between experimental (ε_y) and theoretical (ε_y^t) yield deformations for epoxy polymers EP-1 (1) and EP-2 (2) [107]

Intensification of stress relaxation at loading of epoxy polymers under the conditions, analogous to the above-described (Figure 7.34, curves 2 – 4), was detected [110]. The authors [110] have explained the observed effect by partial break of chemical bonds. To check this conclusion, the repeated compression tests of samples, loaded before the cold flow plateau and the annealed at $T < T_g$, were performed [107]. The "yield peak" is again observed on $\sigma - \varepsilon$ graph. This may be caused by restoration of unstable clusters, because restoration of broken chemical bonds at $T < T_g$ is of low probability. Thus it should be noted that the "yield peak" lowering as the result of preliminary plastic deformation was also observed for linear amorphous polymers, for example, PC [111], in which the network formed by chemical bonds is admittedly absent.

In accordance with the data obtained [74], height of the "yield peak" ($\Delta\sigma$) for epoxy polymers is decreased with T_g increase. This dependence is also fitted in the case of systems, studied in the works [27, 107]. However, dependence $\Delta\sigma(T_g)$ is not the universal one: for EP-1, $\Delta\sigma$ values are much lower than for EP-2 at the same T_g. In the case of $K_{st} > 1.0$ at comparable T_g, $\Delta\sigma$ values are higher than at $K_{st} \leq 1.0$, i.e. at excess of the curing agent greater amount of unstable clusters is formed [27].

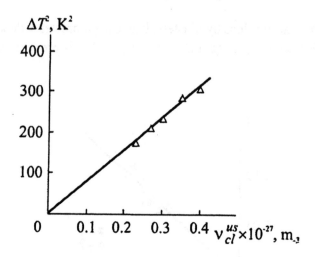

Figure 7.36. Dependence of square temperature interval of glass transition, ΔT_g, on unstable cluster network frequency, v_{cl}^{us}, for epoxy polymer EP-2 [107]

Figure 7.36 shows correlation between the square width of the glass transition temperature interval ΔT_g, determined experimentally, and unstable cluster network frequency v_{cl}^{us}, calculated by the equation as follows [107]:

$$\Delta\sigma = \frac{1.73Eb\left(l_0C_\infty v_{cl}^{us}\right)^{1/2}}{4\pi(1+v)}. \tag{7.32}$$

Comparison of data in Figures 6.14 and 7.36 shows that for polymer networks the factor regulating $\Delta\sigma$ is deviation, K_{st}, from its stoichiometric value equal 1.0.

The behavior of deformable solid subject to mechanical impact is determined by formation and evolution of dissipative structures (DS) providing for the optimal mode of energy dissipation supplied from the outside [108, 112]. In the case of metals, this approach is commonly accepted, though there is no general point of view on the restructuring mechanisms of deformable body [113]. For polymers this problem is not practically studied, despite the fact that the presence of DS in them had already been studied (refer to Section 5.2). Simultaneously, its solution allows approaching deformation of polymers from positions of fundamental physical principles of non-equilibrium thermodynamics [114]. These principles can be applied in the presence of quantitative structural model. Shown below will be the cluster model used for this purpose, in the framework of which DS are personified as the local order zones (refer to Section 5.2).

It has been shown [103] that yielding in amorphous linear polymers is realized as the effective Poisson coefficient, $\nu_y \cong 0.41$, is reached. Accepting this conclusion also correct for polymer networks (EP-1 and EP-2) and using the relation between ν and ν_{cl}, expressed by equation (3.14), values ν_{cl}^t obtained as the yield stress σ_y is reached were calculated [115]. These values are shown in Figure 7.37, accompanied by ν_{cl} values as the function of K_{st} for undeformed epoxy polymers EP-1 and EP-2. Clearly ν_{cl}^t values are independent of K_{st} and coincide for EP-1 and EP-2, which is determined by the primary selection of ν_y. At the same time, they are much lower than ν_{cl}. This means that realization of polymer network yield requires decay of a definite amount of clusters (DS). This situation is diametrically opposite to deformation processes in metals, in which, vice versa, DS (dislocation substructures) formation is observed [108, 114]. The mentioned difference is of the principal character and is stipulated by deviations in ideas on ideal (defect-free) structures of compared classes of materials (refer to Chapter 3).

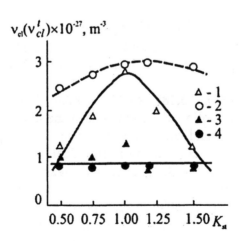

Figure 7.37. Dependencies of the cluster network frequency in non-deformed state ν_{cl} (1, 2) and after the yield ν_{cl}^t (3, 4) on the curing agent:oligomer ratio K_{st} for EP-1 (1, 3) and EP-2 (2, 4) [115]

Expectedly, the mentioned structural changes will determine parameters characterizing the yielding in polymer networks. Let us consider this statement on the example of yielding deformation ε_y. It would appear reasonable that the greater amount of DS is decayed during yielding, the higher is ε_y value. The mentioned amount of DS can be determined as the difference of cluster network frequencies before (v_{cl}) and after (v_{cl}') the yield as follows: $\Delta v_{cl} = v_{cl} - v_{cl}'$, which can easily be obtained from the graphs in Figure 7.37. Figure 7.38 shows correlation $\varepsilon_y(\Delta v_{cl})$, which proves the above suggestion.

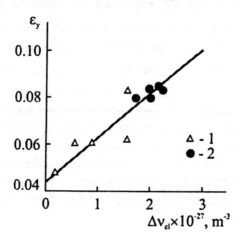

Figure 7.38. Dependence of the yielding deformation, ε_y, on the difference of cluster network frequencies before and after yielding, Δv_{cl}, for epoxy polymers EP-1 (1) and EP-2 (2) [115]

Correlations (5.33), (5.38), (5.45) combined with equation (2.14), considered in Chapter 5, allow determination of the cluster functionality F for EP-1 and EP-2 before and after yielding. The most typical difference of dependencies $F(K_{st})$ under comparison is significant increase of F at reaching the yield stress (Figure 7.39). Simultaneous decrease of v_{cl} and increase of F epoxy polymer deformation before the yield stress (Figures 7.37 and 7.39) means decomposition of unstable clusters possessing low values of F. As a consequence, at σ_y only stable clusters with high F remain. Decomposition of unstable clusters induces mechanical devitrification of the packless matrix, which is the reason explaining the rubber-like behavior of the polymer at the plateau of stimulated rubber-like elasticity (cold flow) (refer to Section 2.1).

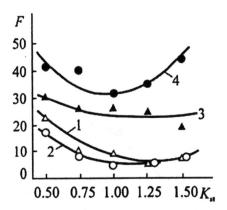

Figure 7.39. Dependencies of functionality F in non-deformed state (1, 2) and after yielding (3, 4) on the curing agent:oligomer ratio K_{st} for epoxy polymers EP-1 (1, 3) and EP-2 (2, 4) [115]

In the framework of the suggested model, let us estimate energy expenses associated with cluster decomposition under the effect of external load. Energy U_{cl} required for association (dissociation) of a couple of segments in the cluster can be calculated by the equation (3.19). Then total energy U_{tot} required for reaching the yield deformation ε_y will be determined as follows [115]:

$$U_{tot} = U_{cl} \Delta v_{cl}. \tag{7.33}$$

On the other hand, U_{tot} value can be determined from σ - ε graph assuming it approximately triangle shaped before the yield stress:

$$U_{tot} = \frac{1}{2}\varepsilon_y \sigma_y. \tag{7.34}$$

Setting equations (7.33) and (7.34) to one another, one can determine theoretical value of the yield stress (σ'_y) and compare it the corresponded experimental ones, σ_y (Figure 7.40). Good conformity between σ'_y and σ_y means that, actually, the yield stress is determined by energy of unstable clusters' decomposition [115].

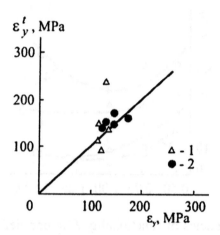

Figure 7.40. Correlation between experimental, σ_y, and calculated by equations (3.19) and (7.34), σ_y^t, values of the yield stress for epoxy polymers EP-1 (1) and EP-2 (2) [115]

As a consequence, the yielding in polymer networks is described in the framework of deformable body synergism, namely, by evolution of dissipative structures. Quantitative DS personification is again obtained in the framework of the cluster model [115].

Let us consider the relation between elastic modules and the yield stress for polymers, which is now suggested to be linear [116]. The results have been obtained [27, 117, 118] indicating linearity of these relations to be a particular case. Figure 7.41 shows the relation between shear modulus G and the yield stress σ_y for two amorphous glassy-like polymers, which is much closer to quadratic than to linear one. In the framework of the cluster model and with respect to interpretation of structural defects, discussed in Chapter 3, the relation between these parameters can be presented as follows (by analogy to equation (3.13)) [117]:

$$\sigma_y = \frac{Gb}{2\pi}\sqrt{\rho_d} \,, \tag{7.35}$$

where b is the Burgers vector; ρ_d is the defects' density.

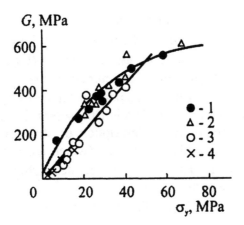

Figure 7.41. Correlation between shear modulus G and yield stress σ_y for PAr (1) and PC (2) [117]

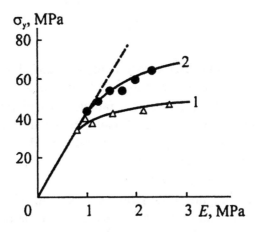

Figure 7.42. Correlation between elasticity modulus E and yield stress σ_y for PHE-Gr-I (1) and PHE-Gr-II (2) composites [118]

Obviously, nonlinearity of the correlation $\sigma_y(G)$ shown in Figure 7.41 is caused by σ_y dependence on two parameters (G and ρ_d). This relation can become linear in particular cases only: $G = $ const, $\rho_d = $ const or when G and ρ_d variation indemnify one another [117].

It has been shown [27] that at extreme variation of the function $E(K_{st})$ the value of σ_y for epoxy polymers is approximately constant. In this case, decrease of E is indemnified by extreme increase of v_{cl} (or ρ_d, refer to equation (3.1)).

Finally, for dispersion-filled composites – polyhydroxyester-graphite (PHE-Gr), a high deviation of σ_y/E relation from linearity already at low volumetric concentration of the filler, φ_n, is indicated (Figure 7.42) [118]. This is stipulated by a strong 'agitation' of structure of the polymeric matrix with φ_n increase (refer to Section 6.6) and corresponded decrease of ρ_d [118].

The nonlinearity under consideration possesses the simple fundamental explanation. As mentioned above, description of structure and properties of thermodynamically non-equilibrium objects (which are polymers) requires, at least, two order parameters. That is why correct description of σ_y using a single parameter (E or G) is the particular case.

For calculation of the yield deformation, ε_y, the following expression was obtained [119]:

$$\varepsilon_y = \frac{\sqrt{2}\ln\left(1/f_f\right)b\left(l_0 C_\infty v_{cl}\right)^{1/2}}{4\pi(1+v)}. \tag{7.36}$$

Correlation (7.36) indicates rather complex dependence of ε_y on three groups of factors. The first of these groups includes molecular characteristics b, l_0 and C_∞; the second group consists of parameters f_c and v determined by the fluctuation free volume of the polymer. Finally, parameter v_{cl} represents purely structural characteristic indicating the local ordering degree of the polymer amorphous state structure. Parameter v_{cl} generally determines temperature dependence of ε_y and its correspondence to structural changes during thermal processing of polymer. Figure 7.43 shows comparison of experimental $\varepsilon_y(T)$ dependencies and theoretical ones, calculated by the equation (7.36), for PAr and PC. It is indicated that calculation by the equation (7.36) does also adequately reflect both the absolute values of ε_y (the difference between theoretical and experimental values is below 10%) and general features of $\varepsilon_y(T)$ dependence. Note also that equation (7.36) contains no adjusting parameters [119].

Let us consider the features of the yielding in amorphous-crystalline polymers. It is known that any property of polymers of this class depends upon the crystallinity degree [120]. It is commonly accepted that the determinative contribution into their properties is made by the crystalline phase. Contrary to the above-considered amorphous glassy-like polymers displaying numerous discrepant concepts of the yielding, its mechanism in amorphous-crystalline polymers is essentially commonly accepted: starting from the basic assumption that the yielding in these polymers represents partial melting and recrystallization of crystalline zones [121]. In the latest works devoted to this problem [83, 123] this point of view on the yielding in amorphous-crystalline polymers remained practically unchanged.

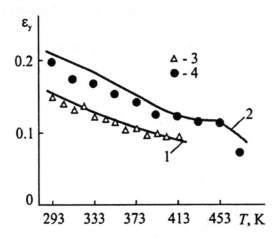

Figure 7.43. Correlation between elasticity modulus E and yield stress σ_y for PHE-Gr-I (1) and PHE-Gr-II (2) composites [118]

Based on this general concept a series of correlations of parameters characterizing the yielding and cold flow processes with characteristics of the crystalline phase of amorphous-crystalline polymers was obtained, among them the relation between the yield stress σ_y and crystallinity degree [83], σ_y and melting heat [122], and σ_y and crystallite size [123]. Hence it has been assumed that yielding and cold flow are independent of the existence of non-crystalline zones. However, recently the works appeared pointing out underestimation of the crystalline zone effect on mechanical properties of amorphous-crystalline polymers (for instance, [39]).

To the above-mentioned concept dislocation models are also corresponded. Hence, defects of the polymer crystalline phase analogous to dislocations in crystal lattices are usually considered.

Figure 7.44 (curves 1 and 2) shows dependencies of σ_y on epoxy polymer concentration, C_e, for HDPE composites containing 0.1 wt.% of superfine Fe/FeO mixture and 1 – 5 wt.% of diane epoxy polymer, injected to the composites with the curing agent during mixture preparation. The dependence of crystallinity degree of the composites, K_ρ, calculated from sample density with appropriate correction introduced for additives' content, on C_e is also shown in this Figure. Curve 1 presents results of blow tests, performed by the Charpy technique; curve 2 shows results obtained for film samples subject to quasi-static stretching. Despite the sharp difference in test conditions, both curves of $\sigma_y(C_e)$ dependence display analogous extreme increase of σ_y with C_e. Of special importance is the fact that such variation of σ_y proceeds with monotonous K_ρ decrease, which avoids the possibility of linear correlation $\sigma_y(K_\rho)$ [124].

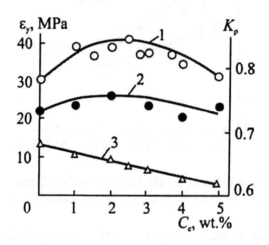

Figure 7.44. Dependence of the yield stress σ_y in blow (1) and quasi-static (2) tests, and crystallinity degree K_ρ (3) on epoxy polymer concentration C_e in HDPE [124]

Since any kinds of admixtures are rejected from crystalline zones during crystallization and grouped on their borders [125], studies of non-crystalline zones structure variation was carried out by measuring their gas permeabilities by oxygen and nitrogen. Hence, it has been assumed that gas transfer proceeds through non-crystalline zones exclusively. Figure 7.45 shows dependencies of normalized gas permeability coefficient P/α_{n-c} (where α_{n-c} is the volumetric concentration of non-crystalline zones) on C_e. Comparing graphs in Figures 7.44 and 7.45 one can observe that σ_y increase corresponds to P/α_{n-c} decrease, i.e. the decrease of relative free volume f_f in non-deformed samples of studied composites. Such relation between σ_y and P/α_{n-c} suggests (combined with the graph $K_\rho(C_e)$, Figure 7.44) that structural conditions of non-crystalline zones in amorphous-crystalline polymers are sufficient for determination of the yielding parameters.

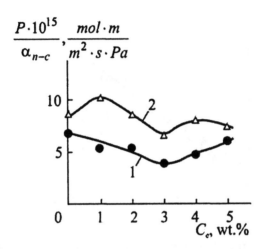

Figure 7.45. Dependencies of gas permeability coefficient P normalized by volumetric content of non-crystalline zones α_{n-c} by oxygen (1) and nitrogen (2) on epoxy polymer concentration C_e in HDPE [124]

It is shown [126] that dependencies $\sigma_y(K_p)$ for five different polyethylenes are approximated well by lines, different for different test temperatures. However, extrapolation of these graphs to finite non-zero K_p values at $\sigma_y = 0$ assumes the presence of, at least, one more factor affecting the yielding of polyethylenes.

Dependencies $\sigma_y(v_{cl})$ for these polyethylenes are also approximated by lines passing though the origin of coordinates at different inclinations, A_v. It has been found that the effect of crystallinity degree on A_v with increase of the absolute value of K_p is of approximate quadratic type. The above-presented results allow plotting the total graph showing dependence of the yield stress for studied polyethylenes on both crystallite stability and structure of non-crystalline zones (Figure 7.46). Clearly for five polyethylenes at different test temperatures a linear dependence of σ_y on the complex structural characteristic $v_{cl} K_p^2$ is observed. For every polyethylene, the value of v_{cl} was varied by changing test temperature under the supposition that K_p is constant in the current temperature range. The latter assumption correlates with experimental results for polyethylenes [39].

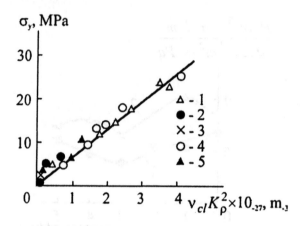

Figure 7.46. Correlation between the yield stress σ_y and complex parameter $v_{cl} K_\rho^2$ for HDPE (1 – 3), LDPE (4) and HDPE + Z composite (5) [126]

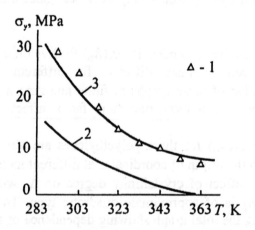

Figure 7.47. Dependencies of the yield stress σ_y on test temperature T for HDPE: 1 – experimental data; 2 – calculation for non-crystalline zones; 3 – calculation with respect to crystalline zones' contribution [127]

As a consequence, the yielding phenomenon in amorphous-crystalline polymers is affected by both the crystalline phase and non-crystalline zones. However, the former effect is much stronger, especially for high-crystalline polymers.

These statements were proved [127] using dislocation analogies, discussed in detail in Chapter 3 (refer also to equation (7.35)). Figures 7.47 and 7.48 show temperature dependencies of σ_y, both obtained experimentally and calculated from this equation, for HDPE and polypropylene (PP), respectively. Expectedly, σ_y values calculated by the cluster network frequency of non-crystalline zone (curves 2) are much lower than the experimental ones. This is explained by association of the yield in amorphous-crystalline polymers with both non-crystalline and crystalline zones [123, 126].

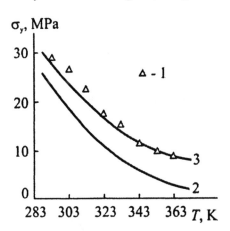

Figure 7.48. Dependencies of the yield stress σ_y on test temperature T for polypropylene (PP): 1 – experimental data; 2 – calculation for non-crystalline zones; 3 – calculation with respect to crystalline zones' contribution [127]

For amorphous-crystalline polymers, the contribution of crystalline zones into σ_y value can be estimated as follows. The density of linear defects, ρ_d^c, corresponded to crystalline zones is defined as the length of macromolecules in these zones (compare with the equation (2.7)) [127]:

$$\rho_d^c = \frac{K}{S}. \tag{7.37}$$

Now substituting ρ_d^c to the equation (7.35) the contribution of crystalline zones into the yield stress can be calculated. Figures 7.47 and 7.48 do also present the comparison of experimental σ_y values and the ones calculated with respect to the total length of defects $(\rho_d + \rho_d^c)$ indicating good conformity of these σ_y values.

Thus the model suggested [127], based on the principally new concept of the structural defect, allows quantitative estimation of contributions from both crystalline and non-crystalline zones to the yielding of amorphous-crystalline polymers.

Similar ideas are used [128] for quantitative description of the yielding for a series of branched polyethylenes (BPE). The parameter K value was presented [84] by the sum as follows:

$$K = \alpha_c + \alpha_b, \tag{7.38}$$

where α_c and α_b are parts of chain units in the proper crystallites and anisotropic interface zones, respectively.

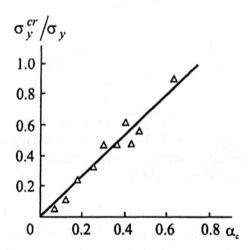

Figure 7.49. Dependence of the relative contribution of crystalline zones, σ_y^{cr}/σ_y, into the yield stress on the part of proper crystallites α_c for a series of BPE [128]

Assessments made by equations (7.35) and (7.37) under the conditions $K = \alpha_c$ indicate the contribution of the crystalline phase to the yield stress (σ_y^{cr}) to be always lower than the experimental values of σ_y. Figure 7.49 shows the dependence σ_y^{cr}/σ_y (the relative contribution of the crystalline phase to σ_y) on α_c. Clearly this dependence is linear, and at $\alpha_c = 0$ the trivial result $\sigma_y^{cr} = 0$ is obtained. Note also this extrapolation suggests $\sigma_y \neq 0$ at $\alpha_c = 0$. At high α_c, contribution of the crystalline phase dominated, and at $\alpha_c \cong 0.75$ $\sigma_y^{cr}/\sigma_y = 1$.

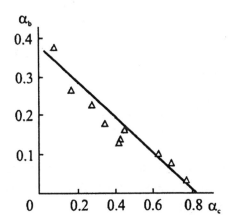

Figure 7.50. Correlation between relative parts of interface zones, α_b, and proper crystallites, α_c, for a series of BPE [128]

Figure 7.50 shows the dependence $\alpha_b(\alpha_c)$ for the same ten BPE. The data in the Figure indicate linear decrease of α_b with α_c increase. Such variation of α_b and simultaneous increase of σ_y^{cr}/σ_y (Figure 7.49) with α_c increase assume a decrease of the local order degree, which determines contribution of non-crystalline zones to σ_y with increase the crystallinity degree. Moreover, expectedly the dependence in Figure 7.50 indicates general concentration of the local order areas of non-crystalline zones of polyethylenes in interface anisotropic zones [129]. At $\alpha_c \cong 0.82$, $\alpha_b = 0$ that correlates with the zero contribution of non-crystalline zones into σ_y, obtained from the data of Figure 7.50 at approximately the same value of α_c.

Figure 7.51 shows comparison of the contributions of crystalline (α_c) and non-crystalline (α_b) zones - experimental and calculated by equation (7.35), into

σ_y for the series of BPE. Clearly theoretical and experimental values fit one another well, except for BPE possessing the highest α. However, the experimental graph $\sigma_y(\alpha_c)$ of the latter BPE displays the overestimated yield stress [84].

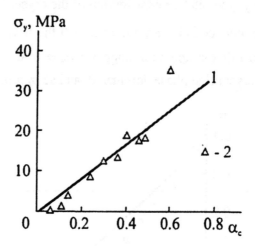

Figure 7.51. Dependence of the yield stress σ_y on the relative part of proper crystallites σ_c for a series of BPE: 1 – experimental data; 2 – calculation [128]

It has been indicated [84, 130] that decline of stress outside the yield stress area of polyethylenes is the more stronger expressed, the higher α_c value is. As indicated above for amorphous polymers, the current stress decline (the "yield peak") is stipulated by decomposition of clusters and is the more so clearer, the higher density of the mentioned clusters is [107]. By analogy to this mechanism one can assume that the mentioned stress decline in amorphous-crystalline polymers will be so higher, the greater the part of crystallites subject to mechanical disordering (partial melting) in the yielding is. The mentioned part of the crystallites, χ_{cr}, can be determined by the difference as follows [131]:

$$\chi_{cr} = \chi - \alpha_{am-ph},\qquad(7.39)$$

where χ ios the part of the polymer subject to elastic deformation, determined from the equation (7.21); α_{am-ph} is the part of the amorphous phase.

Figure 7.52 shows the relation between χ_{cr} and α_c, which indicates χ_{cr} increase with α_c. At low α_c values (≤ 0.37) crystallites are not disordered at all

($\chi_{cr} = 0$). As a consequence, the "yield peak" in the stress-deformation curves is absent, and the graphs obtain the shape typical of rubbers [84, 130]. As a consequence, more intensive stress decrease, $\Delta\sigma$, after the yield stress is induced by χ_{cr} increase [128].

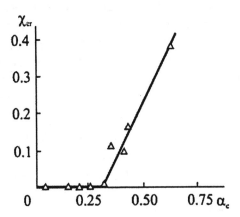

Figure 7.52. Correlation between relative parts of crystallites disordered during the yielding (χ_{cr}) and the proper ones (α_c) for the series of BPE [128]

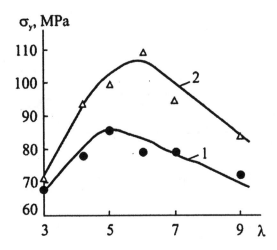

Figure 7.53. Dependencies of the yield stress σ_y on the extrusion stretch λ for UHMWPE-A1 (1) and UHMWPE-bauxite (2) componors [132]

Structural analysis of the yielding in extruded UHMWPE-Al and UHMWPE-bauxite componors was performed in the work [132]. Figure 7.53 shows dependencies σ_y on extrusion stretch λ, which indicate the extreme change of σ_y with λ. Hence, despite lower filler content in the composite, absolute values of σ_y for UHMWPE-bauxite componors are higher than for UHMWPE-Al ones. As mentioned above, presently it is assumed that σ_y is the function of either crystallinity degree, K, [122] or the size of crystallites, L_{200}, [123]. The contributions of these factors can hardly be separated, because their usual variation with the structural state is symbate [83]. Comparison of dependencies $\sigma_y(\lambda)$ (Figure 7.53), $K(\lambda)$, and $L_{200}(\lambda)$ (Figure 7.54) for studied componors indicate that they are not symbate. The more so, absolute values of L_{200} at lower σ_y values are generally higher for UHMWPE-Al than for UHMWPE-bauxite. To put it differently, dependencies $\sigma_y(\lambda)$ for componors cannot be explained in the frames of traditional ideas [122, 123]. For this purpose, the equation (7.35) is used, in which the value of G for oriented polymers was estimated as follows [133]:

$$G = \frac{E}{4},$$

(7.40)

where E is the Young modulus.

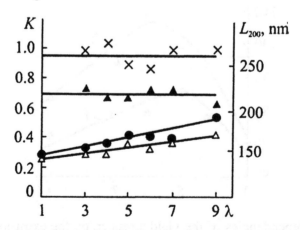

Figure 7.54. Dependencies of crystallinity degree K (1, 2) and the size of crystallites L_{200} (3, 4) on the extrusion stretch λ for UHMWPE-Al (1) and UHMWPE-bauxite (2) componors [132]

The yield stress values, experimental (σ_y) and calculated by the above-mentioned method (σ'_y), for extruded UHMWPE and componors derived from it are compared in Figure 7.55. Clearly quite good conformity of the theory and experiment is observed, though regular increase of σ'_y values in relation to appropriate σ_y values is obtained. This is explained by purely methodical reasons. The absence of the "yield peak" (the stress maximum) on $\sigma - \varepsilon$ graphs of extruded componors induces determination of σ_y as the cross point of tangent lines to elasticity and cold flow segments. Such composition is quite frequently used [134], but always gives underestimated σ_y values.

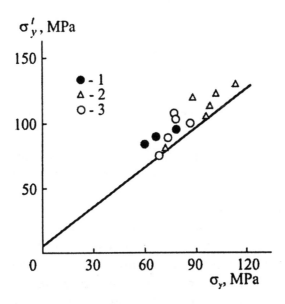

Figure 7.55. Correlation between experimental, σ_y, and calculated by equation (7.35), σ'_y, of the yield stress for UHMWPE (1), UHMWPE-A1 (2) and UHMWPE-bauxite (3) [132]

As a consequence, the results [132] indicate unsatisfactory description of the yielding of oriented filled amorphous-crystalline polymers in the framework of the models associating it with melting (disordering) of the crystalline phase. At the same time, the concept described in [127], which considers the yielding as the consequence of crystallites' disordering and the loss of their mechanical stability in the mechanical stress field, allows its adequate description [132].

In Section 2.2 the correspondence between macroscopic thermodynamic hierarchic and cluster models was considered. The applicability of this interpretation to description of the yielding of a broad selection of amorphous glassy-like and amorphous-crystalline polymers was also shown [135]. As shown above, the yield can be considered as the loss of stability by the polymer structure in the field of mechanical stresses; the resistance measure to this process is the yield deformation, ε_y. It has been indicated [136] that typical lifetime of supermolecular structures (t^{im}) is bound to the Gibbs function of the self association of the cluster structure (ΔG^{im}) at $T = T_g - \Delta T$ as follows:

$$t^{im} \sim \exp(-\Delta G^{im}/RT). \qquad (7.41)$$

Assuming $t^{im} \sim t_y$ (where t_y is the time for reaching σ_y) and taking into account that $\varepsilon_y \sim t_y$, one can write down as follows [135]:

$$\varepsilon_y \sim \exp(-\Delta G^{im}/RT). \qquad (7.42)$$

Note that correctness of the relation (7.42) applied to polymers means regulation of the yielding by thermodynamic stability of supersegmental structures in them.

Figure 7.56 shows dependence of ε_y on $\exp(-\Delta G^{im}/RT)$, corresponded to the relation (7.42), for the polymers considered in the work [135]. Despite a definite (and expected) scattering of data, obviously, all the results are grouped into two branches, each corresponded to different supersegmental structures: the right branch is fitted by data on polymers with negative ΔG^{im} values, and the left branch displays positive ΔG^{im}. The latter group consists of amorphous-crystalline polymers with the packless matrix, devitrified at the test temperature (PTFE, HDPE, LDPE and PP). The cluster model postulates thermofluctuation type of the cluster formation and their decay at $T \geq T_g$. That is why the presence of clusters in devitrified amorphous phase of the mentioned amorphous-crystalline polymers is of absolutely different origin, namely, it is stipulated by strain of amorphous chains during crystallization. This effect induces practical realization of the condition $\Delta G^{im} > 0$, which in the case of low-molecular substances relates to hypothetical, not really existing "overheated liquid →solid" transitions [137].

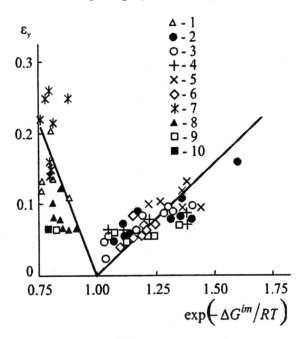

Figure 7.56. Correlation between the yield deformation ε_y and parameter (-$\Delta G^{im}/RT$) for polymers with devitrified (1) and glassy-like (2) of the packless matrix [135]

Another feature proving the fact of existence of the left branch in the graph in Figure 7.56 shall also be noted. Presently, ε_y increase with temperature for polyethylenes and decrease for amorphous glassy-like polymers is commonly known [138]. Clearly this experimental fact is properly explained by the presence of two branches in the graph in Figure 7.56. Subsequently, this represents one more proof for existence of supersegmental structures, quasi-balanced with "free" segments. This quasi-balance is characterized by ΔG^{im} so that in the present case $\Delta G^{im} > 0$.

Expectedly, $\varepsilon_y = 0$ at $\exp(-\Delta G^{im}/RT) = 1$ or $\Delta G^{im} = 0$. The latter condition is reached at $T = T_g$, when the yield deformation is always zero. Note also that the data for amorphous-crystalline polymers with the glassy-like packless matrix (PA-6, PET) fit the right branch of the graph in Figure 7.56. This indicates that the existence of supersegmental structures ($\Delta G^{im} > 0$) is induced directly by glass transited amorphous phase rather than crystallinity. At $T < T_g$ viscosity of the amorphous phase is sharply increased and strain of amorphous chains is unable to perform its action shifting parts of macromolecules. As a consequence, the local order formation is of thermofluctuation type [135].

The results of [135] can give one more, at least fractional, explanation of the "cell effect". As shown above [139], there is an approximate relation between ε_y and Grüneisen parameter γ as follows:

$$\varepsilon_y = \frac{1}{2\gamma}.$$

(7.43)

Using this correlation and the graphs in Figure 7.56, one can easily show that $|(\Delta G^{im}|$ and, consequently, φ_{cl} decrease (Figure 5.10) leads to γ increase characterizing the anharmonicity of intermolecular bonds. This parameter indicates the rate of intermolecular interaction weakening under external (for example, mechanical [139]) influence on the polymer: the higher γ is, the higher the weakening rate of intermolecular interaction is, all other factors being the same. To put it differently, the higher $|(\Delta G^{im}|$ and φ_{cl} are, the lower γ is and the higher resistibility of the polymer to external impact is. Figure 7.57 shows dependence of γ on the mean number of segments in a single cluster, n_{cl}, that strictly illustrates all the above-said [135].

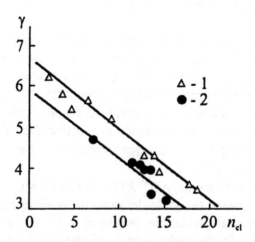

Figure 7.57. Dependence of Grüneisen parameter (γ) on the mean number of segments in a single cluster (n_{cl}) for PC (1) and PAr (2) [135]

Presently, it is commonly accepted that the yield stress (σ_y) is independent of the polymer molecular weight (MW) [140, 141]. This point was revised within the cluster model [142]. Figure 7.58 shows dependencies of the yield stress σ_y and elasticity modulus E on aging duration t_{age} of poly(arylate sulfone) (PAS) film samples [143]. Clearly both σ_y and E are decreased linearly with t_{age} increase. Since the mean molecular weight $\overline{M}_w$ is decreased simultaneously [143], there are grounds for associating all these changes with one another.

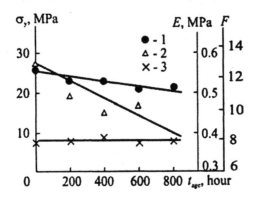

Figure 7.58. Dependencies of the yield stress σ_y (1), elasticity modulus E (2) and functionality of clusters F (3) on aging duration, t_{age}, at $T_{age} =$ 448 K for PAS [142]

Dependencies of entanglement cluster network frequency ν_{cl} and σ_y on $\overline{M}_w$ for aged PAS samples are shown in Figure 7.59. Both parameters decrease symbately with $\overline{M}_w$, MW values corresponding to the bending on the S-shaped property-MM curve [143]. These data give an opportunity to state that correlation $\sigma_y(\overline{M}_w)$ does exist, which is stipulated by lowering of the local order as MW decreases. The following note on this point shall be made. The technique [144] presumes ν_{cl} determination from data on the cold flow plateau. To put it differently, the density of clusters linked by "tie chains" only (the analogue of tie chains linking crystallites in amorphous-crystalline polymers) will be determined. Break of such "tie chains" as the result of thermal degradation during heat aging separates the cluster (or a pair of segments in its composition) from the general deformable structure of the polymer. That is why if even chains with low MM form local order zones (clusters), the absence of "tie chains" eliminates the effect of such clusters on mechanical properties of the polymer (including parameters σ_y and E). In this connection, of interest is dependence $F(t_{age})$ shown in Figure 7.58. As indicated in Section 2.1, test temperature increase causes symbate decrease of v_{cl} and F. To put it differently, the temperature increase varies the number of segments in the cluster (equal $F/2$) not changing total number of the latter. In the case of thermal aging of PAS as v_{cl} is 1.5-forl decreased (Figure 7.59), the number of segments in each cluster remains practically unchanged. This observation suggests either decrease of the number of clusters in the polymer or increase of the number of clusters not linked to one another and, thereby, not included in macromolecular entanglement network, and, as a consequence, not contributing in formation of mechanical properties of the polymer.

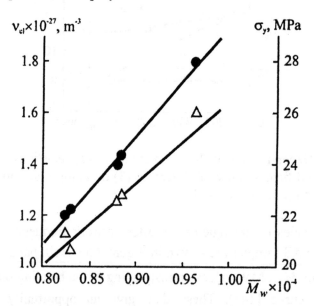

Figure 7.59. Dependencies of entanglement cluster network frequency v_{cl} (1) and the yield stress σ_y (2) on the mean molecular weight $\overline{M}_w$ for aged PAS samples [142]

Thus these results suggest a change in "effective" frequency (i.e. participating in formation of mechanical properties of the polymer) of entanglement cluster network with MM variation in the bending area of the S-shaped property-MM curve and, consequently, existence of σ_y dependence on MM [142].

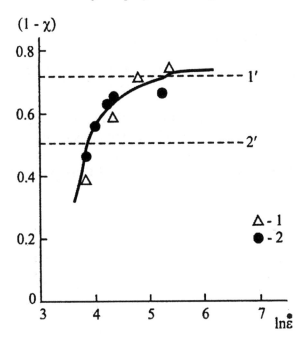

Figure 7.60. Dependence of the part of plastic deformed zones $(1 - \chi)$ on the deformation rate $\overset{\bullet}{\varepsilon}$ presented in logarithmic coordinates for HDPE (1) and PAr (2). 1' - crystallinity degree for HDPE; 2' - the part of packless matrix for PAr [147]

At present, the yield stress increase in polymers with the deformation rate $\overset{\bullet}{\varepsilon}$ is known well [145, 146]. This effect can be described in the framework of the cluster model using two approaches. The first approach [147] is based on application of the fractal concept of polymer plasticity [102]. Using equation (7.21) it is possible to estimate the relative part of elastically deformed polymer $(1 - \chi)$ and thus φ_{cl} value as the function of $\overset{\bullet}{\varepsilon}$ basing on the above-considered condition $(1 - \chi) = \varphi_{cl}$ [103]. Figure 7.60 shows dependence of $(1 - \chi)$ on $\overset{\bullet}{\varepsilon}$ for HDPE and PAr. As mentioned above, the yielding of amorphous polymers includes devitrification of the packless matrix, the relative part of which for non-deformed polymer is indicated by horizontal dashed line in Figure 7.60. Data in this Figure show that as the entire packless matrix is devitrified at the lowest deformation rate $\dot{\varepsilon}$ = 48 s^{-1} of the blow test, just a part of it is devitrified at $\dot{\varepsilon}$ increase. This part equals χ and is so lower, the higher $\dot{\varepsilon}$ is. Since glass transition is a relaxation process [23], this effect was expected.

After that σ_y was estimated as the function of $\dot{\varepsilon}$ theoretically [147] using the equation (7.35). Figure 7.61 gives comparison of experimental and theoretical (calculated in the mentioned manner) values of the yield stress for HDPE and PAr, from which good conformity of these parameters proving the suggested model is observed.

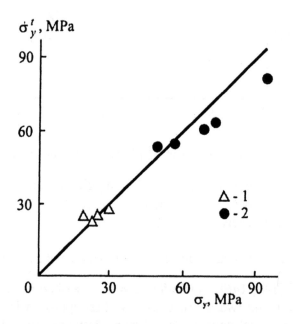

Figure 7.61. Correlation between experimental (σ_y) and calculated by equation (7.35) (σ_y^t) yield stress values for HDPE (1) and PAr (2) [147]

The second approach [148] suggests time dependence of the order parameter $\varphi_{cl}(t)$, the value of which can be determined from the equation (7.16). This calculation can be executed defining t as duration of the linear part of the load – time $(P - t)$ graph in blow tests and accepting φ_{cl}^0 equal the relative part of clusters in quasi-static tensile tests [131]. Figure 7.62 shows dependence of

$\overset{\bullet}{\varphi_{cl}}(t)$ on $\ln \overset{\bullet}{\varepsilon}$ that indicates φ_{cl} increase with the deformation rate, i.e. decrease of the experiment time scale.

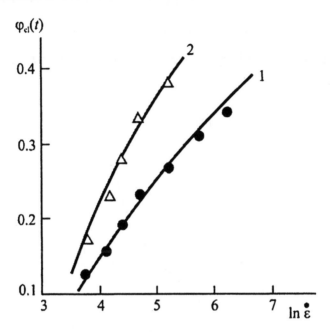

Figure 7.62. Dependencies of the relative part of clusters $\varphi_{cl}(t)$ on the deformation rate $\overset{\bullet}{\varepsilon}$ in logarithmic coordinates for HDPE (1) and PP (20 [148]

The contribution of crystalline zones σ_y^c into the yield stress of HDPE and PP can be determined from equations (7.35) and (7.37). It has been shown [148] that in blow tests the shear modulus of these polymers is independent of $\overset{\bullet}{\varepsilon}$. That is why σ_y^c = const, equal 16.3 MPa for HDPE and 17.6 MPa for PP.

This means that σ_y increase with $\overset{\bullet}{\varepsilon}$ is completely determined by the contribution of non-crystalline zones (σ_y^{n-c}) of the polymers into the yield stress. Parameter σ_y^{n-c} value can be calculated from equation (3.15) with the supposition that the number of segments in closely packed zones ($n_{c.p.}$) varies in proportion to φ_{cl} (Figure 7.62). Figure 7.63 shows dependence of σ_y^{n-c} on the deformation rate $\dot{\varepsilon}$ for HDPE (for PP, the same picture is observed). It is indicated that at σ_y^c = const, σ_y^{n-c} increases with $\dot{\varepsilon}$, and at $\dot{\varepsilon} \cong 150$ s^{-1} $\sigma_y^{n-c} >$ σ_y^c, i.e. the contribution of crystalline zones into σ_y becomes the dominating parameter [148].

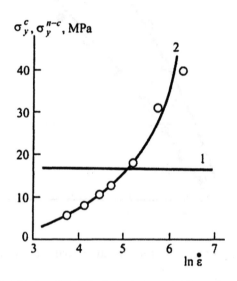

Figure 7.63. Dependencies of contributions of crystalline, σ_y^c, (1) and non-crystalline, σ_y^{n-c}, (2) zones to the yield stress on the deformation rate, $\dot{\varepsilon}$, in logarithmic coordinated for HDPE [148]

Figure 7.64 gives comparison of experimental and theoretical (calculated as the sum $\sigma_y^c + \sigma_y^{n-c}$) dependencies of the yield stress on $\dot{\varepsilon}$ for HDPE and PP. Clearly good correspondence of these parameters is observed. This proves the yielding model for amorphous-crystalline polymers, suggested in [148].

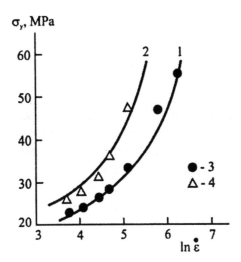

Figure 7.64. Comparison of experimental (1, 2) and theoretical (3, 4) dependencies of the yield stress σ_y on the deformation rate $\dot{\varepsilon}$ in logarithmic coordinates for HDPE (1, 3) and PP (2, 4) [148]

Finally, the following thing shall be mentioned. As follows from the results in the work [148], σ_y increases at simultaneous decrease of E and constancy of G that contradicts to previously suggested proportionality between E and σ_y [116]. The absence of such proportionality is already suggested by the equation (3.12), because parameter v is the function of several structural factors (equation (3.14)). The same conclusion follows from equations (3.15) and (5.41). That is why postulated proportionality of E and σ_y [116] is a particular case only, correct for a structure either invariable or changing monotonously by a specific law.

7.9. COLD FLOW STRESS

General qualitative description of the processes proceeding in the cold flow of amorphous glassy-like polymers can be found in [107]. These processes can be identified as motion of stable clusters interlinked by 'tie chains' in the packless devitrified matrix. High viscosity of this matrix is one of the reasons for transition to the turbulent mode [107, 149]. Similar qualitative deformation model for amorphous glassy-like polymers was suggested by Bekichev and Bartenev [150, 151].

The important additional information on the cold flow can be obtained using ideas of the fractal analysis and irreversible aggregation models (refer to Sections 5.1 and 5.6). Turbulent type of the cold flow is proven in the work [149]. It has been shown [108, 152] that though from geometrical positions multiple slip is the simplest way of plastic yield in metals, it is energetically unprofitable and gives the lowest dissipation efficiency of elastic energy. Limited slip system yielding accompanied by rotations at different structural levels is more effective. This is corresponded to the general statement about higher efficiency of energy dissipation at whirling. That is why in the case of quite high plastic deformations in solids, transition to the turbulent mode of plastic yield is always realized [152].

At high stresses exceeding some critical value, the energy received by deformed medium is accumulated in local highly non-equilibrium zones [113], distributed by the fractal type that provides for effective energy transfer. This process is studied in detail for metals, but the question is arisen about transfer of this statement on plastic yield of amorphous polymers. In this case, principal difficulties are not in the difference of structures of the mentioned materials (polycrystalline for metals and amorphous for polymers), but in their deeper difference, precisely, in the diametrical opposition of their ideal structures [127]. This opposition, discussed in detail in Chapter 3, also implies difference in definition of the plastic yielding mechanism: formation of a dissipative structure (dislocation substructure) for metals and annihilation of already existing dissipative structures (local order zones) for polymers.

It is common knowledge [152] that one of the fundamental properties of turbulent motions is fractality of turbulent structures, which are described using "discontinuous fractals" (refer to Section 5.2). For these fractals, there is a relation between the number of active whirls at the n-th fragmentation step $\langle N_n \rangle$ and typical scale of the turbulent structures L_n, determined from the equation (5.46). It has been suggested [149] to consider $\langle N_n \rangle$ as v_{cl} and apply l_{st} for the scale. In this case, the equation (5.46) is reduced to as follows [149]:

$$v_{cl} \sim l_{st}^{-d_f},\qquad (7.44)$$

where d_f is the fractal dimensionality of the structure.

Figure 7.65 shows the relation between v_{cl} and $l_{st}^{-d_f}$ for PC and PAr, found linear and, as a consequence, correlating with the equations (5.46) and (7.44). Thus parameter v_{cl} can be taken for the number of active whirls at the segmental level [149].

A general equation for estimation of fractal liquid viscosity $\eta(l)$ has been suggested [153]:

$$\eta(l) \sim \eta_0 l^{2-d_f},\qquad (7.45)$$

where l is the typical scale of the yielding; η_0 is a constant.

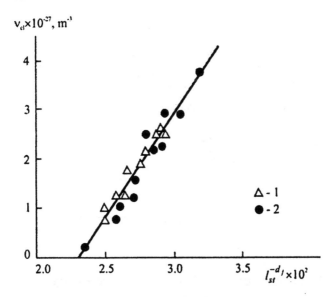

Figure 7.65. Correlation between parameter $l_{st}^{-d_f}$ and entanglement cluster network frequency v_{cl} for PC (1) and PAr (2) [149]

Supposedly, for amorphous polymers, the stress at stimulated rubber elasticity plateau (cold flow), σ_B, will be so higher, the higher viscosity of the polymer, $\eta(l_{st})$, is. In this case, the following condition shall be met [149]:

$$\sigma_B \sim l_{st}^{2-d_f} . \tag{7.46}$$

Figure 7.66 shows linear dependence $\sigma_B(l_{st}^{2-d_f})$ for PC and PAr. As a consequence, the relation (7.46) is met, at least, for the polymers under consideration in [149]. Since the relation (7.45), from which (7.46) was deduced, was obtained for fractal turbulent liquid, linearity of dependence $\sigma_B(l_{st}^{2-d_f})$ is the proof of the turbulent type of cold flow in amorphous polymers [149].

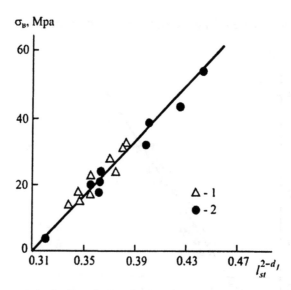

Figure 7.66. Dependence of the cold flow plateau stress σ_B on parameter $l_{st}^{2-d_f}$ for PC (1) and PAr (2) [149]

The above-considered cold flow model of amorphous glassy-like polymers allows two basic conclusions. Firstly, the reasons for transition to the turbulent mode shall be associated with high viscosity of the packless matrix. As a consequence, the part of the latter inflected by a cluster during its displacement appears in the effect zone of the next displacing cluster, is also inflected by it and so on, which leads to the turbulent mode of the cold flow. Obviously, the higher the part of clusters (or v_{cl}) is, the better turbulence of the process is expressed, which is illustrated by the dependence in Figure 7.65.

Secondly, preservation of the stable cluster network with high frequency ν_{cl} at this part of $\sigma - \varepsilon$ graph explains high values of σ_B comparable to the yield stress value [149].

It is common knowledge [138] that a macroscopic polymeric sample is capable of standing quite high stresses after reaching some molecular mass (MM_{cr}) only. The value of MM_{cr} is determined by formation of macromolecular entanglement network able to transmit load by the entire polymeric sample. Thus the entanglement network represents something like a carcass of bonds in metal-insulator alloy [154]. As shown in Chapter 1, two types of macromolecular entanglement network are formed in glassy-like polymers: the "hooking" network and the cluster network. It is discussed [155] which of two above-mentioned macromolecular carcasses determines the stress transmission in macromolecular samples of glassy-like polymers.

The ability to conduct current at a definite conductivity level, g, is obtained by the metal-insulator alloys after reaching the percolation threshold, i.e. when conductive bonds form a continuous percolation carcass [156]. As mentioned above, macroscopic polymer samples become capable of transmitting stress, when a continuous carcass of macromolecular entanglements is formed in them. This obvious analogy allows application of the modern physical conductivity models to non-oriented systems for the purpose of describing dependence of the cold flow plateau stress, σ_B, on frequency of macromolecular entanglement carcass in amorphous polymers. Clearly [157] conductivity depending upon the length scale L - $g(L)$ - is described by the following relation:

$$g(L) \sim L^{\beta}, \qquad (7.47)$$

where the exponent β is determined as follows [158]:

$$\beta = d_f - 2 - \theta. \qquad (7.48)$$

In this equation d_f is the fractal dimensionality of the polymer structure; θ is the index from the equation of distance (r) dependent diffusion coefficient D [158]:

$$D(r) \sim r^{-\theta}. \qquad (7.49)$$

The structure of amorphous glassy-like polymers can be simulated as the totality of a great number of Witten-Sander clusters (refer to Section 5.6), for which the following connection between fractal, d_f, and spectral, d_s, dimensionalities is true [159]:

$$d_s = \frac{2d_f}{1+d_f}. \tag{7.50}$$

Combined equations (7.47) – (7.51) allow obtaining of the simple expression as follows [155]:

$$g(L) \sim L^{-1}. \tag{7.52}$$

For polymers, the natural choice is L determination as the chain segment length between neighbor fixation points of it (chemical cross-link points for polymer networks and physical entanglement points for linear polymers) [158]. The chain segment length between clusters, L_{cl}, can be determined from the equation (5.36), As a consequence and with respect to the above-mentioned analogy, it can be presented as follows [155]:

$$\sigma_B \sim L_{cl}^{-1}. \tag{7.53}$$

Figure 7.67 shows dependence $\sigma_B(L_{cl}^{-1})$ corresponded to the relation (7.53) for PC and PAr. Clearly a good linear correlation is obtained that proves the above-suggested analogy. It is typical that at $L_{cl}^{-1} = 0$ (that corresponds to $\nu_{cl} = 0$ or rubber-like polymer [24]) the dependence $\sigma_B(L_{cl}^{-1})$ is extrapolated to a non-zero value, $\sigma_B \cong 5$ MPa. It is known [53] that at $T \geq T_g$ the macromolecular hooking network displaying frequency $\nu_h < \nu_{cl}$ is the unique carcass of entanglements in the polymer. This fact explains the presence of low but finite σ_B values at $L_{cl}^{-1} = 0$ (or $\nu_{cl} = 0$).

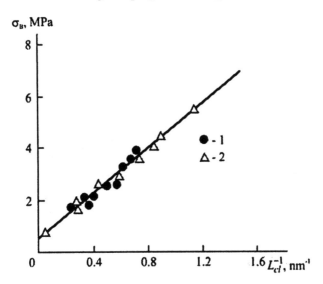

Figure 7.67. Dependence of the stress of stimulated rubbery state (cold flow) plateau σ_B on length of the chain segment between clusters L_{cl} for PC (1) and PAr (2) [155]

Thus the suggested analogy between conductivity and stress transfer in non-oriented structures (amorphous polymers) allows explanation of temperature dependence of σ_B. Analytically, the dependence $\sigma_B(L_{cl}^{-1})$ can be presented by the empirical equation as follows [155]:

$$\sigma_B \approx 5 + 44(L_{cl}^{-1}), \text{ MPa}, \tag{7.54}$$

where L_{cl} dimensionality is nm.

Note that using ideas of the macromolecular hooking carcass the analogous dependence cannot be obtained, because the frequency v_h (and consequently, the length of the chain segment between hookings, L_h) is independent of test temperature [155].

Referring to Section 5.6, the structure of amorphous glassy-like polymers can be simulated as a totality of multiple Witten-Sander (WS) clusters. These clusters possess a compact central zone, which is associated with the "cluster" notion in the simulation [53]. To avoid readings, in the present Section by the "cluster" term this very compact zone of the local order is meant. The molecular friction coefficient, ξ_0, of every particle of the radius a in the cluster is determined as follows [160]:

$$\xi_0 = 6\pi\eta_0 a, \tag{7.55}$$

where η_0 is viscosity of the medium, in which the particle is moving.

For the cluster composed of n_{cl} particles (in the current case, statistical segments), the friction stress σ_{fr}^0 can be expressed by the following equation [160]:

$$\sigma_{fr}^0 = \xi_0 c n_{cl}^{1/d}, \tag{7.56}$$

where d is the cluster dimensionality; c is the coefficient determined from the following expression [160]:

$$c = \frac{1}{a\rho^{1/d}}, \tag{7.57}$$

where ρ is the polymer density.

Substitution of equations (7.55) and (7.57) to the expression (7.56) gives the following [160]:

$$\sigma_{fr}^0 = 6\pi\eta_0 \left(\frac{n_{cl}}{\rho}\right)^{1/d}, \tag{7.58}$$

i.e. the friction stress is independent of the size of particles forming the cluster (in the case under consideration, the cross-section of the macromolecule).

To determine macroscopic stress of stimulated elasticity, σ_B, σ_{fr}^0 shall be multiplied by the number of clusters, N_{cl}, per specific volume, which is determined from the equation (5.27). As follows from this equation and data in Figures 2.1 and 2.2, N_{cl} value is practically independent of the test temperature. Since σ_B is calculated per the sample cross-section and N_{cl} is taken per specific volume, obviously the value $N_{cl}^{2/3}$ shall be used in calculations. Moreover, since in the present simulation displacement of not the entire WS aggregate but of its compact part only is considered, dimensionality of the central compact zone of WS aggregate is accepted for d [161], i.e. $d = 3$, but not the entire WS aggregate dimensionality equal ~2.5 [160].

Based on the above-discussed ideas, one can present the following expression determining σ_B [161]:

$$\sigma_B = 6\pi\eta_0 \left(\frac{n_{cl} N_{cl}^2}{\rho} \right)^{1/2}. \tag{7.59}$$

Figure 7.68 shows dependence of σ_B experimental values on parameter $n_{cl}^{1/3}$ for PS. Clearly this dependence is linear, passes through the origin of coordinates and is approximated by the following empirical equation [161]:

$$\sigma_B = 1.4 \times 10^7 n_{cl}^{1/3}, \text{ Pa.} \tag{7.60}$$

The comparison of equations (7.59) and (7.60) indicates that the graph in Figure 7.68 suggests the condition $\eta_0 = \text{const}$ and the absolute value of η_0 can be determined from the mentioned combined equations. The calculation gives the following result: $\eta_0 = 0.69 \times 10^7$ Pa·s.

The alternative value η_0 can be estimated from the equation [28] as follows:

$$\eta_0 = \frac{\sigma_B}{\dot{\varepsilon}}. \tag{7.61}$$

The assessment by equation (7.61) gives $\eta_0 = 0.82 \times 10^9$ Pa·s for $T = 293$ K. Moreover, as follows from this equation, η_0 is not constant and will be decreased with T increase that contradicts to experimental data from Figure 7.68.

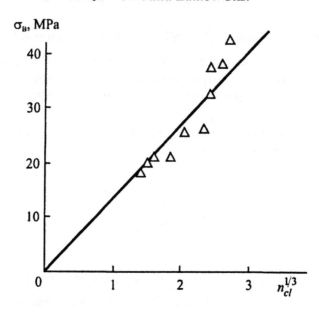

Figure 7.68. Dependence of stimulated rubbery state plateau stress σ_B on parameter $n_{cl}^{1/3}$ for PC [161]

The above-mentioned deviation between η_0 values by two orders of magnitude, determined by the mentioned methods, can easily be explained by in the framework of the model [107]. Clearly calculation by the equation (7.61) is performed under the suggestion that the polymer is present in the glassy state, and the model [107] assumes that displacement of clusters proceeds in mechanically devitrified packless matrix. For the latter case, elasticity modulus E of devitrified polymer can be estimated from the equation for the rubbery state [55]:

$$E = kT\nu_h, \qquad (7.62)$$

where k is the Boltzmann constant; ν_h is the macromolecular hooking network frequency, because in devitrified state clusters are absent.

For such devitrified matrix, the stress σ can be estimated from the following fractal correlation [162]:

$$\sigma = \frac{E}{4.5}\left(\lambda^2 - \lambda^{-2.5}\right), \tag{7.63}$$

where λ is the stretch degree accepted [161] equal 1.20, i.e. corresponded to the point on the graph σ - ε directly after the yield stress.

Calculation by the equation (7.61) using $\sigma \cong 0.33$ MPa (equation (7.63)) gives $\eta_0 = 0.66\times10^7$ Pa·s that correlates well with the estimation by equations (7.59) and (7.60). This correspondence illustrates that the cold flow of the polymer outside the yield stress is possible under the condition of devitrified packless matrix only. Actually, calculation by the equation (7.59) using $\eta_0 = 0.82\times10^9$ Pa·s that correlates well with the estimation by equations (7.59) and (7.60). This correspondence illustrates that the cold flow of the polymer outside the yield stress is possible under the condition of devitrified packless matrix only. Actually for a glassy-like packless matrix, calculation by the equation (7.59) using $\eta_0 = 0.66\times10^7$ Pa·s gives $\sigma_B \cong 1.75$ GPa, which is physically meaningless. This value exceeds theoretical strength of PC and thus the polymer shall be destroyed before occurrence of the cold flow, i.e. before occurrence of macroscopic yield and mechanical devitrigication of the packless matrix.

For two polymers (HDPE and PAr), theoretical estimations of σ_B were performed [161] by the equation (7.59) under the condition η_0 = const = 0.69×10^7 Pa·s. The calculation gave the following σ_B values: 18.0 MPa for HDPE and 47.8 MPa for PAr, which correlate well (within 10% inaccuracy) with the experimental data.

Thus combination of the cluster model of the polymer amorphous state structure and the model of WS aggregates friction at translational motion [160] allows both qualitative and quantitative description of the solid-phase polymers behavior on the cold flow plateau. Hence the cluster model explains feature of the polymer behavior in the mentioned area of σ - ε graph, which cannot be described in the framework of other models [163].

7.10. LOCAL PLASTICITY

Amorphous glassy-like polymers are capable of being deformed by such localized mechanisms of plastic deformation as crazes [164] and shear bands [165]. This circumstance can be considered as the feature of structural inhomogeneity of these polymers subject to the impact of force fields. Nevertheless, under definite conditions continual concepts of the polymer mechanics [166] are widely used for description of behavior of these polymers at deformation. This raises the question about the level of properties, at which the polymer may be considered still as a homogeneous solid. As shown [167] (refer to equation (2.12)), inhomogeneity of the polymer structure determines its affinity to crazing: the higher inhomogeneity is, the easier polymers are crazed. Thus there appears a possibility to obtain direct connection between the polymer structure and its affinity to crazing. In the work [168] it was made performed using the fluctuation theory [167].

In the framework of equation (2.12) the volume V_0 free from hookings is considered as a random value that excludes its connection with the polymer structure. In the framework of the cluster model, the volume V_0 can be represented as a cubic volume of the packless matrix (by definition the volume free from cluster network cross-link points), limited by eight clusters located in the cube corners. Its volume, V_{str}, can be easily determined as the value reverse to the number of clusters N_{cl} per specific volume of the polymer (refer to equation (5.27)).

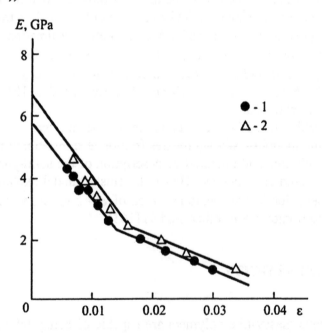

Figure 7.69. Dependencies of elasticity modulus E on deformation ε for PC (1) and PAr (2) [12]

The value of V_{str}, obtained in this manner, can be compared with V_0, the volume necessary for formation of a craze cavity, estimated from the equation (2.12). Let us discuss the physical meaning of parameters used in this equation. The crazing stress, σ_c, can be interpreted differently. Clearly there are stresses of craze nucleation, destruction, etc. [169, 170]. For σ_c the maximum basic stress of the craze nucleation, σ_{yy}, determined by the technique [170], was chosen [168]. Such selection is stipulated by consideration of cavity and, as a consequence, craze nucleation.

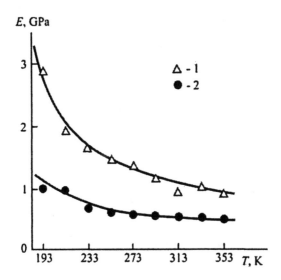

Figure 7.70. Temperature dependencies of elasticity modulus E, determined by tangent of linear area of P-t diagram (1) and deformation $\varepsilon = \varepsilon_c$ by impact resilience method (2) for PAr [168]

More detail consideration is also required for experimental determination of elasticity modulus E, which as mentioned in Chapter 2 is approximately equal volumetric modulus B from equation (2.12) (at $\nu \cong 0.33$, equation (2.13)). It is known [12, 171, 172] that for blow tests of polymers a strong dependence of E on deformation ε is observed, induced by anharmonicity of intermolecular bonds. For PC and PAr, this dependence is shown in Figure 7.69. As we are talking about nucleation of crazes, the selection of E at deformation of the craze nucleation, ε_c, will be correct. For PC and PAr, ε_c can be accepted equal ~0.02 [173]. Figure 7.70 shows comparison of E values, determined by the slope of the linear part of load-time ($P - t$) graph in blow tests and estimated by the elastic recoil method [12] at $\varepsilon \cong 0.02$. Obviously, the former are much (by 1.5-fold, approximately) higher than the latter. This deviation will affect accuracy of V_0 assessment by the equation (2.12). Finally, as before, T_g was accepted for T_0 [168].

Comparison of temperature dependencies of V_0 and V_{str} for PC and PAr are shown in Figures 7.71 and 7.72, respectively. Calculations of V_0 by equation (2.12) using E values, estimated by the slope of the linear part of $P - t$ graph, give overestimated values of this parameter compared with V_{str}. Nevertheless, it shall be noted that in this case also V_0 and V_{str} display values of the same order of magnitude and identical shapes of temperature dependencies. In accordance with the above-mentioned reasons, much better correspondence is obtained under application of E values at $\varepsilon = \varepsilon_c$. In this case, maximal deviation does not exceed ~22%.

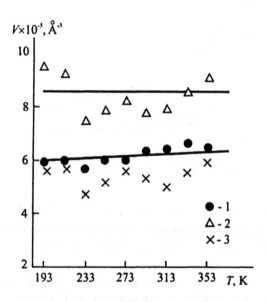

Figure 7.71. Temperature dependencies of volumes V_{str} (1) and V_0 (2, 3) for PC [168]

Typically, V_{str} value does not change with temperature, thus representing the polymer property. Figure 7.73 shows correlations between v_{cl} and F for PC and PAr, well approximated by straight lines. This means that v_{cl} increase with temperature decrease (Figure 2.1) proceeds by addition of segments to already existing ones, but not by formation of new clusters. Extrapolation of $v_{cl}(F)$

graphs to $F = 2$ gives a finite value, which is the characteristic of the "hooking" network with the frequency $\sim 0.4 \times 10^{27}$ m^{-3}, which is close to experimental values of ν_h for these polymers [174].

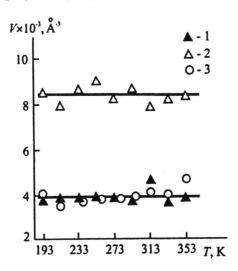

Figure 7.72. Temperature dependencies of volumes V_{str} (1) and V_0 (2, 3) for PAr [168]

Essentially, analogous interpretation of the interconnection between crazing processes and the entanglement ("hookings") network has been suggested by Michler [175]. However, contrary to the above-mentioned model, it is of speculative and qualitative character.

As a consequence, the value of V_{str} due to its correspondence to V_0 characterizes affinity of the polymer to crazing. It is illustrated by comparison of V_0 for PC and PAr ($\sim 6.0 \times 10^3$ Å^3 and $\sim 3.9 \times 10^3$ Å^3, respectively) and V_0 for polystyrene (PS) obtained in [167] ($\sim 450 \times 10^3$ Å^3). Clearly PS possesses much higher tendency to crazing, than PC and PAr, which was numerously proved by experiments [176, 177]. Note also that V_0 decrease does also reduce the critical level of sizes, above which continual concepts can be applied to the polymer. In practice PC deformation is realized by the diffusion method [178] and PS by localized "rough" shear bands [165].

Donald and Kramer [179 – 181] have suggested micromechanical models of crazing in amorphous polymers, where realization probability of one elastic deformation mechanism or another is determined by the structural factor, which is macromolecular hooking network frequency ν_h. As demonstrated by this model, ν_h increase intensifies the shift mechanism and appropriate suppression of the crazing one. Kramer [182] has suggested the following formula for effective surface energy Γ:

$$\Gamma = \gamma_B + \frac{1}{4}d_{ent}\nu_h U_b,$$ (7.64)

where γ_B is the van-der-Waals surface energy; d_{ent} is the distance between entanglement network points; U_b is the energy of chemical bond break in the backbone.

The increase of ν all other conditions being the same determines Γ increase, which in its turn implies fibrillation stress increase and, as a consequence, prevents craze formation. Henkee and Kramer [183] show also that formation of fibrils in the craze requires "configurative loss" of entanglements, which can be realized by two methods: break of macromolecules or their slipping and further yielding from the entanglement [184].

The advantage of the above-considered model by Donald and Kramer is its connection of the craze structure and macromolecular entanglement network density, which is the important structural parameter of block polymers (it should be noted that both ν_h as the structural characteristic and crazing as the mechanism of elastic deformation are typical of the polymer state of the substance, and thus the interconnection between them suggests itself). Predictive validity of the mentioned model is increased by correlations between entanglement network parameters and molecular characteristics of polymers, observed in the works [174, 185]. Limitations of the alternative of macromolecular entanglement ("hooking") network, chosen by the authors of the model, are equally obvious. It is common knowledge [141, 186] that ν is estimated from the test results of polymers above T_g and is accepted equal the value, obtained in this manner (i.e. the constant), in the whole temperature interval of existence of the glassy state, i.e. at $T \leq T_g$.

Clearly the scheme of explanation of the crazing mechanism temperature dependence was limited by the selected structural model of the polymer. It is assumed that the "geometrical loss" of entanglements is realized by macromolecule break at relatively low (about room) temperature, and at higher temperature by their "extrication" (chain yield from the entanglement). To explain high-temperature shear – crazing transition Donald [141] has suggested the following model, based on the schematic temperature dependence of yield (shear) stress and crazing, shown in Figure 7.74.

For high-molecular PS "extrication" of macromolecules is hindered. That is why the crazing stress σ_c is reduced with temperature slower than for low-molecular PS (refer to equation (7.64)). If the yield stress σ_y is higher than σ_c, than crazing is the dominating process and, vice versa, at $\sigma_y < \sigma_c$, the yield dominate. It is assumed [141] that σ_y depends on the polymer molecular mass. Since the stretch degree of fibrils in the craze, λ_c, depends on molecular mass of the chain segment between "hookings", M_1, the concept of macromolecule "extrication" with temperature was suggested for explanation of λ_c increase with T [187, 188]. For this concept it is assumed that the coefficient of friction between macromolecules, ξ_0, is high at relatively low temperatures, and reduced load breaks the backbone. While temperature is increased, ξ_0 decreases and the load allowing slipping of the macromolecule becomes lower than the chemical bond strength, which determines the advantage of the first process. Slipping of macromolecules in a "pipe" leads to "extrication" or η and subsequent σ_c decrease. In the framework of this concept experimental results on temperature dependencies of λ_c and Γ were explained [91, 188].

Nevertheless, the above-considered concept induces draw some objections. The most important (from the authors' point of view) of them are considered below [189].

1) It is common knowledge [190, 191] that the crazing stress is the function of temperature and increases monotonously with T decrease, three-fold increasing with T decrease from 300 to 100 K [191]. If the "geometrical loss" of the entanglement already at room temperature proceeds by the mechanism of macromolecules break exclusively, then the reason for the mentioned increase of σ_c is not clear.

2) The relation between σ_y and σ_c (Figure 7.74) and so much their absolute values cannot be accepted as the unique index of realization simplicity of one mechanism or another. It has been shown [129, 192] that HDPE modification and cross-linking induce $2 - 3$-fold increase of shear intensity (that can be easily traced by sizes of "shear lips" on the surfaces of samples' destruction), hence σ_y is not decreased but even increased by $5 - 10\%$. Thus simultaneous suppression of crazing is observed [129].

3) Some doubts are raised on the independence of σ_y of the polymer molecular mass, postulated by Donald [141] (refer to Section 7.8).

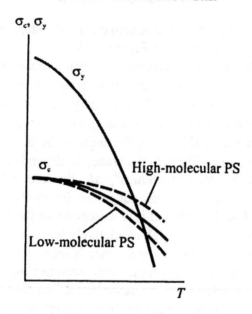

Figure 7.74. Schematic image of temperature dependencies of shear yield and crazing stresses (σ_y and σ_c, respectively) on the example of PS [141]

4) Macromolecule slipping in the "pipe" under the effect of load and, as a consequence, its escape from the entanglement may hardly be considered to be the one-direction process. A speculative three-stage model as follows has been suggested for this phenomenon: first a local balance without slipping is set; secondly, the chain is stretched; and finally, a statistical coil configuration due to micro-Brownian motion is restored. The three stages mentioned were suggested for polymeric melt. For the glassy-like polymer Donald [141] considered two initial stages only and assumed the third one absent due to appearance of the stretched chain. However, it should not be forgotten that the point of discussion is the active zone at the polymer - craze interface, where increased molecular mobility can take place due to closeness of a free surface and/or mechanical devitrification of the polymer [104]. Note also that formation of macromolecular entanglements during "healing" of polymers by means of diffusion is known well even for the stressed state [193, 194].

5) Assessment of σ_c temperature dependence for PS under blow loading conditions has indicated [195] its decrease from 25 to 18 MPa with temperature increase in the range of 193 – 353 K. The more so, at $T \cong 313$ K a maximum of σ_c is observed, corresponded to β-transition in PS [196]. Berger and Kramer

[187] have estimated the "extrication" time for macromolecules, τ_{extr}, equal $\sim 10^2 \div 10^3$ s. PS sample deformation time in blow tests is by $5 - 6$ orders of magnitude shorter that makes "extrication" the process of low probability. Nevertheless, craze formation energy in the temperature interval of $313 - 353$ K for PS is much lower than at $T < 353$ K [195].

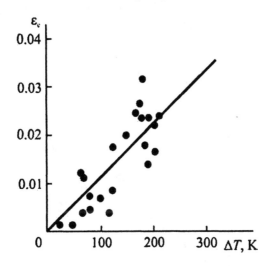

Figure 7.75. Dependence of crazing deformation ε_c on the temperature difference $\Delta T = T_g - T$ [189]

6) Finally, the main reason for development of the alternative interpretation of the temperature dependence of crazing and shear mechanisms is non-correspondence of results from works by Kambour *et al.* [173, 197, 198] to the Donald model [141]. The authors of the works [173, 197, 198] have obtained empirical linear correlations between crazing parameters (crazing deformation, ε_c; maximum main stress, σ_{yy}; hydrostatic stretch, P_s) and the basic parameters of block polymers (T_g and cohesion energy density, W_c), as well as between σ_y and E. The correlations are as follows:

$$\varepsilon_c \sim \frac{W_c(T_g - T)}{E}, \tag{7.65}$$

$$\varepsilon_c \sim \frac{W_c(T_g - T)}{\sigma_y}, \tag{7.66}$$

$$\sigma_{yy} \sim W_c(T_g - T), \tag{7.67}$$

$$P_s \sim W_c(T_g - T). \tag{7.68}$$

The analogy between equation (5.71) and correlations (7.65) - (7.68) shall be noted. This presents additional proofs for the benefit of the alternative interpretation, suggested in [189].

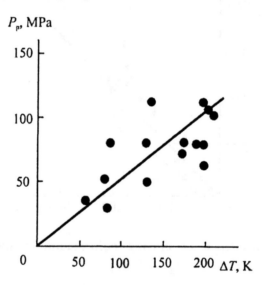

Figure 7.76. Dependence of hydrostatic stress P_s on the temperature difference $\Delta T = T_g - T$ [189]

Presently, the well-known are correlations between E and W_c [76, 199], E and σ_y [116], and W_c and T_g [200]. That is why in the first approximation the correlations (7.65) - (7.68) reflect the dependence of ε_c (or σ_{yy} and P_s) on the temperature difference $\Delta T = T_g - T$ or, to put it differently, on test temperature approach to T_g. To prove this approximation, Figures 7.75 and 7.76 show correlations $\varepsilon_c(\Delta T)$ and $P_s(\Delta T)$, plotted by data of the works [173, 198], respectively. Though these graphs display higher dispersion than the correlations (7.65) and (7.68), the tendency of ε_c and P_s increase with ΔT is obvious. Therefore, the reasons for increase of dispersion of the data are based on deviation of results from the straight line for correlations $E(W_c)$ and $T_g(W_c)$, physical bases for which are discussed in detail in works [76, 199, 200]. Obviously the increase of parameters ε_c, σ_{yy} and P_s means the progressing difficulty of crazing realization and at equal ΔT all polymers must display equal probability of crazes' formation. In the framework of the model, described in [179, 180], this condition should mean equal ν_h values. However, as mentioned above, the "hooking" network is "frozen" in the whole interval of temperature of the glassy-like state existence, i.e. ν_h is constant. This inconsistency is improved in the concept [141] by introduction of chain slipping and their escape from the entanglement with temperature increase.

Let us now consider alternative interpretation of the temperature dependence of elastic deformation mechanisms [189], based on the ideas of the cluster model. It is common knowledge [201] that the critical stretch of polymers λ_{cr} (as well as λ_c for fibrils in crazes or λ_{dz} for deformation zones (DZ) [202]) is determined by the length of macromolecule segment between macromolecular entanglement network points, l_{ent}. Qualitatively, this correlation postulates that l_{ent} increase induces growth of λ_{cr}. Thus deformation ε_c in crazes of DZ can be determined as follows [203]:

$$\varepsilon_c = \ln\left(\frac{r_c + \delta_c}{r_c - \delta_c}\right), \tag{7.69}$$

where r_c and δ_c are length and disclosure of the craze (or DZ), respectively; $\lambda_c = 1 + \varepsilon_c$ [138].

Figure 7.77 shows dependence of λ_{dz} on M_{cl} displaying qualitative correspondence to the concepts [201]: λ_{dz} increases with M_{cl} [204].

Plummer and Donald [188] have measured stretches for crazes and DZ (λ_c and λ_{dz}, respectively) in PC, Temperature dependencies of which are shown in Figure 7.78 (curves 1 and 2). Parameter λ_c begins increasing sharply at $T \cong 383$ K (compare with T_g' for PC, Figure 2.1), whereas parameter λ_{dz} remains almost unchanged. Theoretical value of the critical stretch in crazes, λ_c, and DZ, λ_{dz}, can be calculated from the following equation [179, 180]:

$$\lambda_c\,(\lambda_{dz}) = \frac{l_{eng}}{R_{cl}}. \tag{7.70}$$

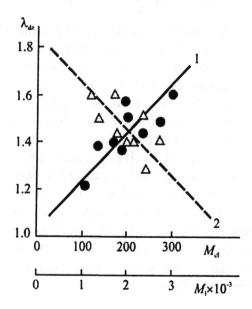

Figure 7.77. Dependence of the critical stretch in DZ, λ_{dz}, on molecular mass of the chain segment between clusters, M_{cl} (1), and hookings, M_h (2), for PAS films, formed from different diluters [204]

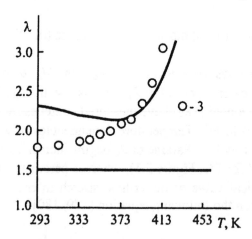

Figure 7.78. Experimental dependencies of the critical stretch in crazes, λ_c (1), and DZ, λ_{dz} (2), on test temperature, T, for PC [188]; 3 – calculation by the equation (7.70) [189]

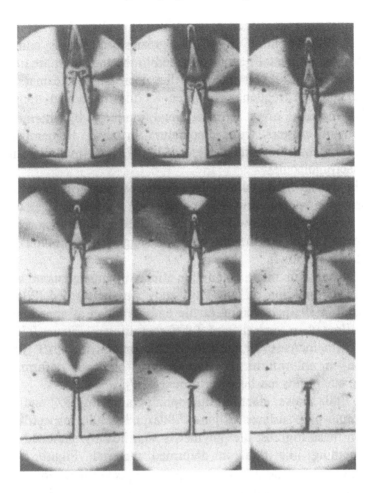

Figure 7.79. Optical microphotograph showing DZ development in PAS film formed from solution in chloroform with sample deformation increase. Magnification 18× [205]

Points in Figure 7.78 show calculation results of λ_c and λ_{dz} for PC. Therefore, their comparison with the experimental data by Plummer and Donald [188] is quite amazing. For PC, calculated and experimental λ_c values in the crazing zone (as mentioned above, the start of this zone corresponds to T_g') coincide well, which quire unexpected paying attention to the fact that they were obtained in different laboratories and for different PC trademarks. In the zone of DZ realization ($T \leq 343$ K) λ_{dz} values correlate quite well by both the shape of temperature dependence and absolute values (deviation does not exceed 15%). But the more so, of the greatest interest is comparison of theoretical and experimental results, obtained in the transition area from DZ to crazes ($T = 333 - 373$ K). It indicates that decrease of the macromolecular entanglement network frequency, required for the mentioned transition, and, consequently, increase of λ is determined uniquely by thermofluctuation decomposition of clusters with temperature increase. Remember that for explanation of this transition, the concepts [91, 187, 188] involve the increase of slipping of macromolecules.

Thus the cluster model allows the alternative interpretation of changes in elastic deformation mechanisms and parameters regulating it at temperature variation for amorphous glassy-like polymers. Both qualitative and quantitative correspondence of results of the interpretation suggested to previously obtained experimental data is shown.

Let us discuss in more detail the effect of entanglement network on formation and deformation of DZ. This study was performed [204] for PAS films possessing a sharp undercut, at the end of which at stretching a DZ is formed. The films were formed from 9 different diluters. Figure 7.79 show DZ development with increase of the sample deformation for PAS film, formed from solution in chloroform. Fractal analysis of local deformations and degradation of such films has been performed [205].

Figure 7.80 shows electron microphotographs of DZ and elastically deformed sample part surfaces (Figure 7.80a and 7.80b, respectively). As is observed from these Figures, the polymer is significantly stretched in DZ: the globules something like float in deformed material. Figure 7.77 shows correlation between λ_{dz} and M_{cl} displaying the expected behavior: λ_{dz} increase with M_{cl} [204]. The dependence $\lambda_{dz}(M_{ent})$, also shown in this Figure, displays the opposite tendency. This allows a supposition that in the glassy state of PAS parameters of DZ are regulated by the entanglement cluster network. For the same stretched PAS samples without undercuts, comparison of λ_{dz} values with the experimental destruction stretch, λ_{exp}, indicated very close absolute values and tendencies in parameter changes (Table 7.5).

Clearly [179] both crazes and DZ are formed in locally oriented polymer and differ by the presence of cavities in the former ones. The presence of cavities in crazes determines "geometrical losses" of entanglements [179], due to which the stretch of fibrils in crazes exceeds λ_{exp}. The equality of λ_{exp} and λ_{dz} indicates the absence of entanglement losses in DZ. Thus the polymer stretch in

DZ does not practically differ from stretch of the same sample without undercut, except for its localization at the undercut apex – the stress concentrator [204].

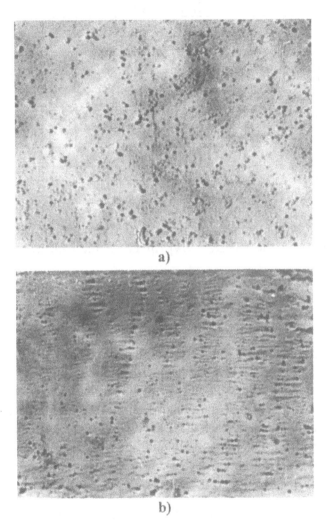

a)

b)

Figure 7.80. Electron microphotographs of DZ (a) and elastically deformed sample (b) surfaces for PAS film, formed from solution in chloroform. Magnification 5625× [204]

Table 7.5

Experimental and calculated stretch of PAS films formed from different diluters [204]

Diluter	λ_{dz}	λ_{exp}	λ_1	λ_2
Chlorobenzene	1.62	-	1.98	3.07
Tetrahydrofuran	1.58	1.55	1.60	2.07
N,N-dimethylformamide	1.58	1.62	1.36	1.75
Dichloroethane	1.55	-	1.38	1.79
1,4-Dioxane	1.50	-	1.90	2.47
N,N-dimethylacetamide	1.44	1.46	1.79	2.33
Methylene chloride	1.43	1.57	1.49	1.94
Chloroform	1.29	1.49	1.20	1.56
1,1,4,4-Tetrachloroethane	1.70	1.73	-	-

There are several methods for theoretical assessment of the stretch, two of which were used in the studies [204]. In the framework of the rubber state concept [206]:

$$\lambda_1 = n^{1/2}, \tag{7.71}$$

where n is the number of independently moving kinetic units of the macromolecule between the entanglement cross points.

Another method assumes that the critical stretch λ_2 can be calculated by the equation (7.70) [179]. Table 7.5 shows λ_1 and λ_2 values, calculated by the mentioned method. All samples display good conformity between experimental and theoretical data. Of special attention is the fact that λ_2 values are approximately 1.5-fold higher than λ_{dz} ones. This can be associated with the type of plasticity localization zone: for crazes $\lambda_2/\lambda_c \cong 1$, and for DZ it equals ~1.4 [179].

The above-shown data on PAS deformation in DZ allow a conclusion that the cluster model adequately describes both macroscopic and local deformation processes [204].

Local plasticity zone (either crazes of shear yield zones, the so-called "shear lips") represent the most important element of deformed polymers under quasi-friable (quasi-viscous) degradation. Type of the local plasticity zone determines the degradation type: if at the apex of the defect a craze is formed, then degradation of the polymeric sample is quasi-brittle, and in the case of local shear yield zone – quasi-viscous [166]. Transition of the deformation mechanism from crazing to shear yield is usually considered to be the brittle-viscous process [138]. Usually, increase of sizes of both craze [207] and shear yield zone [208] causes to viscosity increase at polymer degradation, characterized by critical coefficient of the stress intensity, K_{Ic}, or critical rate of deformation energy release, G_{Ic}. Essentially, the unique theoretical method of the local plasticity zone assessment, r_p, is the Dagdale-Barenblatt equation [166]:

$$r_p = \frac{\pi}{8} \frac{K_{Ic}^2}{\sigma_y^2}. \qquad (7.72)$$

Though equation (7.72) gives good correspondence to experimental data for both crazes [207] and local shear zones [88], it does not bind parameter r_p to any structural characteristics of polymers. Obviously, this circumstance prevents forecasting of r_p value and, correspondingly, parameters of quasi-brittle (quasi-viscous) degradation of polymers. In some works [209, 210] the percolation theory is used for obtaining such interconnection, which assumes formation of an infinite cluster bracing the system at the percolation threshold [211]. Formation of the percolation cluster in the polymer structure at temperature decrease to T_g using the cluster model was discussed above in detail (refer to Section 5.5). If one assumes r_p to be the length of the cluster network correlation, it can be written down as follows [211]:

$$r_p \sim \left(\varphi_{cl} - \varphi_{cl}^{cr}\right)^{\nu_p}, \qquad (7.73)$$

where φ_{cl}^{cr} is the percolation threshold; ν_p is a critical index.

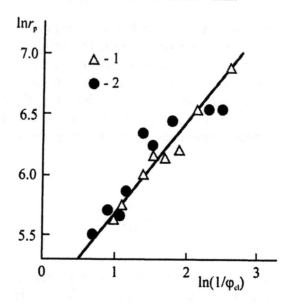

Figure 7.81. Correlation between local plasticity zone length, r_p, and relative part of clusters, φ_{cl}, corresponded to the relation (7.73), in double logarithmic coordinates for HDPE (1) and PS (2) [209]

As shown in Section 5.5, for the polymer structure, the percolation threshold on the temperature axis is T_g (or T_{melt}). As approaching it, φ_{cl} is reduced and may become indefinitely small. That is why in the first approximation it is accepted [209, 210] that $\varphi_{cl}^{cr} = 0$. Figure 7.81 shows dependence $r_p(\varphi_{cl}^{-1})$ in double logarithmic coordinates for HDPE and PS, corresponded to the correlation (7.73). Clearly data for both studied polymers fit the same linear dependence, which allows determination of v_p value equal ~0.76. This result correlates well with the classical v_p value equal ~0.80 [211].

Let us consider physical meaning of the critical index v_p. Percolation cluster is a fractal objects with dimensionality d_f, for which correlation (5.76) is true. Thus the index v_p becomes clearly physically meaningful. Estimation of d_f by the equation (5.76) gives the value ~2.63 that correlates with d_f values, obtained by other methods [100].

It should be noted that parameter v_p characterizes the part of the percolation system, which surrounds the percolation carcass [212]. As mentioned in Section 5.5, as applied to polymers this means that parameter v_p characterizes the packless matrix surrounding the cluster network (the percolation carcass), in which the whole fluctuation free volume of the polymer is concentrated [53]. As a consequence, plasticity zone formation is, expectedly, bound to the packless component of the polymer structure and its free volume. It should also be noted that combination of expressions (5.76) and (7.73) represents the typical example of necessary application of, at least, two order parameters to description of structure and properties of polymers. In the case under consideration these parameters are φ_{cl} and d_f [210].

Thus for the first time structural meaning of the local plasticity zone length in polymers was cleared out [209, 210]. The use of the percolation theory has indicated that independently of the polymer class and the type of local plasticity zone its length is the length of the polymer structure correlation, deformed before occurrence of inelastic deformations. Such approach gives a possibility of forecasting r_p value in the framework of the model suggested and, consequently, forecasting of polymer properties at quasi-brittle (quasi-viscous) degradation.

7.11. DEGRADATION STRESS

At present the approach is known well [138], in which deformation-strength behavior of polymers is associated with the macromolecular entanglement network existing in them. As a consequence, it can be assumed that the strength of non-oriented glassy-like polymers depends on the number of entanglements per specific area in direction transverse to the load application. Bersted [213] has deduced the following expression for the degradation stress σ_d:

$$\sigma_d = \frac{Q\left(T,\overset{\bullet}{\varepsilon}\right)}{2}\left(\frac{2}{3}\right)^{5/6} N_A^{1/3}(E_0 S U_b)^{1/2}\left[\frac{\rho}{M_{cr}}\left(1-\frac{M_{cr}}{\overline{M}_n}\right)\right]^{5/6}, \qquad (7.74)$$

where Q is an adjustment coefficient considering the difference between the test temperature and the one, at which mobility of macromolecules is completely suppressed; N_A is the Avogadro number; E_0 is the longitudinal elasticity modulus of the crystal lattice; S is the cross-section of the macromolecule; U_b is the energy of covalent bond break; ρ is the polymer density; M_{cr} is the critical molecular mass, at reaching which a network of macromolecular entanglements is formed in the polymer; $\overline{M}_n$ is the mean molecular mass.

The value of M_{cr} is determined as follows [185]:

$$M_{cr} \cong 2M_{ent}. \qquad (7.75)$$

Substituting a series of values for E_0, S and U_b, Narisawa [138] has reduced the equation (7.74) (as applied to calculations of σ_d for PE and PET) to the following form:

$$\sigma_d = 1.4 \times 10^5 \left[\frac{\rho}{M_{cr}} \left(1 - \frac{M_{cr}}{\overline{M}_n} \right) \right]^{5/6}, \text{ Pa}. \qquad (7.76)$$

From the totality of the above-mentioned parameters, the highest dispersion is displayed by E_0. Even for polyethylene, the polymer the most well studied in this relation, values of E_0 obtained by different methods varies within the range from 40 to 400 GPa [215]. That is why the constant before the square brackets in the equation (7.76), used in the work [214], was applied as the adjustment parameter equaled 1.17×10^5 for both polymers under consideration, PC and PAr. The following circumstance shall be mentioned. The value of M_l for the "hooking" network is determined at $T > T_g$ and thus is accepted constant at temperatures below T_g. Since the degradation stress is the temperature function [138], Bersted [213] has considered it with the help of coefficient Q, which is the implicit technique. For the entanglement cluster network, M_{cl} is the function of temperature increasing with it (refer to equation (2.2)). As a consequence, in this interpretation σ_d decrease with temperature growth is stipulated by M_{cl} increase that seems much more natural. For the strength characteristic, the value of the true degradation stress, σ_d^{true}, was accepted [214].

Comparison of temperature dependencies of σ_d^{true} for PAr, obtained by the experiment and theoretically by equation (7.76), is shown in Figure 7.82. Clearly from this Figure, substitution of M_{cl} values of the cluster network into the equation (7.76) gives quite accurate presentation of the experimental dependence $\sigma_d^{true}(T)$, whereas substitution of M_h values corresponded to the "hooking" network, is inconsistent with both absolute values of σ_d^{true} and the temperature dependence [214]. It should be noted that coincidence of σ_d^{true} values for M_h (M_{cl}) ones, determined by both techniques, is observed at $T = 493$ K only, i.e. at the glass transition temperature of PAr [214]. This is explained by the fact that at T_g the cluster network decays, and the single "hooking" network remains in the polymer.

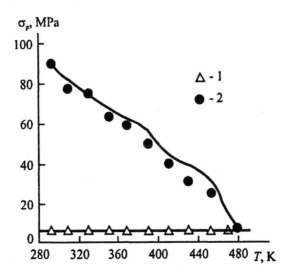

Figure 7.82. Dependencies of the degradation stress, σ_d^{true}, on the test temperature, T, calculated for the "hooking" (1) and cluster (2) network. The Upper dependence is composed by experimental data [214]

For PC, calculation was performed in the reverse manner [214], e.g. based on experimental values of σ_d^{true} the value of M_h was calculated by the equation (7.76). Figure 7.83 compares M_h values, calculated in this manner, and appropriate M_{cl} values, calculated by equations (2.1) and (2.2). Data from this Figure indicate that parameters of the cluster network do correspond to the experimental stretch strength of PC by both absolute values and the temperature dependence.

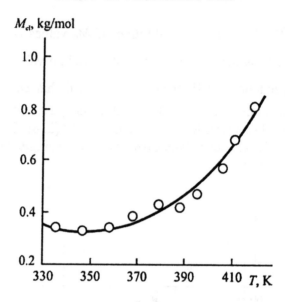

Figure 7.83. Temperature dependencies of molecular mass, M_{ent}, of the chain segment between entanglements for PC (points mark M_{ent} calculated by equation (7.76); curve shows M_{cl} calculation by equations (2.1) and (2.2)) [214]

As a consequence, the strength level observed for amorphous glassy-like polymers (about several ten MPa) can be explained by the presence of the entanglement cluster network of high frequency, which is approximately by two orders of magnitude higher than that of the "hooking" network [214]. Analogous results were obtained [216] using the model by Mikos and Peppas [217], which indicated that the strength of amorphous-crystalline polyethylenes at blow destruction is determined by ν_{cl} value. Moreover, as shown in the previous Section on the example of PAS, the critical stretch λ_2 is also regulated by the cluster network parameters [204].

7.12. BLOW VISCOSITY

At present it commonly accepted [218] that the main source of high blow viscosity of amorphous-crystalline polymers with the amorphous phase devitrified at the test temperature (similar to polyethylenes) is the structural composite, mentioned above. However, this conclusion possesses rather speculative bases and is not proved by any quantitative assessments, which is stipulated by the absence of appropriate models. The cluster model and the fractal analysis allow filling this blank in the knowledge on behavior of mechanically loaded amorphous-crystalline polymers. This has been implemented [219, 220] on the example of HDPE samples, which is typical representative of the class of polymers under consideration.

In the equation (7.21) parameter χ characterizes the part of polymer not participating in plastic but subject to elastic deformation. For amorphous-crystalline polymer, this part includes devitrified amorphous phase and a part of the crystalline phase, subjected to partial mechanical disordering [208]. To put it differently, parameter χ characterizes structural state of deformed polymer. For HDPE considered in works [219, 220], the crystallinity degree $K_\rho = 0.687$ and, as a consequence, the part of amorphous phase, φ_a, equals $1 - K_\rho = 0.313$. As indicated in assessments, performed by the equation (7.21), parameter χ for HDPE depending on the undercut length and test temperature varies within the range of $\sim 0.40 - 0.83$, i.e. exceeds φ_a. Since the yield in polymers is realized in closely packed zones (crystallites and clusters), some part of the crystalline phase shall be necessarily disordered, and its value χ_{cr} can be determined from the equation (7.39). Such disordered component of HDPE structure can also be the effective dissipater of the blow energy. Thus the following structural components can be suggested for the role of parameter determining high plasticity of HDPE: devitrified packless matrix of the amorphous phase, the part of which equals $\varphi_{p.m.}$; disordered part of the crystalline phase equal χ_{cr}; their sum equal $(\varphi_{p.m.} + \chi_{cr})$.

The part of energy dissipated during the blow loading, η, can be estimated by the equation (5.52), in which index β_d represents the fractional part of the degradation surface fractal dimensionality d_d, which, in its turn, is determined by equations (5.53) and (5.54) for quasi-brittle and quasi-viscous degradation types, respectively. These types of degradation are limited by the condition $v = 0.35$ [108]. As shown in Section 5.2, for polymers the self-similarity coefficient, Λ, equals the characteristic relation C_∞ (for HDPE, $C_\infty = 6.8$ [68, 185]).

Figure 7.84 shows correlations between η and $\varphi_{p.m.}$, χ_{cr} and $(\varphi_{p.m.} + \chi_{cr})$ parameters for HDPE in the temperature range of $293 - 333$ K. Clearly increase of any structural components among the mentioned ones induces η increase; however, the relation $1 : 1$ was obtained for $(\varphi_{p.m.} + \chi_{cr})$ only. This correlation of full energy dissipation is typical of rubber, which is presented by both devitrified packless matrix $(\varphi_{p.m.})$ and mechanically disordered part of crystallites (χ_{cr}). As a consequence, the mentioned structural components of the deformed state but not the amorphous phase itself, the part of which (clusters) does not participate in the blow energy dissipation, determine energy dissipation in HDPE [220].

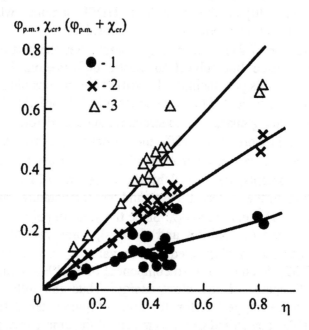

Figure 7.84. Correlation between parts of packless matrix, $\varphi_{p.m.}$ (1), disordered crystallites, χ_{cr} (2), and their sum, ($\varphi_{p.m.} + \chi_{cr}$) (3), and the part of dissipated energy, η, for HDPE samples with a sharp undercut in the temperature interval of 293 – 353 K [219]

It may be expected that, ultimately, parameter η is determined by the polymer structure, the state of which can be characterized by its fractal dimensionality d_f. Figure 7.85 shows dependence $\eta(d_f)$ for HDPE, which was found linear and displayed η increase with d_f. Such type of the dependence was expected, because d_f approaching 3 means approaching the rubber-like state [100], for which $\eta = 1$. It is typical that at $\eta = 0$ dimensionality $d_f = 2.5$, which characterizes the transition border to brittle degradation [108]. The value $\eta = 1$ is reached at $d_f \cong 2.95$ that correlates with the Poisson coefficient, maximum possible for solids: $\nu = 0.475$ [108]. Analytically, the correlation shown in Figure 7.85 can be presented in the following form [220]:

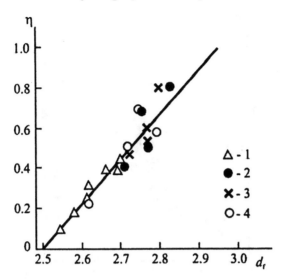

Figure 7.85. Dependence of the part of dissipated energy, η, on the fractal dimensionality of the structure, d_f, for HDPE samples with sharp undercut at $T = 293$ K (1), 313 K (2), 333 K (3), and 353 K (4) [215]

$$\eta \cong 2.1(d_f - 2.5). \qquad (7.77)$$

In this case, the fractal dimensionality of the degradation surface d_d from equation (5.52) and with respect to $\Lambda = C_\infty$ can be presented as follows:

$$d_d = 2 - \frac{\ln(6 - 2d_f)}{\ln C_\infty}. \qquad (7.78)$$

As follows from the equation (7.78), d_d value (and, consequently, blow viscosity A_d) is determined by structural (d_f) and molecular (C_∞) characteristics of HDPE [219].

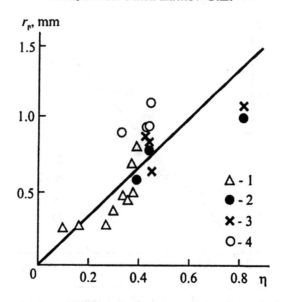

Figure 7.86. Dependence of "shear lips" size r_p on the part of dissipative energy η for HDPE samples with sharp undercut at $T = 293$ K (1), 313 K (2), 333 K (3), and 353 K (4) [219]

It is common knowledge [129] that for HDPE, the main mechanism of blow energy dissipation is the local shear deformation, which follows from linear increase of A_d with r_p ("shear lips"). Based on these ideas a definite correlation between r_p and η should be expected. The curve $r_p(\eta)$ shown in Figure 7.86 proves this supposition. Expectedly, η increases with r_p. At $\eta \cong 1$ parameter r_p equals 1.8 mm, approximately. In the Brown model [221] it is assumed that the brittle-viscous transition is realized under the following condition:

$$r_P = \frac{B}{2},\qquad (7.79)$$

where B is the sample width.

For HDPE samples, the brittle-viscous transition is realized at $v \cong 0.35$ or $\eta \cong 0.42$ (equations (5.4) and (7.77)), corresponded to $r_p \cong 0.85$ mm (Figure 7.86), which is much lower than the value suggested by the criterion (7.79) (at B

= 6 mm). For HDPE, correlation between r_p and η can be presented in the form as follows [220]:

$$r_p \cong 2\eta, \text{ mm,} \qquad (7.80)$$

or with respect to equation (5.4):

$$r_p \cong 4.2(d_f - 2.5), \text{ mm.} \qquad (7.81)$$

It is common knowledge [138] that stimulation of molecular mobility induces energy dissipation and polymer viscosity increase at degradation. As indicated in Section 7.6, molecular mobility can be characterized by the fractal dimensionality of the chain segment between molecular entanglement points D ($1 < D \leq 2$). Variation of the fractional part of D (i.e. $D - 1$) from 0 to 1 gives the full possible spectrum of molecular mobility of this chain segment. Figure 7.87 shows the correlation between η and $(D - 1)$, which possesses the expected type and, analytically, is presented as follows [220]:

$$\eta \cong D - 1. \qquad (7.82)$$

With respect to equation (5.4), it is obtained [220]:

$$D \cong 2.1(d_f - 4.25), \qquad (7.83)$$

that for two border cases gives: $D = 1.0$ at $d_f = 2.5$ and $D = 1.945$ at $d_f = 2.95$ [220].

For polymers, the equation (5.52) can be reduced with respect to correlations $\Lambda = C_\infty$ and $\eta = D - 1$ as follows [219]:

$$d_d = 2 - \frac{\ln(2 - D)}{\ln C_\infty}. \qquad (7.84)$$

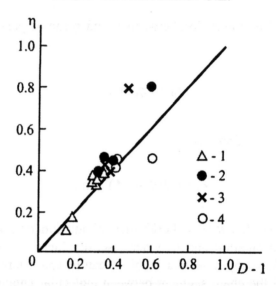

Figure 7.87. Correlation between the part of dissipated energy η and fractional part of the fractal dimensionality of the chain segment between entanglements D for HDPE samples with sharp undercut at $T =$ 293 K (1), 313 K (2), 333 K (3), and 353 K (4) [219]

This equation is remarkable, because it binds fractal dimensionality of the degradation surface d_d to two specific factors, most important for polymers: chain mobility and rigidity. Moreover, the present correlations allow association of microscopic (D and d_f) and macroscopic (r_p) parameters of the structure and degradation process with the polymer properties. For instance, equations (7.80) and (7.82) can be generalized as follows [220]:

$$r_p = 2(D - 1), \text{ mm.} \tag{7.85}$$

The equation (7.85) suggests that the fractal dimensionality d_d, determined at the microlevel by correlation (7.84), can be associated with d_d value at the microlevel by appropriate use of r_p value. In this interpretation, fractal dimensionality of degradation surfaces of HDPE sample can be attributed quite definite meaning in the material science. If it is assumed that schematically the "shear lips" can be presented similar to the upper insertion in Figure 7.88, it is suitable to model them by elements of Koch figures [222], shown in the lower insertion of the Figure. In this case, fractal dimensionality of the Koch figure is determined as ln4/ln3 = 1.263, i.e. as the relation of hyperbolic logarithms of the number of elements after transformation and before it. It is obvious from schematic cross-section of the degradation surface that since the size of "shear lips", r_p, can be arbitrary, the number of elements is fractional. Then by analogy with the Koch figures, macroscopic fractal dimensionality of the degradation surface, d_{pm}, is determined as follows [223]:

$$d_{pm} = \frac{\ln(B + 2r_p)}{\ln B},$$ (7.86)

where coefficient 2 at r_p reflects the presence of two "shear lips" on the degradation surface of HDPE.

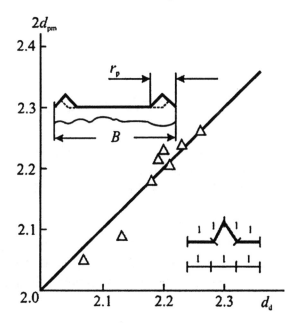

Figure 7.88. Correlation between fractal dimensionalities of degradation surfaces at the macrolevel, $2d_{pm}$ (equation (7.86)), and microlevel, d_d (equation (7.84)), for HDPE samples with a sharp undercut at $T = 293$ K. Insertions: upper – schematic cross-section of the sample with "shear lips"; lower – Koch curve elements illustrating determination of "shear lips" d_{pm} [219]

Figure 7.88 shows comparison of fractal dimensionalities of the degradation surfaces, d_d and d_{pm}, calculated by equations (7.84) and (7.86), respectively. This means conformity of fractal dimensionalities of the degradation surfaces at micro- and macrolevel in the case of HDPE blow degradation.

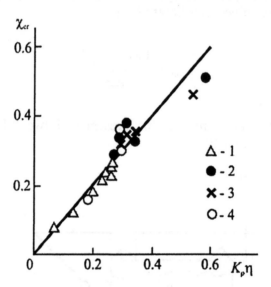

Figure 7.89. Correlation between the part of disordered crystallites χ_{cr} and multiplication $K_\rho \eta$ for HDPE samples with a sharp undercut at T = 293 K (1), 313 K (2), 333 K (3), and 353 K (4) [219]

Further on, let us consider interconnection between initial and deformed HDPE structure. Figure 7.89 shows practically full coincidence of χ_{cr} and $K_\rho \eta$ values, i.e. as suggested before, the relation χ_{cr}/K_ρ is the parameter that determines η value. Following from the data in Figures 7.84 and 7.89, it can be presented [219]:

$$\chi_{cr} + \varphi_{p.m.} = \frac{\chi_{cr}}{K_\rho}, \qquad (7.87)$$

or

$$\chi_{cr} = \frac{\varphi_{p.m.} K_\rho}{1 - K_\rho}. \qquad (7.88)$$

As a consequence, increase of the undercut length leads to mechanical glass transition in the amorphous phase of HDPE, and simultaneously with $\varphi_{p.m.}$ decrease under the condition K_p = const χ_{cr} decreases. Both these effects lead to decrease of the part of dissipated energy, decrease of the "shear lips" size and reduction of blow viscosity A_p. Essentially, the equation (7.88) demonstrates the interconnection between the initial HDPE structure and after deformation.

The results shown indicate that the packless matrix of devitrified amorphous phase and the part of the crystalline phase, disordered during deformation, are structural components determining the blow energy dissipation and, consequently, blow viscosity of amorphous-crystalline polymers. Fractal analysis allows correct quantitative description of processes proceeding at blow loading of HDPE. Existence of an interconnection between characteristics of the initial polymer structure and changes in it proceeding during deformation is of importance [219, 220].

Dependence of blow viscosity, A_p, on entanglement cluster network frequency, v_{cl}, for amorphous glassy-like polymers is briefly discussed by the authors of [142]. Figure 7.90 shows dependence of blow viscosity, A_p, on entanglement cluster network frequency, v_{cl}, for aged PAS samples in double logarithmic coordinates. Linearity of it allows obtaining of the following empirical correlation between A_p and v_{cl} [142]:

$$A_p \sim v_{cl}^{3.7}. \tag{7.89}$$

Clearly power correlation of equation (7.89) suggests extremely strong dependence of A_p on v_{cl}. This observation is proved by the experiment. Decrease of v_{cl} by, approximately, 1.5 times leads to A_p decrease by more than an order of magnitude [143]. The reason for such abrupt decrease of A_p is weakening of the shear mechanism (which is the effective dissipater of the blow energy) with v_{cl} decrease [183]. Analogous correlation between A_p and v_{cl} was also obtained for polyethylenes [192]. Equation (7.89) combined with data in Figure 7.59 does also explain strong dependence of PAS blow viscosity on molecular mass of samples [224]. It should be noted that in the case under consideration A_p decrease is accompanied by reduction of σ_y that contradicts to the basic statements of the model [141].

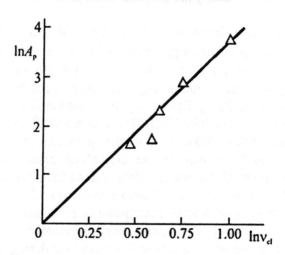

Figure 7.90. Dependence of blow viscosity, A_p, on entanglement cluster network frequency, v_{cl}, in double logarithmic coordinates for aged PAS samples [142]

7.13. GAS PERMEABILITY

The free volume theory is successfully used for interpretation of diffusion processes in polymers above the glass transition temperature. However, its application to the glassy-like state meets some definite difficulties [225]. Therefore, analysis of polymer diffusion properties below T_g (T_{melt}) in the framework of two interconnected interpretations has been performed [226, 227]: the fluctuation free volume model and the cluster model.

Figure 7.91 shows dependence of the gas permeation coefficient per nitrogen, P_{N_2}, on relative free volume, f_t, determined from the equation (2.33). Parameter P_{N_2} is considered without coefficient 10^{17} (refer to Table 2.6), i.e. in fact, in dimensionless units. Clearly this dependence is approximated well by a straight line that testifies about close connection of gas transfer processes and the free volume in polymeric materials [75].

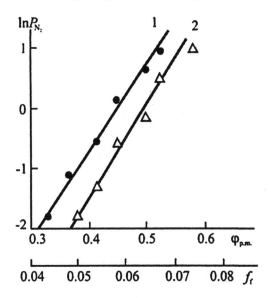

Figure 7.91. Dependencies of gas permeability coefficients per nitrogen, P_{N_2}, on the part of packless matrix, $\varphi_{p.m.}$ (1), and fluctuation free volume, f_f (2), for HDPE + Z composites. Parameter P_{N_2} is considered in dimensionless units (with no regard to coefficient 10^{17}, refer to Table 2.6) [227]

As shown in Chapter 2 (refer to Table 2.6), injection of 0.05 wt.% of Z into HDPE induces extreme decrease of P_{N_2}. Table 7.6 shows some structural characteristics for three most typical HDPE + Z composites. Clearly comparison of Tables 2.6 and 7.6 indicates that at practically constant crystallinity degree, K, the gas permeability coefficient per nitrogen for HDPE + Z composite containing 0.05 wt.% of Z is 17-fold lower, approximately, than for initial HDPE, and K decrease for HDPE + Z composite containing 1.0 wt.% of Z causes not increase but decrease of P_{N_2}. That is why explanation of the effect observed shall be searched for in structural changes of the non-crystalline zone. It is known [228] that transfer processes in amorphous-crystalline polymers are usually realized via the amorphous phase. It may be assumed that these processes are realized in the packless matrix, because the whole fluctuation free volume is concentrated in it directly. Dependence $P_{N_2}(\varphi_{p.m.})$ in logarithmic coordinates was also found linear (Figure 7.91). This implies correctness of the above-suggested structural interpretation of diffusion processes in polyethylenes [226].

Table 7.6
Crystallinity degree K and size of crystallites L_{002} for HDPE + Z composites [227]

Z content, wt.%	K	L_{002}, nm
0	0.68	28.0
0.05	0.65	25.0
1.00	0.52	26.5

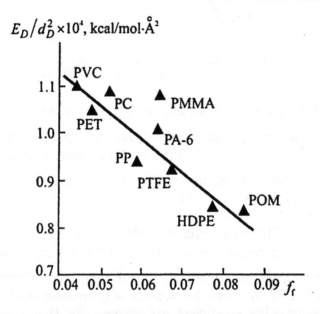

Figure 7.92. Dependence of the relation between diffusion activation energy to square diameter of diffusing molecule, E_D/d_D^2, on the relative fluctuation free volume, f_f, by data from [225] at $T = 293$ K [227]

As shown above [225], there is a definite correlation between the relation of the activation energy of diffusion to the square diameter of diffusing particles, E_D/d_D^2, and the free volume. Linear decrease of E_D/d_D^2 with f_f increase was observed [225]. Figure 7.92 shows dependence of the ratio E_D/d_D^2, accepted from the data of [225], on f_c value, estimated by the equation

(2.33), at $T < T_g$. Clearly this dependence is linear and expectedly displays the following type: f_f increase induces decline of E_D/d_D^2. The data correspond to temperature equal 293 K. Taking into account linear correlation $f_f(\varphi_{p.m.})$ (Figure 5.21), one can state that analogous correlation does also exist between E_D/d_D^2 and $\varphi_{p.m.}$ [227].

The cluster model allows consideration of non-crystalline (amorphous and interphase) zones as a microcomposite consisting of clusters and the packless matrix [229], gas transfer processes realizing in the latter structural component only [230]. The interpretation mentioned allows application of the gas permeability model for composites, suggested by Nielsen [231], to structural analysis of the gas permeability of amorphous-crystalline polymers. In the framework of this model, the relative permeability coefficient for composites is presented as follows:

$$\frac{P_{comp}}{P_{pol}} = \frac{P_{ip}}{P_{pol}\varphi_{fl}^{dp} + P_{ip}\left(1 - \varphi_{fl}^{dp}\right)}\left(\frac{\Phi_{l-ip}}{\tau^*}\right) + \left(\frac{\Phi_{pol} + \Phi_{l-pm}}{\tau}\right), \qquad (7.90)$$

where P_{comp}, P_{pol}, and P_{ip} are permeability coefficients of the composite, the polymer matrix and the interphase layers, respectively; φ_{fl} is the volumetric part of the filler; dp is the index for discrete particles of the filler equal 1/3; Φ_{l-ip} and Φ_{l-pm} are volumes of liquid dissolved in the interphase layer and polymeric matrix, respectively; Φ_{pol} is the volumetric part of the polymeric matrix; τ^* and τ are the sinuosity coefficients showing increase of the penetrating agent path compared with the sample thickness for the interphase layer and the polymeric matrix, respectively.

Let us consider the methods for estimating parameters in the equation (7.90) in relation to the structure of HDPE + Z composites. The analogue of the composite in the suggested simulation are non-crystalline zones, through which the transfer processes proceed; thus the analogue of P_{comp} is the permeability coefficient of HDPE + Z composite (P), and for P_{pol} the analogue is the same parameter of the packless matrix of the amorphous phase, P_{ap}. For φ_{fl}, φ_{cl} shall be accepted. It has also been suggested [230] that $\Phi_{l-ip} \cong \varphi_{ip}$, $\Phi_{l-pm} = 0$, and due to the above-mentioned ideas Φ_{pol} is accepted equal the relative part of the packless matrix, $\varphi_{p.m.}$

Table 7.7
Structural parameters and permeability coefficients for HDPE + Z composites [230]

Z content, wt.%	0	0.05	1.0
φ_{cl}	0.12	0.18	0.13
$\varphi_{p.m.}$	0.20	0.06	0.35
φ_{ip}	0	0.11	0.03
τ	1.06	-	1.02
τ^*	-	1.06	-
$P_{ap} \times 10^{-17}$, mol/m·s·Pa	4.55	-	2.33
$P_{ip} \times 10^{-17}$, mol/m·s·Pa	-	0.303	-
$P^e \times 10^{-17}$, mol/m·s·Pa	0.860	0.056	0.816
$P_{par}^t \times 10^{-17}$, mol/m·s·Pa	0.910	0.172	0.826
$P_{subs}^t \times 10^{-17}$, mol/m·s·Pa	9.17	0.59	4.57

The value of τ (or τ^*) can be determined from the equation as follows [231]:

$$\tau = 1 + \frac{W}{2L}\varphi_{cl},\tag{7.91}$$

where W and L are the thickness and the length of filler particles (or clusters), respectively.

For clusters $W \cong L$; therefore, [230] we obtain:

$$\tau = \tau^* = 1 + 0.5\varphi_{cl}.\tag{7.92}$$

Since it is assumed [129] that clusters are basically concentrated in partly oriented interphase zones, the value τ^* was determined from the equation (7.92), and τ was accepted equal one (Table 7.7).

Further on, let us consider the assessment of permeability coefficients for structural components of HDPW + Z composites. As for the initial HDPE $\varphi_{ip} = 0$ [129], the equation (7.90) is reduced to the following form [230]:

$$\frac{P}{P_{ap}} = \varphi_{p.m.}\tag{7.93}$$

The equation (7.93) allows determination of P_{ap} for HDPE, the value of which is shown in Table 7.7. For HDPE + Z composite containing 1.0 wt.% of Z, P_{ip} estimation has given a negative value, which is physically meaningless. That is why P_{ap} value, determined for the mentioned composite, was found approximately two-fold lower than for the initial HDPE. Using this P_{ap} value and equation (7.90), P_{ip} value was calculated [230] for HDPE + Z composites containing 0.05 wt.% of Z. The results shown in Table 7.7 indicate approximately an order of magnitude lower P_{ip} compared with previously calculated P_{ap} values. This result was expected due to two factors: partial ordering of interphase zones [232] and concentration of closely packed in them [129].

Calculation of theoretical permeability coefficient values, P^t, involved two models: parallel and subsequent transfer of the penetrating compound through amorphous and interphase zones. In the first case, P_{par}^t (for parallel transfer) is determined as follows [231]:

$$P_{par}^t = P_{ap}\varphi_{p.m} + P_{ip}\varphi_{ip}.$$ (7.94)

For subsequent transfer, the expression is as follows [231]:

$$\frac{1}{P_{subs}^t} = \frac{\varphi_{p.m.}}{P_{ap}} + \frac{\varphi_{ip}}{P_{ip}}.$$ (7.95)

Table 7.7 shows comparison of P_{par}^t and P_{subs}^t values with experimental ones, P^e. As indicated the subsequent model gives overestimated P_{subs}^t values, whereas P_{par}^t displays quite good coincidence with the experiment.

One more structural difference of high importance in the analysis of gas permeability for composites and amorphous-crystalline polymers shall be noted. For composites it is assumed [231] that the permeability coefficient for filler – polymeric matrix interfaces is much higher than the appropriate value for the bulky polymeric matrix. That is why decline of the relative part of these layers induces decrease of P_{comp} for the composite. For amorphous-crystalline polymers, due to the above-mentioned reasons, the opposite effect is observed (refer to Table 7.7).

As a consequence, the above-mentioned results of the structural analysis of amorphous-crystalline HDPE + Z composites' gas permeability allow the conclusions as follows. For two structural components, through which penetrating gas is transferred (amorphous and interphase zones), permeation degree of the former one is much higher (by an order of magnitude, approximately) than of the latter. Practical absence of interphase zones in the initial HDPE and their low content in HDPE + Z composite containing 1.0 wt.% of Z determine high P values for the polymers under consideration. Vice versa, high concentration of interphase zones in HDPE + Z composite containing 0.05 wt.% of Z determines decline of P by more than an order of magnitude. Probably, the observation that the transfer process in films from studied polymers is realized by the parallel mechanism is stipulated by low thickness of these films [230].

The presence of a jump of the diffusion coefficient, D, at the melting temperature, T_{melt}, of polyethylenes is the clearly determined experimental fact [233]. A theoretical model describing the temperature dependence of D has been suggested [233]. It presents two different equations for temperature intervals $T < T_{melt}$ and $T > T_{melt}$. It should be noted that the present model does not consider structural differences of the amorphous phase of polyethylenes in the amorphous-crystalline ($T < T_{melt}$) and viscoelastic ($T > T_{melt}$) states. Description of the dependence $D(T)$ and the jump of D at T_{melt}, in particular, in the framework of the cluster model, which takes into account the above-mentioned structural differences, has been presented [234].

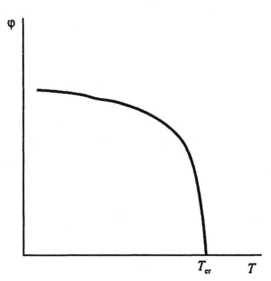

Figure 7.93. Schematic presentation of temperature dependence of the order parameter φ [212]

Let us discuss physical grounds for existence of D jump at T_{melt} in the suggested interpretation. As supposed in the model suggested [233], structure of the amorphous phase is identical for the amorphous-crystalline polyethylene and its melt. In this case, the role of the crystalline phase is reduced to D decrease proportional to the crystallinity degree. Moreover, it has been assumed [233] that the limiting stage of gas diffusion in the amorphous-crystalline polyethylenes is the transfer through the places of crystallites' contact that should be considered a rather arbitrary suggestion. It is common knowledge [235] that transition of the crystalline phase into amorphous one is the most closely part of non-crystalline zones in polyethylenes (refer to Table 7.7). Thus there are no reasons for expecting both direct contact of crystallites and increased porosity in places of such contacts, if they take place.

In the framework of the interpretation suggested, structures of the amorphous phase in amorphous-crystalline polyethylene and its melt differ by the presence of local order zones (clusters) in the former one, through which gas transfer processes do not proceed, too. Occurrence of these zones at $T \le T_{melt}$ is caused by crystallization and chain stretch in amorphous interlayers, induced by it (refer to Section 2.5). Some part of the local order zones is also preserved at $T > T_{melt}$, but in this case, the transfer will be of the dynamic origin and short-living [38, 236]. The jump of D at T_{melt} in the frames of the most general physical ideas is stipulated by the specificity of the order parameter variation at critical temperature T_{cr} (for polyethylenes, $T_{cr} = T_{melt}$). Figure 7.93 schematically shows the temperature dependence of the order parameter $\varphi(T)$, which indicates extremely fast φ increase in a narrow temperature interval $T \le T_{melt}$ [21]. As shown in Section 5.5, the dependence $\varphi_{cl}(T)$ can be described using the thermal cluster concept, in which φ_{cl} is determined by the equation (5.74) at $\beta = 0.4$ [234]. For temperature interval of about 5 K corresponded to D jump [233] and $T_{melt} = 403$ K [38], $\varphi_{cl} = 0.176$ is obtained, which is the direct reason for the jump mentioned [234].

Much simpler and speculative illustration of the physical meaning of D jump at $T = T_{melt}$ is given by the following empirical equations [237]:

$$D = D_0' e^{13\varphi_{p.m.}}, \qquad (7.96)$$

$$D = D_0'' e^{62.5 f_f}, \qquad (7.97)$$

where D_0' and D_0'' are constants.

Though equations (7.96) and (7.97) were deduced for description of oxygen diffusion in a series of amorphous and amorphous-crystalline polymers [237], nevertheless, similarity of diameters of O_2 and CH_4 molecules allows their use for preliminary analysis of CH_4 diffusion. For example, for $T = T_{melt} - 5$ K in accordance with the model [53] the value of $\varphi_{p.m.}$, given as $(1 - \varphi_{cl})$, equals ~0.827; at $T > T_{melt}$ equation (5.7) gives $\varphi_{p.m.} \cong 0.933$. For these states, calculation by the equation (7.96) gives the difference in $\lg D$ equal 0.61, approximately. Calculation by the equation (7.97) for $f_c = 0.091$ at $T = T_{melt} - 5$

K and for $f_c = 0.133$ at $T > T_{melt}$ gives the difference $\lg D \cong 0.60$ for the same states. These results correlate well with the experimentally obtained value of $\lg D$ equal ~0.63 [233].

Comparison of temperature dependencies of the diffusion coefficient for HDPE and its melt, calculated by the equation (7.96) and obtained experimentally [233], is shown in Figure 7.94. Unexpectedly good (with regard to the equation (7.96) [237]) conformity between theoretical and experimental results was obtained; therefore, constant value of D at $T > T_{melt}$ is stipulated by the condition $\varphi_{p.m.} = const = 0.933$ accepted for this temperature interval [234]. It is also important that one and the same constant D_0' equal 3.65×10^{-11} m^2/s is used for both temperature intervals.

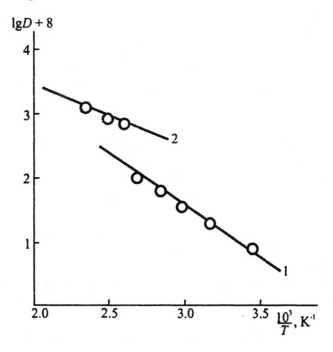

Figure 7.94. Temperature dependence of the diffusion coefficient D for methane in HDPE (1) and its melt (2). Lines show data from [233], points mark calculation results by equation (7.96) [234]

As a consequence, the cluster model gives an opportunity to explain behavior of the diffusion coefficient in the framework of the entire structural approach for temperatures both above and below T_{melt}. The interpretation suggested by the authors of [234] explains D jump at T_{melt} as the result of local order zones' (clusters') formation (decomposition), impermeable for the gas transfer process. The interpretation under consideration gives both qualitative and quantitative description of dependence $D(T)$ in the whole temperature interval studied.

REFERENCES

1. Nilsen L.E., *Mechanical Properties Of Polymers And Composites*, New York, Marcel Dekker, Inc., 1974, 309 p.
2. Popova S.S., Budtov V.P., Ryabikova V.M., and Khudobina G.V., *Analysis Of Polymerizing Plastics*, Leningrad, Khimia, 1988, 304 p. (Rus)
3. Mashukov N.I., Mikitaev A.K., Gladyishev G.P., Belousov V.N., and Kozlov G.V., *Plasticheskie Massy*, 1990, No. 11, pp. 21 – 23. (Rus)
4. Mashukov N.I., Serdyuk V.D., Belousov V.N., Kozlov G.V., Ovcharenko E.N., and Gladyishev G.I., *Izvestiya AN SSSR, Ser. Khim.*, 1990, No. 8, pp. 1915 - 1917. (Rus)
5. Volyinskaya A.V., Godovsky Yu.K., and Papkov V.S., *Vysokomol. Soedin.*, 1979, vol. **A21**(5), pp. 1059 – 1063. (Rus)
6. Lipatov Yu.S., Privalko V.P., Demchenko S.S., Besklubenko Yu.D., Titov G.V., and Shumsky V.F., *Vysokomol. Soedin.*, 1986, vol. **A28**(3), pp. 573 – 579. (Rus)
7. Kozlov G.V., Dolbin I.V., and Zaikov G.E., In Coll.: *Proc. V All-Russian Seminar "Modeling of Non-Equilibrium Systems"*, Krasnoyarsk, KGU, 2002, pp. 96 – 97. (Rus)
8. Aharoni S.M., *Macromolecules*, 1985, vol. **18**(12), pp. 2624 - 2630.
9. Di Benedetto A.T. and Trachte K.L., *J. Appl. Polymer Sci.*, 1970, vol. **14**(11), pp. 2249 - 2262.
10. Godovsky Yu. K., *Heat Physics Of Polymers*, Moscow, Khimia, 1982, 280 p. (Rus)
11. Sanditov D.S. and Kozlov G.V., *Fizika I Khimia Stekla*, 1995, vol. **21**(6), pp. 547 - 576. (Rus)
12. Kozlov G.V. and Sanditov D.S., *Nonlinear Effects And Physicomechanical Properties Of Polymers*, Novosibirsk, Nauka, 1994, 261 p. (Rus)
13. Slutsker A.I. and Filippov V.E., *Vysokomol. Soedin.*, 1988, vol. **A30**(11), pp. 2386 - 2390. (Rus)
14. Kozlov G.V., Burya A.I., Sviridenok A.I., and Zaikov G.E., *Doklady NAN*

Belarusi, 2003, vol. **47** (in press). (Rus)

15. Kozlov G.V., Beloshenko V.A., Kuznetsov E.N., and Lipatov Yu.S., *Doklady NAN Ukrainy*, 1994, No. 12, pp. 126 - 128. (Rus)

16. Vainstein B.K., *X-ray Diffraction On Chain Molecules*, Moscow, Izd. AN SSSR, 1963, 372 p. (Rus)

17. Serdjuk V.D., Kozlov G.V., Mashukov N.I., and Mikitaev A.K., *J. Mater. Sci. Techn.*, 1997, vol. **5**(2), pp. 55 - 60.

18. Irzhak V.I., Rozenberg B.A., and Enikolopyan N.S., *Polymer Networks. Synthesis, Structure, Properties*, Moscow, Nauka, 1976, 248 p. (Rus)

19. Timm D.C., Ayorinde A.J., and Foral R.F., *Brit. Polymer J.*, 1985, vol. **17**(2), pp. 227 - 232.

20. Chang T.-D. and Brittain J. O., *Polymer Eng. Sci.*, 1982, vol. **22**(18), pp. 1221 - 1227.

21. Kozlov G.V., Beloshenko V.A., Stroganov I.V., and Lipatov Yu.S, *Doklady NAN Ukrainy*, 1995, No. 10, pp. 117 - 118. (Rus)

22. Aliguliev R.M., Khiteeva D.M., and Oganyan V.A., *Vysokomol. Soedin.*, 1988, vol. **B30**(4), pp. 268 – 272. (Rus)

23. Sanditov D.S. and Bartenev G.M., Physical Properties Of Non-oriented Structures, Novosibirsk, Nauka, 1982, 256 p. (Rus)

24. Beloshenko V.A., Kozlov G.V., and Lipatov Yu.S., *Fizika Tverdogo Tela*, 1994, vol. **36**(10), pp. 2903 - 2906. (Rus)

25. Beloshenko V.A., Pakter M.K., Beresnev B.I., Zaika T.P., Slobodina V.G., and Shepel V.M., *Mekhanika Kompositnyikh Materialov*, 1990, vol. **26**(2), pp. 195 - 199. (Rus)

26. Beloshenko V.A., Kozlov G.V., and Varyukhin V.N., *Fizika I Tekhnika Vysokikh Davlenii*, 1994, vol. **4**(2), pp. 70 - 74. (Rus)

27. Beloshenko V.A. and Kozlov G.V., *Mekhanika Kompozitnyikh Materialov*, 1994, vol. **30**(4), pp. 451 - 454. (Rus)

28. Kalinchev E.L. and Sakovtseva M.B., Properties And Processing Of Thermoplastics, Leningrad, Khimia, 1983, 288 p. (Rus)

29. Flory P.J., *J. Chem. Phys.*, 1977, vol. **66**(12), pp. 5720 - 5729.

30. Stoyanov O.V. and Deberdeev R.Ya., *Vysokomol. Soedin.*, 1987, vol. **B29**(1), pp. 22 - 24. (Rus)

31. Bartenev G.M., Shut N.I., and Kaspersky A.V., *Vysokomol. Soedin.*, 1988, vol. **B30**(5), pp. 328 - 332. (Rus)

32. Hg S.C., Hosea T.Y.C., and Goh S.H., *Polymer Bull.*, 1987, vol. **18**(2), pp. 155 - 158.

33. Stehling F. and Mandelkern L., *Macromolecules*, 1970, vol. **3**(2), pp. 242 - 252.

34. Popli R., Glotin M., Mandelkern L., and Benson R.S., *J. Polymer Sci.: Polymer Phys. Ed.*, 1984, vol. **24**(3), pp. 407 - 448.

35. Ashcraft C.R. and Boyd R.H., *J. Polymer Sci.: Polymer Phys. Ed.*, 1976, vol. **14**(11), pp. 2153 - 2193.

36. Axelson D.E., Mandelkern L., Popli R., and Mathieu P., *J. Polymer Sci.: Polymer Phys. Ed.*, 1983, vol. **21**(10), pp. 2319 - 2335.

37. Popli R. and Mandelkern L., *Polymer Bull.*, 1983, vol. **9**(2), pp. 260 - 267.

38. Bernstein V.A. and Egorov V.M., *Differential Scanning Calorimetry In Physicochemistry Of Polymers*, Leningrad, Khimia, 1990, 256 p. (Rus)

39. Wedgewood A.R. and Seferis Y.C., *Pure Appl. Chem.*, 1983, vol. **55**(5), pp. 873 - 892.

40. Belousov V.N., Kozlov G.V., and Mashukov N.I., *Doklady Adyigeiskoi (Cherkesskoi) Mezhdunar. AN*, 1996, vol. **2**(1), pp. 74 - 82. (Rus)

41. Boyer R.F., *Polymer Eng. Sci.*, 1968, vol. **8**(3), pp. 161 - 185.

42. Kargin V.A., Andrianova G.P., and Kardash G.G., *Vysokomol. Soedin.*, 1967, vol. **A9**(2), pp. 267 - 287. (Rus)

43. Egorov E.A. and Zhizhenkov V.V., *Vysokomol. Soedin.*, 1982, vol. **A24**(1), pp. 67 - 72. (Rus)

44. Egorov E.A., Zhizhenkov V.V., Marikhin V.A., Myasnikova L.P., and Popov A., *Vysokomol. Soedin.*, 1983, vol. **25**(4), pp. 693 - 701. (Rus)

45. Krigbaum W.R., Roe R.-Y., and Smith K.Y., *Polymer*, 1964, vol. **5**(3), pp. 553 - 542.

46. Papkov S.P., *Vysokomol. Soedin.*, 1977, vol. **A19**(1), pp. 3 - 18. (Rus)

47. Tanabe Y., Strobl G.R., and Fischer E.W., *Polymer*, 1986, vol. **27**(8), pp. 1147 - 1153.

48. Mansfield M.L., *Macromolecules*, 1987, vol. **20**(6), pp. 1384 - 1393.

49. Averkin B.A., Egorov E.A., Zhizhenkov V.V., Slutsker A.I., Stepanov A.B., and Timofeev V.S., *Vysokomol. Soedin.*, 1989, vol. **A31**(10), pp. 2173 - 2177. (Rus)

50. Kozlov G.V., Beloshenko V.A., and Varyukhin V.N., *Ukrainskyi Fizicheskyi Zhurnal*, 1996, vol. **41**(2), pp. 218 - 221. (Rus)

51. Wunderlich B., *J. Chem. Phys.*, 1960, vol. **61**(8), pp. 1052 - 1056.

52. Boyer R.F., *J. Macromol. Sci. - Phys.*, 1973, vol. **7B**(3), pp. 487 - 501.

53. Kozlov G.V. and Novikov V.U., *Uspekhi Fizicheskikh Nauk*, 2001, vol. **171**(7), pp. 717 - 764. (Rus)

54. Bartenev G.M. and Zelenev Yu.V., *Physics And Mechanics Of Polymers*, Moscow, Vysshaya Shkola, 1983, 391 p. (Rus)

55. Bartenev Г. M. and Frenkel S.Ya., *Polymer Physics*, Leningrad, Khimia, 1990, 432 p. (Rus)

56. Kozlov G.V., Temiraev K.B., Shetov R.A., and Mikitaev A.K., *Materialovedenie*, 1999, No. 2, pp. 34—39. (Rus)
57. Boyd R.H., *Polymer*, 1985, vol. 26(8), pp. 1123 - 1133.
58. Lamarre L. and Sung C.S.P., *Macromolecules*, 1983, vol. 16(11), pp. 1729 - 1736.
59. Yu W.-C. and Sung S.P., *Macromolecules*, 1988, vol. 21(2), pp. 365 - 371.
60. Robertson R.E., Simha R., and Curro J.G., *Macromolecules*, 1984, vol. 17(4), pp. 911 - 919.
61. Kozlov G.V., Shustov G.B., and Zaikov G.E., *Vestnik KBGU. Khim. Nauki*, 2003, pp. 52 – 57. (Rus)
62. Jean Y.C., Sandreczki T.C., and Ames D.P., *J. Polymer Sci.: Part B: Polymer Phys.*, 1986, vol. 24(6), pp. 1247 - 1258.
63. Deng Q., Zandiehnadem F., and Jean Y.C., *Macromolecules*, 1992, vol. 25(3), pp. 1090 - 1095.
64. Bantle S., Hasslin H.W., ter Meer H.-U., Schmidt M., and Burchard W., *Polymer*, 1982, vol. 23(12), pp. 1889 - 1893.
65. Kozlov G.V., Beloshenko V.A., and Lipskaya V.A., *Ukrainsky Fizichesky Zhurnal*, 1996, vol. 41(2), pp. 222 - 225. (Rus)
66. Kozlov G.V., Beloshenko V.A., and Shogenov V.N., *Fizikokhimicheskaya Mekhanika Materialov*, 1999, vol. 35(5), pp. 105 - 108. (Rus)
67. SHogenov V.N., Belousov V.N., Potapov V.V., Kozlov G.V., and Prut E.V., *Vysokomol. Soedin.*, 1991, vol. A33(1), pp. 155 - 160. (Rus)
68. Graessley W.W. and Edwards S.F., *Polymer*, 1981, vol. 22(10), pp. 1329 - 1334.
69. Arzhakov S.A. and Kabanov V.A., *Vysokomol. Soedin.*, 1971, vol. B13(5), pp. 318 - 319. (Rus)
70. Arzhakov S.A., Bakeev N.F., and Kabanov V.A., *Vysokomol. Soedin.*, 1973, vol. A15(5), pp. 1154 - 1167. (Rus)
71. Novikov V.U. and Kozlov G.V., *Analysis Of Polymer Degradation In the Framework Of The Fractal Concept*, Moscow, Izd. MGOU, 2001, 135 p. (Rus)
72. Kozlov G.V., Mikitaev A.K., and Novikov V.U., *Izv. AN. Mekhanika Tverdogo Tela*, 2000, No. 1, pp. 93 - 96. (Rus)
73. Askadsky A.A. and Matveev Yu.I., *Chemical Structure And Physical Properties Of Polymers*, Moscow, Khimia, 1983, 248 p. (Rus)
74. Filyanov E.M., *Vysokomol. Soedin.*, 1987, vol. A29(5), pp. 975 - 981. (Rus)
75. Sanditov D.S., Kozlov G.V., Belousov V.N., and Lipatov Yu.S., *Ukrain. Polymer. J.*, 1992, vol. B(3-4), pp. 241 - 258.
76. Mikitaev A.K., Kozlov G.V., and Shogenov V.N., *Plast. Massy*, 1985, No.

2, pp. 32 - 33. (Rus)

77. Kozlov G.V., SHogenov V.N., Kharaev A.M., and Mikitaev A.K., *Vysokomol. Soedin.*, 1987, vol. **B29**(4), pp. 311 - 314. (Rus)

78. Graessleu W.W., *Macromolecules*, 1980, vol. **13**(2), pp. 372 - 376.

79. Sanditov D.S. and Kozlov G.V., *Fizika I Khimia Stekla*, 1993, vol. **19**(4), pp. 593 - 601. (Rus)

80. Fischer E.W. and Dettenmaier M., *J. Monocryst. Solids*, 1978, vol. **31**(1-2), pp. 181 - 205.

81. Serdjuk V.D., Burya A.I., Sanditov D.S., and Kozlov G.V., *In Proc. 2nd Intern. Conf. "Research and Development in Mechanical Industry"*, RaDMI, Vrnjačka Banja, Yugoslavia, 2002, vol. **2**, pp. 1068 - 1072.

82. Andrews E.H., *Pure Appl. Chem.*, 1974, vol. **39**(1-2), pp. 179 - 194.

83. Popli R. and Mandelkern L., *J. Polymer. Sci.: Part B: Polymer Phys.*, 1987, vol. **25**(3), pp. 441 - 483.

84. Peacock A.J. and Mandelkern L., *J. Polymer. Sci.: Part B: Polymer Phys.*, 1990, vol. **28**(11), pp. 1917 - 1941.

85. Pakhomov P.M., Korsukov V.E., Shablyigin M.V., and Novak I.I., *Vysokomol. Soedin.*, 1984, vol. **A26**(6), pp. 1288 - 1293. (Rus)

86. Manevich L.I., Oshmyan V.G., Gai M.I., Akopyan E.L., and Enikolopyan N.S., *Doklady AN SSSR*, 1986, vol. **289**(1), pp. 128 - 131. (Rus)

87. Levene A., Pullen W.J., and Poberts J., *J. Polymer Sci. Part A-2*, 1965, vol. **3**(2), pp. 697 - 701.

88. Kozlov G.V., Serdjuk V.D., and Milman L.D., *Vysokomol. Soedin.*, 1993, vol. **B35**(12), pp. 2049 - 2050. (Rus)

89. Dotsenko V.S., *J. Phys. C: Solid State Phys.*, 1985, vol. **18**(15), pp. 6023 - 6031.

90. Kozlov G.V., Aloev V.Z., Bazheva R.Ch., and Zaikov G.E., In Coll.: *Articles Of IV All-Russian Sci.-Res. Conf. On "New Chemical Technologies: Production And Application"*, Russia, Penza, 2002, p. 59 - 61. (Rus)

91. Plummer C.J.G. and Donald A.M., *Macromolecules*, 1990, vol. **23**(12), pp. 3929 - 3937.

92. Kozlov G.V. and Novikov V.U., *Prikladnaya Fizika*, 1997, No. 1, pp. 85 - 93. (Rus)

93. Kozlov G.V., Novikov V.U., and Mikitaev A.K., *Materialovedenie*, 1997, No. 6 – 7, pp. 3 - 6. (Rus)

94. Novikov V.U. and Kozlov G.V., *Materialovedenie*, 2000, No. 1, pp. 2 - 12. (Rus)

95. Bergman D.J. and Kantor Y., *Phys. Rev. Lett.*, 1984, vol. **53**(6), pp. 511 - 514.

96. Webmann I., In Coll.: *Fractals In Physics*, Ed. Pietronero L. and Tosatti E., Amsterdam, North-Holland, 1986, 670 p.
97. Song H.H. and Roe R.-J., *Macromolecules*, 1987, vol. **30**(9), pp. 2723 - 2732.
98. Nemilov S.V., *Fizika I Khimia Stekla*, 1987, vol. **13**(6), pp. 801 – 809. (Rus)
99. Shklovsky B.I. and Efros A.L., *Uspekhi Fizicheskikh Nauk*, 1975, vol. **117**(3), pp. 401 – 436. (Rus)
100. Kozlov G.V. and Novikov V.U., *Synergism And Fractal Analysis Of Polymer Networks*, Moscow, Klassika, 1998, 112 p. (Rus)
101. Aleksanyan G.G., Berlin Al.Al., Goldansky A.V., Grineva N.S., Onishchuk V.A., Shantarovich V.P., and Safonov G.P., *Khimicheskaya Fizika*, 1986, vol. **5**(9), pp. 1225 - 1234. (Rus)
102. Balankin A.S. and Bugrimov A.L., *Vysokomol. Soedin.*, 1992, vol. **A34**(10), pp. 135 - 139. (Rus)
103. Balankin A.S., Bugrimov A.L., Kozlov G.V., Mikitaev A.K., Sanditov D.S., *Doklady AN*, 1992, vol. **326**(3), pp. 463 - 466. (Rus)
104. Andrianova G.P. and Kargin V.A., *Vysokomol. Soedin.*, 1970, vol. **A12**(1), pp. 3 - 8. (Rus)
105. Kachanova I.M. and Roitburd A.L., *Fizika Tverdogo Tela*, 1989, vol. **31**(4), pp. 1 - 9. (Rus)
106. Hartmann B., Lee G.F., and Cole R.F., *Polymer Eng. Sci.*, 1986, vol. **26**(8), pp. 554 - 559.
107. Kozlov G.V., Beloshenko V.A., Gazaev M.A., and Novikov V.U., *Mekhanika Kompozitnyikh Materialov*, 1996, vol. **32**(2), pp. 270 - 278. (Rus)
108. Balankin A.S., *Synergism Of Deformable Body*, Moscow, MO SSSR, 1991, 404 p. (Rus)
109. Shogenov V.N., Kozlov G.V., and Mikitaev A.K., *Vysokomol. Soedin.*, 1989, vol. **A31**(8), pp. 1766 - 1770. (Rus)
110. Kartsovnik V.I., Volkov V.N., and Rozenberg B.A., *Vysokomol. Soedin.*, 1977, vol. **B19**(4), pp. 280 - 284. (Rus)
111. Matsuoka S. and Bair H.E., *J. Appl. Phys.*, 1977, vol. **48**(10), pp. 4058 - 4062.
112. Papin V.E., Likhachev V.A., and Grinyaev Yu.V., *Structural Levels Of Deformation Of Solids*, Novosibirsk, Nauka, 1985, 226 p. (Rus)
113. Ivanova V.S. and Vstovsky G.V., In Coll.: *Wrap-up Of Science and Technology. Metal Science And Thermal Processing*, 1990, vol. **24**, Moscow, VINITI, pp. 43 – 98. (Rus)
114. Balankin A.S., *Pis'ma V ZhTF*, 1990, vol. **16**(7), pp. 14 – 20. (Rus)

115. Kozlov G.V., Beloshenko V.A., and Varyukhin V.N., *Prikladnaya Mekhanika I Tekhnicheskaya Fizika*, 1996, vol. 37(3), pp. 115 – 119. (Rus)
116. Brown N., *Mater. Sci. Eng.*, 1971, vol. 8(1), pp. 69 - 73.
117. Serdjuk V.D., Kosa P.N., and Kozlov G.V., *High Pressure Physics And Technology*, 1995, vol. 5(3), pp. 37 – 42. (Rus)
118. Abaev A.M., Beloshenko V.A., Kozlov G.V., and Mikitaev A.K., *High Pressure Physics And Technology*, 1998, vol. 8(2), pp. 102 – 109. (Rus)
119. Kozlov G.V., Sanditov D.S., Serdjuk V.D., and Lipatov Yu.S., *Doklady NAN Ukrainy*, 1993, No. 9, pp. 144 - 147. (Rus)
120. Wunderlich B., *Macromolecular Physics*, Vol. 3, New York, Academic Press, 1980, 484 p.
121. Kargin V.A. and Sogolova T.I., *Zh. Fiz. Khim.*, 1953, vol. 27(7), pp. 1039 – 1049. (Rus)
122. Gent A.N. and Madan S., *J. Polymer Sci.: Part B: Polymer Phys.*, 1989, vol. 27, pp. 1529 - 1542.
123. Bernstein V.A., Sirota A.G., Egorova L.M., and Egorov V.M., *Vysokomol. Soedin.*, 1989, vol. A31(4), pp. 776 - 779. (Rus)
124. Kozlov G.V., Shustov G.B., and Mashukov N.I., In Coll.: *Proc. Intern. Sci.-Res. Conf. On "Modern Materials And Technologies"*, Russia, Penza, 2002, pp. 20 – 22. (Rus)
125. Schultz J.M., *Polymer Eng. Sci.*, 1984, vol. 24(10), pp. 770 - 785.
126. Mashukov N.I., Belousov V.N., Kozlov G.V., Ovcharenko E.N., and Gladyishev G.P., *Izv. AN SSSR, Ser. Khim.*, 1990, No. 9, pp. 2143 - 2146. (Rus)
127. Kozlov G.V., Shustov G.B., Aloev V.Z., and Ovcharenko E.N., In Coll.: *Proc. II All-Russian Sci.-Appl. Conf. On „Innovations In Machine Building"*, Penza, PGU, 2002, pp. 21 – 23. (Rus)
128. Mashykov N.I., Serdjuk V.D., Kozlov G.V., Ovcharenko E.N., Gladyishev G.P., and Vodakhov A.B., *Stabilization And Modification Of Polyethylene By Oxygen Acceptors* (Preprint), Moscow, ICP AS USSR, 1990, 64 p. (Rus)
129. Kennedy M.A., Peacock A.J., and Mandelkern L., *Macromolecules*, 1994, vol. 27(19), pp. 5297 - 5310.
130. Kosa P.N., Serdjuk V.D., Kozlov G.V., and Sanditov D.S., *High Pressure Physics And Technology*, 1995, vol. 5(4), pp. 70 – 81. (Rus)
131. Aloev V.Z., Beloshenko V.A., Abazekhov M.M., *Vestnik KBGU. Ser. Fiz. Nauki*, 2000, Iss. 4, p. 50 -51. (Rus)
132. Owen A.J. and Ward I.M., *J. Mater. Sci.*, 1971, vol. 6(6), pp. 485 - 489.

133. Argon A.S. and Bessonov M.I., *Phil. Mag.*, 1977, vol. **35**(4), pp. 917 - 933.
134. Kozlov G.V., Shustov G.B., Zaikov G.E., Burmistr M.V., and Korenyako V.A., *Voprosy Khimii I Khimicheskoi Tekhnologii*, 2003, No. 1, pp. 68 – 72. (Rus)
135. Gladyishev G.P., *Thermodynamics And Macrokinetics Of Natural Hierarchic Processes*, Moscow, Nauka, 1988, 290 p. (Rus)
136. Gladyishev G.P. and Gladyishev D.P., *Zh. Fiz. Khim.*, 1994, vol. **68**(5), pp. 790 – 792. (Rus)
137. Narikawa I., *Strength Of Polymeric Materials*, Moscow, Khimia, 1987, 400 p. (Rus)
138. Sanditov D.S., Kozlov G.V., and Sanditov B.D., *Fizika I khimia Stekla*, 1996, vol. **22**(6), pp. 683 - 693. (Rus)
139. Zhorin V.A., Malkin A.Ya., and Enikolopyan N.S., *Vysokomol. Soedin.*, 1979, vol. **A21**(4), pp. 820 - 824. (Rus)
140. Donald A.M., *J. Mater. Sci.*, 1985, vol. **20**(7), pp. 2630 - 2638.
141. Kozlov G.V., Mokaeva K.Z., Gazaev M.A., and Mikitaev A.K., *Manuscript deposited to VINITI RAS*, Moscow, August 09, 1995, No. 2407-V95. (Rus)
142. Shogenov V.N., Kharaev A.M., and Guchinov V.A., In Coll.: *Polycondensation Processes And Polymers*, Ed. A.K. Mikitaev, Nalchik, KBGU, 1988, pp. 14 – 21. (Rus)
143. Belousov V.N., Kozlov G.V., Mikitaev A.K., and Lipatov Yu.S., *Doklady AN SSSR*, 1990, vol. **313**(3), pp. 630 – 633. (Rus)
144. Chou S.C., Robertson K.D., and Rainey J.H., *Exp. Mech.*, 1973, vol. **13**(10), pp. 422 - 430.
145. Beatty C.L. and Weaver J.L., *Polymer Eng. Sci.*, 1978, vol. **18**(14), pp. 1109 - 1116.
146. Kozlov G.V., Milman L.D., and Mikitaev A.K., *Manuscript deposed to VINITI RAS*, Moscow, February 26, 1997, No. 623-V97. (Rus)
147. Kozlov G.V., *Manuscript deposed to VINITI RAS*, Moscow, November 01, 2002, No. 1884-V2002. (Rus)
148. Gazaev M.A., Kozlov G.V., Milman L.D., and Mikitaev A.K., *High Pressure Physics And Technology*, 1996, vol. **6**(1), pp. 76 – 81. (Rus)
149. Bekichev V.I., *Vysokomol. Soedin.*, 1974, vol. **A16**(7), pp. 1479 - 1485. (Rus)
150. Bekichev V.I., *Vysokomol. Soedin.*, 1974, vol. **A16**(8), pp. 1745 - 1747. (Rus)
151. Balankin A.S., *Pis'ma V ZhTF*, 1991, vol. **17**(6), pp. 84 – 89. (Rus)

430   *Kozlov G.V. and Zaikov G.E.*

152. Goldstein R.V. and Mosolov A.B., *Doklady AN*, 1992, vol. **324**(3), pp. 576 – 580. (Rus)
153. Stanley H., I Coll.: *Fractals in Physics*, Ed. Pietronero L. and Tosatti E. Amsterdam, North-Holland, 1986, 670 p.
154. Kozlov G.V., Burya A.I., Sviridenok A.I., and Zaikov G.E., *Doklady NAI Belarusi*, 2003, vol. **47**(1), pp. 62 – 64. (Rus)
155. Gefen Y., Aharony A., and Alexander S., *Phys. Rev. Lett.*, 1983, vol. **50**(1), pp. 77 - 80.
156. Rammal R. and Toulouse G., *J. Phys. Lett.*, 1983, vol. **44**(1), pp. L13 - L22.
157. Alexander S., Laermans C., Orbach R., and Rosenberg H.M., *Phys. Rev. B.*, 1983, vol. **28**(8), pp. 4615 - 4619.
158. Meakin P., Majid I., Havlin S., and Stanley H.E., *J. Phys. A.*, 1984, vol. **17**(18), pp. L975 - L981.
159. Chen Z.-Y., Deutch J.M., and Meakin P., *J. Chem. Phys.*, 1984, vol. **80**(6), pp. 2982 - 2983.
160. Kozlov G.E., Shustov G.B., and Zaikov G.E., *Izv. VUZov, Severo-Kavkazsk. Region, Estestv. Nauki*, 2003, No. 2, p. 58. (Rus)
161. Balankin A.S., *Fizika Tverdogo Tela*, 1992, vol. **34**(4), pp. 1245 - 1258. (Rus)
162. Haward R.N., *Macromolecules*, 1993, vol. **26**(22), pp. 5860 - 5869.
163. Kambour R.P., *J. Polymer Sci.: Macromol. Rev.*, 1973, vol. **7**, pp. 1 - 154.
164. Kramer E.J., *J. Polymer Sci.: Polymer Phys. Ed.*, 1975, vol. **13**(2), pp. 509 - 525.
165. Bucknell C.B., *Toughened Plastics*, London, Applied Science, 1977, 332 p.
166. Fellers J.F. and Huang D.S., *J. Appl. Polymer Sci.*, 1979, vol. **23**(8), pp. 2315 - 2326.
167. Kozlov G.V., Burya A.I., and Zaikov G.E., *Hungarian Chemical J.*, 2003 (in press).
168. Fraser R.A.W. and Ward I.M., *J. Mater. Sci.*, 1977, vol. **12**(2), pp. 459 - 468.
169. Ishikawa M., Ogawa H., and Marisawa I., *J. Macromol. Sci.-Phys.*, 1981, vol. **19B**(3), pp. 41 - 443.
170. Kozlov G.V., Shetov R.A., and Mikitaev A.K., *Vysokomol. Soedin.*, 1987, vol. **A29**(5), pp. 1109 - 1110. (Rus)
171. Kozlov G.V. and Mikitaev A.K., *Vysokomol. Soedin.*, 1987, vol. **B29**(7), pp. 490 - 492. (Rus)
172. Kambour R.P., *Polymer Commun.*, 1984, vol. **25**(12), pp. 357 - 360.

173. Wu S., *J. Polymer Sci.: Part B: Polymer Phys.*, 1989, vol. **27**(4), pp. 723 - 741.

174. Michler G.H., *Macromol. Chem. Macromol. Symp.*, 1991, vol. **45**(1), pp. 39 - 54.

175. Plati P.E. and Williams J.G., *Polymer Eng. Sci.*, 1975, vol. **15**(6), pp. 470 - 477.

176. Kozlov G.V., Shogenov V.N., and Mikitaev A.K., In Coll.: *Polycondensation Processes And Polymers*, Ed. V.V. Korshak, Nalchik, KBGU, 1984, pp. 35 – 44. (Rus)

177. Li J.C.M. and Wu J.B.C., *J. Mater. Sci.*, 1976, vol. **11**(2), pp. 445 - 457.

178. Donald A.M. and Kramer E.J., *J. Polymer Sci.: Polymer Phys. Ed.*, 1982, vol. **20**(4), pp. 899 - 909.

179. Donald A.M. and Kramer E.J., *J. Polymer Sci.: Polymer Phys. Ed.*, 1982, vol. **20**(5), pp. 1129 - 1141.

180. Donald A.M. and Kramer E.J., *Polymer*, 1982, vol. **23**(3), pp. 461 - 465.

181. Kramer E.J., *Polymer Eng. Sci.*, 1984, vol. **24**(10), pp. 741 - 769.

182. Henkee C.S. and Kramer E.J., *J. Polymer Sci.: Polymer Phys. Ed.*, 1984, vol. **22**(4), pp. 721 - 737.

183. Yang A.C.-M., Kramer E.J., Kuo C.C., and Phoenix S.L., *Macromolecules*, 1986, vol. **19**(7), pp. 2010 - 2019.

184. Aharoni S.M., *Macromolecules*, 1983, vol. **16**(9), pp. 1722 - 1728.

185. Lin Y.-H., *Macromolecules*, 1987, vol. **20**(12), pp. 3080 - 3083.

186. Berger L.L. and Kramer E.J., *Macromolecules*, 1987, vol. **20**(6), pp. 1980 - 1985.

187. Plummer C.J.G. and Donald A.M., *J. Polymer Sci.: Polymer Phys. Ed.*, 1989, vol. **27**(2), pp. 325 - 336.

188. Kozlov G.V., Beloshenko V.A., and Lipatov Yu.S., *Intern. J. Polymer Mater.*, 1998, vol. **39**(2), pp. 201 - 212.

189. Haward R.N., Murphy B.M., and White E.F.T., *J. Polymer Sci. Part A-2*, 1971, vol. **9**(5), pp. 801 - 814.

190. Hoare J. and Hull D., *J. Mater. Sci.*, 1975, vol. **10**(11), pp. 1861 - 1870.

191. Mashukov N.I., Gladyioshev G.P., and Kozlov G.V., *Vysokomol. Soedin.*, 1991, vol. **A33**(12), pp. 2538 - 2546. (Rus)

192. Jud K., Kausch H., and Williams J.G., *J. Mater. Sci.*, 1981, vol. **16**(1), pp. 204 - 210.

193. Kausch H., *Pure Appl. Chem.*, 1983, vol. **55**(5), pp. 833 - 844.

194. Kozlov G.V. and Mikitaev A.K., *Izv. SKNTs VSh. Estestvennye Nauki*, 1987, No. 3(59), pp. 66 – 69. (Rus)

195. Ward I.M., *Mechanical Properties Of Solid Polymers*, London, Wiley –

Interscience Ltd., 1973, 357 p.

196. Kambour R.P. and Gruner C.L., *J. Polymer Sci.: Polymer Phys. Ed.*, 1978, vol. 16(4), pp. 703 - 716.

197. Kambour R.P., *Polymer Commun.*, 1983, vol. 24(10), pp. 292 – 296.

198. Willbourn A.H., *Polymer*, 1976, vol. 17(11), pp. 965 - 976.

199. Lee W.A. and Sewell J.H., *J. Appl. Polymer Sci.*, 1968, vol. 12(6), pp. 1397 - 1409.

200. Termonia Y. and Smith P., *Macromolecules*, 1988, vol. 21(7), pp. 2184 - 2189.

201. Donald A.M. and Kramer E.J., *J. Mater. Sci.*, 1981, vol. 16(10), pp. 2967 - 2976.

202. Baskes M.I., *Eng. Fractal Mech.*, 1974, vol. 6(1), pp. 11 - 18.

203. Kozlov G.V., Beloshenko V.A., Shogenov V.N., and Lipatov Yu.S., *Doklady NAN Ukrainy*, 1995, No. 5, pp. 100 - 102. (Rus)

204. Shogenov V.N., Burya A.I., Shustov G.B., and Kozlov G.V., *Physicochemical Mechanics Of Materials*, 2000, vol. 36(1), pp. 51 - 55. (Rus)

205. Haward R.N. and Thackray G., *Proc. Roy. Soc. London*, 1968, vol. A302(1471), pp. 453 - 472.

206. Kozlov G.V., Šetov R.A., and Mikitaev A.K., *Plaste und Kautschuk*, 1988, Bd. 35(7), S. 261 - 263.

207. Kozlov G.V., Serdjuk V.D., and Beloshenko V.A., *Mekhanika Kompozitnyikh Materialov*, 1994, vol. 30(5), pp. 691 - 695. (Rus)

208. Kozlov G.V., Afaunov V.V., and Lipatov Yu.S., *Inzhenerno-Fizicheskyi Zhurnal*, 2000, vol. 73(2), pp. 439 - 442. (Rus)

209. Kozlov G.V., Afaunov V.V., and Novikov V.U., *Materialovedenie*, 2000, No. 9, pp. 19 - 21. (Rus)

210. Sokolov I.M., *Uspekhi Fizicheskikh Nauk*, 1986, vol. 150(2), pp. 221 – 256. (Rus)

211. Bobryishev A.N., Kozomazov V.N., Babin L.O., and Solomatov V.I., *Synergism Of Composite Materials*, Lipetsk, NPO ORIUS, 1994, 153 p. (Rus)

212. Bersted B.H., *J. Appl. Polymer Sci.*, 1979, vol. 23(1), pp. 37 - 50.

213. Kozlov G.V., Belousov V.N., and Lipatov Yu.S., *Doklady AN UkrSSR*, 1990, No. 6., pp. 50 - 53. (Rus)

214. Wool R.P. and Boyd R.H., *J. Appl. Phys.*, 1980, vol. 51(10), pp. 5116 - 5124.

215. Mashukov N.I., Serdjuk V.D., Kozlov G.V., and Gladyishev G.P., *Voprosy Oboronnoi Tekhniki*, 1991, Ser. 15, Iss. 3, No. 97, pp. 13 – 15.

(Rus)
216. Mikos A.G. and Peppas N.A., *J. Chem. Phys.*, 1988, vol. **88**(2), pp. 1337 - 1342.
217. Yamamoto I., Miyata H., and Kobayashi T., In Coll.: *"Benibana" Intern. Symp.*, Oct. 8 – 11, 1990, Yamagata, Japan; Abstract, Yamagata, Japan, 1990, pp. 184 - 189.
218. Novikov V.U. and Kozlov G.V., *Materialovedenie*, 1997, No. 4, pp. 6 - 9. (Rus)
219. Kozlov G.V. and Novikov V.U., *Prikladnaya Fizika*, 1997, No. 1, pp.77 – 84. (Rus)
220. Brown H.R., *J. Mater. Sci.*, 1982, vol. **17**(3), pp. 469 - 476.
221. Smirnov B.M., *Uspekhi Fizicheskikh Nauk*, 1986, vol. **149**(3), pp. 177 - 219.
222. Kozlov G.V., Zaikov G.E., and Mikitaev A.K., *Intern. J. Polymer Mater.*, 1998, vol. **4**(1), pp. 41 – 46.
223. Belousov V.N., Kozlov G.V., and Mikitaev A.K., *Plast. Massy*, 1984, No. 6, p. 62. (Rus)
224. Aharoni S.M., *J. Appl. Polymer Sci.*, 1979, vol. **23**(1), pp. 223 - 228.
225. Mashukov N.I., Vasnetsova O.A., Malamatov A.Kh., and Kozlov G.V., *Lakokrasochnye Materialy I Ikh Primenenie*, 1992, No. 1, pp. 16 - 17. (Rus)
226. Afaunov V.V., Mashukov N.I., Kozlov G.V., and Sanditov D.S., *Izv. VUZov. Severo-Kavkazsk. Region. Estestvennye Nauki*, 1999, No. 4(108), pp. 69 – 71. (Rus)
227. Peterlin A., *J. Macromol. Sci.-Phys.*, 1975, vol. **11B**(1), pp. 57 - 87.
228. Kozlov G.V., Belousov V.N., and Mikitaev A.K., *High Pressure Physics And Technology*, vol. **8**(1), pp. 64 - 70. (Rus)Sozanov V.A., and Kozlov G.V., *Vestnik KGBU. Ser. Fiz. Nauki*, 2000, Iss. 4, pp. 48 – 49. (Rus)
229. Afaunov V.V., Mashukov N.I., Sozanov V.A., and Kozlov G.V., *Vestnik KGBU. Ser. Fiz. Nauki*, 2000, Iss. 4, pp. 48 – 49. (Rus)
230. Nielsen L.E., *J. Macromol. Sci. -Chem.*, 1967, vol. **1A**(5), pp. 929 - 942.
231. Joon D.Y. and Flory P.J., *Polymer*, 1977, vol. **18**(5), pp. 509 - 513.
232. Tochin V.A., Shlyakhov R.A., and Sapozhnikov D.N., *Vysokomol. Soedin.*, 1980, vol. **A22**(4), pp. 752 - 758. (Rus)
233. Kozlov G.V. and Zaikov G.E., *Vysokomol. Soedin.*, Ser. B, 2003 (in press). (Rus)
234. Yoon D.Y. and Flory P.J., *Macromolecules*, 1984, vol. **17**(4), pp. 868 - 871.
235. Krisyuk B.E. and Sandakov G.I., *Vysokomol. Soedin.*, 1995, vol. **A37**(4),

pp. 615 - 620. (Rus)
236. Mashukov N.I., *Doklady Adyigeiskoi (Cherkesskoi) Mezhdunar. AN*, 1994, vol. **1**(1), pp. 69 – 75. (Rus)

Appendix I. Interdisciplinary analysis of cluster structure of polymers

Interdisciplinary analysis applied to solving of scientific problems suggests the use of achievements in allied branches of science. Such approach becomes useful due to development of cybernetics, synergism and fractal analysis in the XXth century. As computers appeared, the Millennium of Computation came. Besides, cybernetics [1] as the theory of system control has given premises for development of new scientific direction – the synergism, which is the theory of self-organizing systems. Presently, synergism primarily occurred as one of the scientific directions of theoretical physics based on thermodynamics of non-equilibrium processes has become the interdisciplinary science, which sets universal regularities of self-organization of spatial structures in dynamic systems of different origin. The determining effect on synergism development was caused by fundamental researches in various branches of science: physics, chemistry, mathematics, mechanics, biology and others. The originator of the structure self-organization theory based on thermodynamics of non-equilibrium processes is the Nobel laureate I. Prigozhine. Development of the mentioned branch of thermodynamics has caused better understanding of the second law of thermodynamics and the entropy notion. Setting of universal laws of self-organization in nonlinear systems, which stipulated the interdisciplinary position of the synergism principles, allowed solution of many problems, raised but not solved by cybernetics, in a short time. The synergism based on the physical essence of adoption of systems to an external impact by self-organization of structures, emphasizes that the effects of structure self-organization (ordering) can appear under conditions, non-equilibrium from positions of thermodynamics.

The theory of structure self-organization under non-equilibrium conditions, developed by I. Prigozhine and his school [2 – 6], has determined the new opinion on the natural laws and promoted development of the universal approach to analysis of complex systems, named as "nonlinear dynamics" (synergism) [7 – 11].

Synergism reflects principally new phase of mathematical physics development allowing description of various processes, based on nonlinear connections, proceeding in various models and systems basing on general statements of the interdisciplinary approach with respect to the structure self-organization phenomenon under nonlinear (non-equilibrium) conditions. Using the principles of synergism every investigator (a physicist, chemist, mathematician, biologist and so on) is able to enrich the resources of synergism ideas and methods using the methods of its own branch of science. It is also able to multiply accelerate solutions of the problems in its branch.

The principles of synergism have disclosed the mechanism of structure adapting to an external impact and its universality for systems of both animate and inanimate nature. Adapting of the structure is the process of rebuilding the structure that became unstable and self-organization of a new, more stable one. During restructuring new fractal (multifractal) structures are formed, which cannot be described in the framework of the Euclidean geometry because of their unusual shape. Fractal structures are, for example, a broad range of natural and artificial topological forms, the basic feature of which is the self-similar hierarchic structure, also displayed by amorphous glassy-like polymers [12].

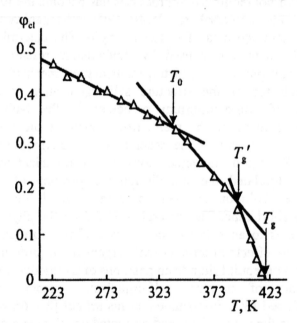

Figure A.1. Temperature dependence of the relative part of clusters φ_{cl} for PC. Arrows mark critical temperatures of bifurcation points [17]

As mentioned above, quantitatively, supermolecular (more precisely, supersegmental) structure of amorphous glassy-like polymers can be described in the framework of the cluster model, based on the local order ideas [13]. The relative part of the local order zones (clusters), φ_{cl}, is the order parameter in the strictest physical meaning of this term [14]. Figure A-1 shows dependence of φ_{cl} on test temperature T for PC, which contrary to the graph in Figure 2.1 is approximated not by a smooth curve, but a broken line, where bending (bifurcation) points reflect the change of the energy dissipation mechanism, associated with the reach of threshold values of φ_{cl}. For example, Figure A.1 indicates T_1 correspondence to the "frost" temperature T_0 [15], T_2 to glass transition temperature of the packless matrix T_g', and T_3 to glass transition temperature of the polymer T_g.

As shown before [16], at transition from the previous point of unstable structure to the next one during self-organization the following universal adapting mechanism is realized:

$$A_m = Z_n / Z_{n+1} = \Delta_i^{1/m} , \tag{A-1}$$

where Z_n and Z_{n+1} are critical values of the control parameter regulating structure formation, and their ratio determines adaptability of the system, A_m, to restructuring; m is the number of restructurings; Δ_i is the measure of structure stability, kept constant at m variation from 1 to m^*. The value $m = 1$ corresponds to the minimum and $m = m^*$ to the maximum of structure adaptability. There is a Table [16] giving values of A_m, m and Δ_i determined from the premium proportion and responding to the spectrum of invariant values of the structure stability measures for the systems of animate and inanimate nature (Table A-1). The use of the Table makes simpler the determination of stability measure and structure adaptability to the external impact, bound by an exponent law.

Taking for Z_n and Z_{n+1} values of φ_{cl} in the mentioned bifurcation points T_0, T_g' and T_g (φ_{cl}' and φ_{cl}^*, respectively) and accompanying them with the data from Table A-1, one can obtain values of parameters A_m, Δ_i and m for PC, shown in Table A-2. Data in this Table indicate systematic decreases of parameters A_m and Δ_i under the condition $m = 1 = \text{const}$, i.e. at minimal structure adaptability to temperature effect. As a consequence, in the framework of synergism the critical temperature T_g' can be characterized as the point of 'ordering – structure degradation' bifurcation, and T_g as 'structure degradation – chaos' one [17].

Table A-1

Self-similarity constants Δ_i of evolving systems, determined by the premium proportion law and corresponded indices m for restructurings of self-controlled systems with internal feedback [16]

i \ m / Δ_i	1	2	4	8	16	32	64	128
1	0.618	0.618	–	–	–	–	–	–	–
2	0.465	0.465	0.682	–	–	–	–	–	–
3	0.380	0.380	0.616	0.785	–	–	–	–	–
4	0.327	0.324	0.569	0.754	0.869	–	–	–	–
5	0.285	0.285	0.534	0.731	0.855	0.925	–	–	–
6	0.255	0.255	0.505	0.711	0.843	0.918	0.958	–	–
7	0.232	0.232	0.482	0.694	0.833	0.913	0.955	0.977	–
8	0.213	0.213	0.461	0.679	0.824	0.908	0.953	0.976	0.988

Table A-2

Critical parameters of cluster supersegmental structure for PC [17]

Temperature interval, K	φ'_{cl}	φ^*_{cl}	A_m	Δ_i	m	m^*
213 ÷ 333	0.528	0.330	0.623	0.618	1	1
333 ÷ 390	0.330	0.153	0.465	0.465	1	2
390 ÷ 425	0.153	0.049	0.323	0.324	1	8

Clearly Δ_i decrease corresponds to the bifurcation critical temperature increase. That is why the increase of critical temperatures with decline of cluster structure stability, Δ_i, should be expected. Figure A.2 shows dependencies of T_0, T_g' and T_g on reverse Δ_i value for PC with corresponded data for polyarylate (PAr, refer to Figure 2.1) applied to it. This correlation is linear and possesses two characteristic points. At $\Delta_i = 1$ linear dependence $T_{cr}\left(\Delta_i^{-1}\right)$ is extrapolated to $T_{cr} = 293$ K, which indicates glassy-like polymer transition to the rubber-like state. Data on Δ_i constant shown in Table A-1 display the minimal values $\Delta_i = 0.213$ at $m = 1$. The graph in Figure A.2 shows the critical temperature $T_{кр} = T_{ll}$, the highest for polymers, determining transition to the unstructured liquid, corresponded to this minimal Δ_i value. For polymers, this means the absence of even dynamic short-living local order [18].

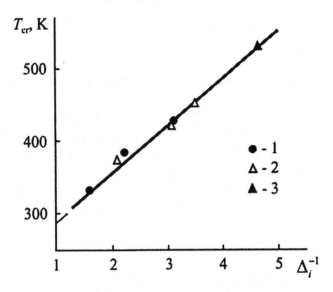

Figure A.2. Dependence of critical temperatures T_{cr} on Δ_i^{-1} value for PC (1) and PAr (2); 3 is T_{ll} value for PC [17]

Thus the above-mentioned results allow the following interpretation of the critical temperatures T_g' and T_g of amorphous glassy-like polymer structure. These temperatures correspond to critical values of the control parameter φ_{cl} at approach of which the subordination principle, one of the mains of synergism, is realized, when the set of variables is subordinate to a single (or several) variable, which is the order parameter. Note also that for PC the number of restructurings $m = 1$ corresponds to the mechanism of the particle-cluster structure formation [19].

Appendix II. Computerized simulation of ordered structures in polymers

It is common knowledge [20 – 22] that computerized simulation is one of the most productive methods of studying molecular processes of the polymer structure formation. Short-chain molecule structure formation processes were studied and molecular mobility and conformation defects in the orientation-ordered structure were analyzed [23]. The microscopic study was performed on 100 chain molecules, each of which consisted of a sequence of CH_2-groups.

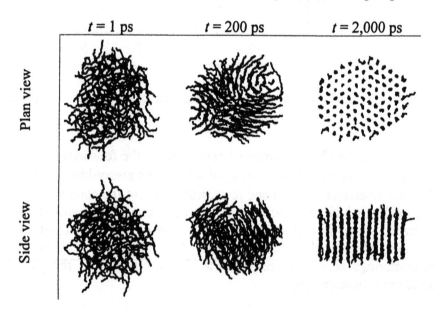

Figure A.3. Photographic images of 00 short-chain molecule configuration at different times: $t = 1, 200$ and $2,000$ ps (from left right-wise) at $T = 400$ K. The upper and the lower figures represent top and side views, respectively [23]

Figure A.3 shows images of the chain configuration at different simulation times t for temperature $T = 400$ K. This Figure shows the following features of the ordered structure formation [23]:

a) at short time periods ($t = 1$ ps) chain molecules possess random conformation;

b) propagation of local oriented-ordered domains in several positions with time is observed ($t = 200$ ps);

c) finally, several domains conjugate into a large domain, and highly ordered structure is formed ($t = 2,000$ ps). In the ordered structure, almost all bonds possess the *trans*-conformation. The connection between the local order and the part of bonds in the *trans*-conformation has been mentioned before and in the framework of the cluster model.

To study the propagation of general orientation order the parameter of it, P_{or}, was calculated as follows [23]:

$$P_{or} = \left\langle \frac{3\cos^2 \psi - 1}{2} \right\rangle_{bond}, \qquad (A.2)$$

where ψ is the angle between two vectors of sub-bonds and $\langle ... \rangle_{bond}$ marks averaging by all couples of sub-bonds. The sub-bond vector is the one connecting centers of two neighbor bonds along the same chain. Parameter P_{or} equals 1.0, when all sub-bonds are parallel, and it is zero when they are randomly oriented.

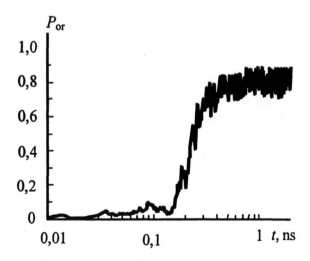

Figure A.4. Dependence of the general orientation order parameter P_{or} on the simulation time t at $T = 400$ K [23]

Figure A.4 shows time dependence of parameter P_{or}. Before $t \cong 150$ ps this parameter is of almost value that means the absence of the general orientation order in this time interval. After $t = 150$ ps P_{or} sharply increases indicating simultaneous sharp increase of the general orientation order after some time period. To compare data of computerized simulation [23] and the cluster model, a simple assessment for polyethylene should be made as follows. According to data in Figure A.4, general order structure parameter P_{or} is maximal equal ~0.8. For amorphous-crystalline polyethylene with the crystallinity degree $K = 0.68$, obviously P_{or} value for crystalline phase $\left(P_{or}^{cr} \right)$ equals one. Further on, the parameter for non-crystalline zones can simply be determined as follows:

$$P_{or} = P_{or}^{cr} K + P_{or}^{n-cr} (1 - K).$$ (A.3)

At the above-mentioned values of P_{or}, P_{or}^{cr} and K directly for non-crystalline zones only, it is obtained $P_{or}^{n-cr} \cong 0.12$ that correlates well with φ_{cl} for HDPE [25].

Further on, the concept of a domain as a yarn of locally orientation-ordered chain molecules, formed by parallel ordering, was introduced [23]. In this work the domain was defined as follows. Two chain molecules correspond to the same domain, if the following two conditions are met: 1) $\left| \vec{r}_c^i - \vec{r}^i \right| < r_0$ and 2) $\alpha_{ij} < \alpha_0$, where $\vec{r}_c^i$ is the vector of the center of mass position for the i-th chain molecule; α_{ij} varying in the interval $0 \le \alpha_{ij} \le \pi/2$ is the angle between the main axis of the lowest moment of inertia of the i-th chain and similar axis of the j-th chain. Clearly such definition represents one of possible variants of the cluster one, present in Chapter 2.

Figure A.5 shows time evolution of the largest domain size, S, at $T = 400$ K. Before $t = 120$ ps only small domains sized below 10 are observed. The largest domain size S starts sharp step-wise increasing at $t = 120$ ps. Note that packing of all chains into a single domain performed under the simulation conditions of the work [23], is determined by two factors as follows: the chains are free, i.e. they are not limited by macromolecular entanglements, and they are too short to form a macromolecular coil [26]. The presence of macromolecular hookings may cause impossibility of the stage (c) realization (Figure A.3), and then $S < 100$.

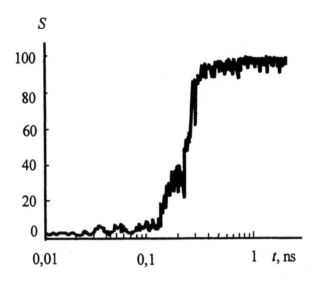

Figure A.5. Dependence of the lowest domain size, S, on the simulation time t at $T = 400$ K [23]

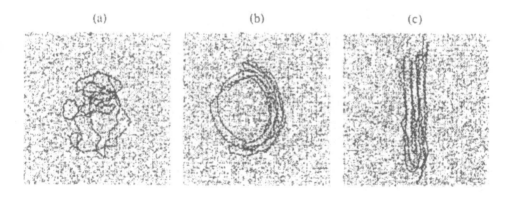

Figure A.6. Configuration of isolated polymeric chain in solution at $T = 350$ K: (a) for t = 0 ps; (b) for $t = 2.5$ ps; (c) for $t = 6$ ps. Black and grew colors mark the polymeric chain and diluter molecules, respectively [27]

  Another result of interest in the present consideration is represented by data of the work [27] on computerized simulation of behavior of isolated polymeric chain in solution. This simulations show that (a) a toroidal structure is formed at high temperature of about 400 K and (b) at lower temperatures of about 350 and 300 K a toroidal structure is formed first, and then it transits to a folded orientation-ordered structure (Figure A.6).

Figure A.6 (c) clearly shows the chain structure allowing formation of the local order zones (clusters) at the contact with other macromolecules, as it is postulated in the cluster model [28].

Appendix III. Computerized forecast of structure and properties of amorphous polymers

---

The immense amount of new polymers is synthesized all over the world every year, and just a minor part of them reaches the industrial production stage. The works on synthesis demand long time and huge assets, which can be reduced by solving the problem of prediction of material properties to be obtained. In particular, one of approaches in this direction [29, 30] allows numerical estimation of some physical parameters basing on the chemical structure of the repeat unit. The authors of [31, 32] have suggested have suggested calculation of mechanical properties of polymers by glass transition temperatures and cohesive energies. A possibility of forecasting mechanical properties basing on the ideas of the cluster model and the fractal analysis has also been suggested [33].

A series of heterochain polyesters, chemical structure of which and some physicochemical characteristics (glass transition temperature T_g, solubility parameter δ_s, cohesive energy density W_c) are shown in Table A.3, were used as the study objects [33]. Films were obtained by the glazing method of 5% polymer solutions in chloroform to horizontal cellophane support. Thus the initial parameters for determination of structure characteristics and properties of studied polyesters, the following ones were accepted: glass transition temperature T_g, cohesive energy density W_c, and skeletal bond length in the backbone l_0. Since structure of the polymer film is significantly dependent on the diluter used and two-component solubility parameter model was selected for estimation of the fractal dimensionality of macromolecular coil [34], then for characterization of polymer – diluter interactions the components δ_f and δ_c of the diluter solubility parameter δ_s were used. The component δ_f characterizes energies of dissipative interactions and dipole bonds' interaction; δ_c characterizes energies of hydrogen bonds' interaction and the interaction between atoms with insufficient and excessive amount of electrons (acceptor-donor couples of atoms). Values of δ_f and δ_c parameters for chloroform are shown in the work [34], and $\Delta\delta_f$ is determined as follows [35]:

$$\Delta\delta_f = |\delta_p - \delta_f|, \tag{A.4}$$

where value of the solubility parameter of polymer δ_p is determined from the known W_c values [29]:

$$\delta_p = W_c^{1/2}.$$ (A.5)

Further on, the fractal dimensionality D of macromolecular coil in solution is determined by the following equation [35]:

$$D = 1.5 + 0.35(\Delta\delta_f)^{2/(1+\delta_c)}.$$ (A.6)

Moreover, in accordance with the technique suggested [33] properties were forecasted by the following equations:
a) for calculation of elasticity modulus E:

$$E = 0.7\left(\frac{S}{C_\infty}\right)^{1/2}, \text{ GPa,}$$ (A.7)

b) for calculation of the Grüneizen parameter γ:

$$\gamma = \frac{1}{2}\left(\frac{d_f}{d - d_f}\right),$$ (A.8)

where d_f is the fractal dimensionality of polymer, calculated by the equation as follows [28]:

$$d_f = 1.5D.$$ (A.9)

Table A-3

Chemical structure and physical characteristics of heterochain polyesters

No.	Chemical structure of repeating unit	T_g, K	$\delta_p, \left(\dfrac{cal}{cm^3}\right)^{1/2}$	$W_c, \dfrac{MJ}{m^3}$
1		446	9.38	369
2		461	9.36	367
3		488	9.52	380
4		491	9.64	389

	Structure			
5	(structure)	508	9.46	375
6	(structure)	518	9.39	369
7	(structure)	449	9.32	364
8	(structure)	548	9.36	367

9		568	9.59	385
10		573	9.43	373

Kozlov G.V. and Zaikov G.E.

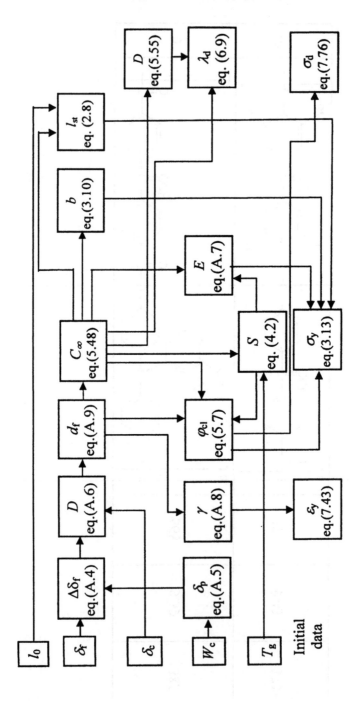

Figure A.7. Block-scheme of computerized forecast

Table A-4

Comparison of experimental and calculated values of mechanical characteristics for the series of heterochain polyesters

No.	E^e, GPa	E^t, GPa	Δ_E, %	σ_y^e MPa	σ_y^t MPa	$\Delta\sigma_y$, %	ε_y^e	ε_y^t	$\Delta\varepsilon_y$, %	σ_d^e, MPa	σ_d^t, MPa	$\Delta\sigma_d$, %	λ_d^e	λ_d^t	$\Delta\lambda_d$, %
1	1.75	1.63	6.9	101	101	0.5	0.100	0.111	9.9	99	115	13.9	1.28	1.80	28.9
2	1.96	1.69	13.7	113	117	3.4	0.108	0.115	6.1	109	123	11.4	1.04	1.73	39.8
3	1.77	1.79	1.1	107	107	3.7	0.090	0.095	5.5	75	91	17.6	1.29	1.41	8.5
4	1.78	1.80	1.1	97	97	0	0.089	0.083	6.7	108	75	30.5	1.49	1.35	9.4
5	1.76	1.86	5.4	105	106	1	0.095	0.103	7.8	101	103	1.9	1.20	1.34	11.1
6	2.08	1.90	8.7	101	120	15.8	0.100	0.111	9.9	121	115	5.0	1.10	1.34	18.3
7	1.74	1.65	5.2	120	107	10.8	0.123	0.119	3.3	119	129	7.8	1.26	1.79	29.6
8	2.19	2.01	8.2	106	129	17.8	0.106	0.114	7.0	127	121	6.3	1.10	1.21	9.1
9	2.28	2.08	8.8	103	107	3.9	0.098	0.088	10.2	103	82	20.4	1.25	1.04	16.8
10	2.18	2.10	3.7	114	122	7.0	0.100	0.107	6.5	103	110	6.4	1.09	1.08	0.9

Notes: Index numbers of polymers correspond to these in Table A-3.
Indices '*e*' and '*t*' mark experimental and theoretical data, respectively.

Figure A.7 shows the block-scheme used for computerized forecast of the basic mechanical properties of polymers (elasticity modulus E, yield stress σ_y and deformation ε_y, stress σ_d and stretch λ_d at degradation), mentioned in Table A.3. Comparison of properties, experimental and calculated by the block-scheme suggested, and their relative inaccuracy Δ as well are shown in Table A.4. Clearly quite good corresponded between theory and experiment is obtained: the average inaccuracy by 5 properties equals ~10%, which equals usual inaccuracy of mechanical tests. For the degradation process theoretical estimation inaccuracy is somewhat above the average value Δ ($\Delta = 12.1\%$ for σ_d and 17% for λ_d). This situation is explained by the effect of possible macroscopic defects on polymer degradation process, which is not considered in the calculated scheme suggested (attention should be paid to the fact that almost all λ_d, calculated theoretically, exceed corresponded experimental values).

Thus the suggested method of mechanical properties forecast using two independent parameters (δ_n and S), which in principle can be calculated by the method of group contributions from chemical structure of polymers only, is based on ideas of both the cluster model and the fractal analysis. The absence of the theory of regular solutions makes application of empirical approximated techniques of parameter D forecast (similar to equation (A.6)) inevitable, which is the source of additional inaccuracy. Nevertheless, even in the simplest (essentially, schematic) alternative, considered above, the forecast accuracy of deformation-strength characteristics is quite high for practical application and close to the accuracy of determination of these parameters in mechanical tests [33].

REFERENCES

1. Wiener N. *Cybernetics Or Control And Communication In The Animal And The Machine*, Sec. Ed., New York, W.H. Freeman and Co., 1961, 343 p.
2. Prigozhin I.M., *From The Existing To The Occurring. Time and Complexity In Physical Sciences*, Moscow, Nauka, 1977, 325 p. (Rus)
3. Glansdorff P. and Prigogine I., *Thermodynamic Theory Of Structure, Stability And Fluctuations*, London, Wiley Interscience, A Division of John Wiley and Sons, Ltd., 1971, 280 p.
4. Nicolis G. and Prigogine I., *Exploring Complexity. An Introduction*, New York, W.H. Freeman and Co., 1989, 342 p.
5. Prigogine I. and Stergers M., *Time, Chaos, Quantum*, Moscow, Progress, 1994, 272 p. (Rus)
6. Prigogine I., *The End Of Determinacy. Time. Chaos And New Laws Of Nature*, Moscow, Publ. Center "Regular and chaotic dynamics", 1999, 215 p. (Rus)
7. Haken H., *Synergetics*, Berlin, Springer-Verlag, 1978, 404 p.
8. Ebeling W., *Strukturbildung bai irreversiblen Prozessen. Eine Einführung in die Theorie disspative strukturen*, Berlin, BSB B.G. Teubner Verlag-Sgesellschaft, 1976, 279 s.
9. Klimontovich Yu.L., *Turbulent Motion And Structure of Chaos. A New Approach To Statistical Theory Of Open Systems*, Moscow, Nauka, 1990, 320 p. (Rus)
10. Akhromeev T.S., Kurdyumov S.P., Malinetsky G.G., and Samarsky A.A., *Non-Stationary Structures And Diffusion Chaos*, Moscow, Nauka, 1992, 281 p. (Rus)
11. Moiseev N.N., *The Algorithm Of Development*, Moscow, Nauka, 1997, 297 p. (Rus)
12. Kozlov G.V. and Novikov V.U., *Synergism And Fractal Analysis Of Cross-linked Polymers*, Moscow, Klassika, 1998, 112 p. (Rus)
13. Kozlov G.V. and Novikov V.U., *Uspekhi Fiz. Nauk*, 2001, vol. **171**(7)pp. 717 – 764. (Rus)
14. Kozlov G.V., Gazaev M.A., Novikov V.U., and Mikitaev A.K., *Pis'ma V ZhTF*, 1996, vol. **22**(16), pp. 31 – 38. (Rus)
15. Kozlov G.V. and Zaikov G.E., In Coll.: *Fractals And Local Order In Polymeric Materials*, Ed. Kozlov G. and Zaikov G., New York, Nova Science Publishers Inc., 2001, pp. 55 - 63.
16. Ivanova V.S., In Coll.: *The 60th Anniversary of A.A. Baikov Institute Of Metallurgy And Material Science, RAS*, Moscow, Eliu, 1998. (Rus)

17. Burya A.I., Ivanova V.S., Novikov V.U., and Kozlov G.V., *Proc. 3rd Intern. Conference "Research and Development in Mechanical Industry. RaDMI – 2003"*, vol. 2 (in press).
18. Kozlov G.V., Shustov G.B., and Zaikov G.E., *Zh. Prikl. Khim.*, 2002, vol. 75(3), pp. 485 – 487. (Rus)
19. Shogenov V.N. and Kozlov G.V., *Fractal Clusters In Physicochemistry Of Polymers*, Nalchik, Poligrafservis i T, 2002, 268 p. (Rus)
20. Cun Feng Fan and Shaw Ling Hsu., *Macromolecules*, 1992, vol. 25(1), pp. 266 - 270.
21. Ladd A.J.C. and Frenkel D., *Macromolecules*, 1992, vol. 25(13), pp. 3435 - 3438.
22. Raaska T., Niemelä S., and Sundholm F., *Macromolecules*, 1994, vol. 27(20), pp. 5751 - 5757.
23. Fujiwara S. and Sato T., *J. Chem. Phys.*, 1999, vol. 110(19), pp. 9757 - 9764.
24. Serdyuk V.D., Kozlov G.V., Mashukov N.I., and Mikitaev A.K., *J. Mater. Sci. Techn.*, 1997, vol. 5(2), pp. 55 - 60.
25. Mashukov N.I., Gladyishev G.P., and Kozlov G.V., *Vysokomol. Soedin.*, 1991, vol. A33(12), pp 2538 – 2546. (Rus)
26. Forsman W.C., *Macromolecules*, 1982, vol. 15(6), pp. 1032 - 1040.
27. Fujiwara S. and Sato T., *Europ. Conf. on Computational Physics*, Aachen, Germany, Sept. 5 – 8, 2001, Book of Abstract. vol. 8, p. A69.
28. Kozlov G.V., Temiraev K.B., Shustov G.B., and Mashukov N.I., *J. Appl. Polymer Sci.*, 2002, vol. 85(6), pp. 1137 - 1140.
29. Askadskii A.A. *Physical Properties Of Polymers. Prediction And Control*, Amsterdam, Gordon and Breach Publ., 1996, 336 p.
30. Askadskii A.A. and Lkinskikh A.F., *Vysokomol. Soedin.*, 1999, vol. A41(1), pp. 83 – 92. (Rus)
31. Shogenov V.N., Kozlov G.V., and Mikotaev A.K., *Vysokomol. Soedin.*, 1989, vol. A31(7), pp. 553 – 557. (Rus)
32. Shogenov V.N., Kozlov G.V., and Mikotaev A.K., *Vysokomol. Soedin.*, 1989, vol. A31(11), pp. 809 – 811. (Rus)
33. Shogenov V.N., Beloshenko V.A., Kozlov G.V., and Varyukhin V.N., *High Pressure Physics And Technology*, 1999, vol. 9(3), pp. 30 – 36. (Rus)
34. Wiche I.A., *Ind. Eng. Chem. Res.*, 1995, vol. 34(2), pp. 661 - 673.
35. Kozlov G.V., Dolbin I.V., Mashukov N.I., Burmistr M.V., and Korenyako V.A., *Voprosy Khimii I Khimicheskoi Tekhnologii*, 2001, No. 6, pp. 71 – 77. (Rus)

SUBJECT INDEX

ABBREVIATIONS

AA	–	reactive monodiamine p,p'-aminoazobensene
APESF	–	aromatic polyestersulfones
DAA	–	p,p'-diaminoazobenzene derivative tt-DAA-DAA with four hydrogen atoms substituted by ethyl groups
DLA	–	diffusion limited aggregation
DS	–	dissipative structure
DZ	–	deformation zone
CEC	–	crystallite with extended chains
WS cluster	–	Witten-Sander cluster
CRZ	–	cooperatively restructuring zones
CFC	–	crystallites with folded chains
IP	–	inclined plate
PA-6	–	polyamide-6
PAr	–	polyarylate
PASF	–	polyarylatesulfone
PBTP	–	poly(butylene terephthalate)
PVC	–	polyvinylchloride
PHE	–	polyhydroxyester
PC	–	polycarbonate derived from bisphenol A
PCA	–	polycaproamide
PMMA	–	poly(methyl methacrylate)
PP	–	polypropylene
PS	–	polystyrene
PSF	–	polysulfone
MFI	–	melt flow index
PTFE	–	polytetrafluoroethylene
HDPE	–	high density polyethylene
LDPE	–	low density polyethylene
PET	–	poly(ethylene terephthalate)
BPE	–	branched polyethylenes
UHMWPE	–	ultrahigh-molecular weight polyethylene
SP-OPD-10/F-1	–	diblockpolyester of oligoformal 2,2-di-(4-oxyphenyl)propane, phenolphthalein and isophthalic

acid dichloranhydride

EP-1   –   amine cross-linking type epoxy polymer derived from bisphenol A diglycidyl ester

EP-2   –   anhydride cross-linking type epoxy polymer derived from bisphenol A diglycidyl ester

EP-3   –   EP-1 epoxy polymer aged under natural conditions during two years with dichlorethylidide

Milton Keynes UK
Ingram Content Group UK Ltd.
UKHW021904071024
449327UK00021B/1613